AF767085

Indoor Aerosol Dynamics

Indoor Aerosol Dynamics

Tengfei (Tim) Zhang
Dalian University of Technology
Dalian, China

Fei Liu
Qingdao University of Technology
Qingdao, China

WILEY

Copyright © 2026 by John Wiley & Sons, Inc. All rights reserved, including rights for text and data mining and training of artificial intelligence technologies or similar technologies.

Published by John Wiley & Sons, Inc., Hoboken, New Jersey.

No part of this publication may be reproduced, stored in a retrieval system, or transmitted in any form or by any means, electronic, mechanical, photocopying, recording, scanning, or otherwise, except as permitted under Section 107 or 108 of the 1976 United States Copyright Act, without either the prior written permission of the Publisher, or authorization through payment of the appropriate per-copy fee to the Copyright Clearance Center, Inc., 222 Rosewood Drive, Danvers, MA 01923, (978) 750-8400, fax (978) 750-4470, or on the web at www.copyright.com. Requests to the Publisher for permission should be addressed to the Permissions Department, John Wiley & Sons, Inc., 111 River Street, Hoboken, NJ 07030, (201) 748-6011, fax (201) 748-6008, or online at http://www.wiley.com/go/permission.

The manufacturer's authorized representative according to the EU General Product Safety Regulation is WILEY-VCH GmbH, Boschstr. 12, 69469 Weinheim, Germany, e-mail: Product_Safety@wiley.com.

Trademarks: Wiley and the Wiley logo are trademarks or registered trademarks of John Wiley & Sons, Inc. and/or its affiliates in the United States and other countries and may not be used without written permission. All other trademarks are the property of their respective owners. John Wiley & Sons, Inc. is not associated with any product or vendor mentioned in this book.

Limit of Liability/Disclaimer of Warranty: While the publisher and the authors have used their best efforts in preparing this work, including a review of the content of the work, neither the publisher nor the authors make any representations or warranties with respect to the accuracy or completeness of the contents of this work and specifically disclaim all warranties, including without limitation any implied warranties of merchantability or fitness for a particular purpose. No warranty may be created or extended by sales representatives, written sales materials or promotional statements for this work. The fact that an organization, website, or product is referred to in this work as a citation and/or potential source of further information does not mean that the publisher and authors endorse the information or services the organization, website, or product may provide or recommendations it may make. This work is sold with the understanding that the publisher is not engaged in rendering professional services. The advice and strategies contained herein may not be suitable for your situation. You should consult with a specialist where appropriate. Further, readers should be aware that websites listed in this work may have changed or disappeared between when this work was written and when it is read. Neither the publisher nor authors shall be liable for any loss of profit or any other commercial damages, including but not limited to special, incidental, consequential, or other damages.

For general information on our other products and services or for technical support, please contact our Customer Care Department within the United States at (800) 762-2974, outside the United States at (317) 572-3993 or fax (317) 572-4002.

Wiley also publishes its books in a variety of electronic formats. Some content that appears in print may not be available in electronic formats. For more information about Wiley products, visit our web site at www.wiley.com.

Library of Congress Cataloging-in-Publication Data Applied for

Hardback ISBN: 9781394399079

Cover Design: Wiley
Cover Image: Courtesy of Tengfei Zhang, Prof., Fei Liu, Assistant Prof.

Set in 9.5/12.5pt STIXTwoText by Straive, Chennai, India

Contents

Preface

Aerosols are tiny solid, liquid, or hybrid particles suspended in a gas medium, and the size of these particles is larger than 1 nm but smaller than 100 μm. The sources of aerosol particles encompass natural and anthropogenic origins, including their respectively induced chemical reactions that form secondary aerosols. Both aerosols of outdoor origin and aerosols emitted indoors can be transported in indoor spaces. Outdoor aerosols can be carried into indoor spaces by means of ventilation or air infiltration. Meanwhile, numerous aerosols are emitted indoors, such as cooking oil fumes, fuel or incense burning aerosols, tobacco smoke aerosols, and even infectious viruses/bacteria. Considering that modern people spend more than 90% of their time indoors, respiratory exposure to indoor aerosols must be given sufficient attention.

Airborne aerosols of certain sizes can be inhaled into the respiratory tract and lungs and deposit inside. Excessive exposure to aerosols may induce a range of negative health effects, including asthma, rhinitis, cancer, cardiovascular diseases, etc. These adverse impacts are mainly due to oxidative stress, inflammation, or damage to the immune system by the deposited aerosols. Indoor aerosols may also facilitate transmission of respiratory infectious diseases such as influenza, COVID-19, MERS, and SARS. The COVID-19 pandemic has resulted in immense losses worldwide. Developing appropriate strategies to prevent or control the indoor transmission of respiratory diseases requires an in-depth understanding of indoor aerosols.

Indoor aerosols constitute a dynamic system, and the aerosol contents vary with the spatial location and time. When we examine indoor aerosol dynamics, we are concerned with the lifecycle of these aerosols from their formation/emission to their airborne spatial motion, settling from the airborne transmission route, and removal in indoor environments. Depending on their aerodynamic sizes, some of the aerosols may remain suspended in the air for a long time. Meanwhile, a portion of the aerosols may deposit on various surfaces, subsequently be resuspended due to disturbances, be exhausted from the rooms by ventilation systems, or be removed by air cleaners. Thus, indoor aerosols evolve throughout their lifecycle.

This book elucidates the mechanical principles of indoor aerosol emission, motion, settling, and removal in indoor spaces. Furthermore, respiratory exposure to indoor aerosols, possible health outcomes, transmission of respiratory infectious diseases, and the engineering control of indoor aerosols are introduced. This volume seeks to address the fundamentals in a clear, logical, step-by-step manner, with the help of concise examples. It can serve as a textbook for postgraduate students who are focusing on the indoor environment, residential health, building ventilation and air cleaning, architectural engineering, infectious disease control, and so on. Researchers who are interested in indoor aerosols may also refer to this book. To guide readers in the practice of numerical modeling, example modeling cases and solution data have been uploaded to the book's webpage as electronic resources. Solution manuals for the calculation problems are also available to course instructors.

Chapters 1 through 6 address the general fundamentals of particle properties and motion mechanics, including physicochemical properties, particle size metrics, force interaction during various types of particle motion, particle adhesion and detachment, particle diffusion, and particle sampling and measurement. Meanwhile, Chapters 7–12 deal with topics that are specific to indoor environments, including typical indoor particle emission sources, indoor particle transport modeling, respiratory particle exposure, common health outcomes after exposure to non-biological aerosols and non-infectious bioaerosols, indoor pathogenic aerosol transmission and infection, indoor particle filtration or air cleaning, and inactivation of indoor microbial aerosols. The topic of particle coagulation and growth is only simply addressed when introducing the condensation particle counter. The particle coagulation is rare in indoor environments due to the relatively low particle concentrations in residential spaces, which do not sufficiently sustain significant coagulation of particles.

This book was finished under the joint efforts of many contributors. Dr. Tengfei (Tim) Zhang initiated the proposal for this book, designed the book contents and structure, thoroughly revised and edited all the manuscripts, and coordinated the overall project. Main contribution credits include Dr. Fei Liu (Chapters 1–5), PhD candidate Jianbo Yang (Chapter 6), PhD candidate Haofu Chen (Chapter 7, with the exception of Section 7.5.2), PhD candidate Zhiyi Zhang (intrusion of molds in Section 7.5.2), PhD candidate Kuibo Wu (Section 8.6), Dr. Xian Li (Chapter 10), Dr. Xiong Mei (Chapter 11), and PhD candidate Di Xiao (Chapter 12). Dr. Zhang is the sole author of the intrusion of pollens in Section 7.5.2, Sections 8.1– 8.5, and Chapter 9. Kuibo Wu assisted in redrawing several figures in Chapter 8, and Di Xiao helped in finalizing the figures and tables in Chapter 9. Dr. Liu prepared the appendix. Ms. Linda Gudeman edited the language and polished the writing of the book as a whole.

We are especially grateful to Summers Scholl, the Executive Editor of Wiley Physical Sciences; Elizabeth Amaladoss, the Managing Editor of Wiley Advanced Chemistry & Chemical Engineering; and their editorial team for providing assistance in publishing this book. Without their hard work, this book could not have been published so smoothly. Although the book has already reached publication, we invite our readers to share their opinions, comments, and suggestions with us. Your input would be extremely helpful to us in correcting any hidden errors and improving the quality of the next edition of this book in a few years.

September 2025

Tengfei (Tim) Zhang
Dalian, China

Fei Liu
Qingdao, China

List of Abbreviations

AC	alternating current
ACGIH	American Conference of Governmental Industrial Hygienists
ACH	air change per hour
ACR	air change rate
AFM	atomic force microscopy
AHAM	Association of Home Appliance Manufacturers
APPJ	atmospheric pressure plasma jet
AQG	air quality guideline
ASHRAE	American Society of Heating, Refrigerating and Air-conditioning Engineers
ATP	adenosine triphosphate
BC	black carbon
BRI	building-related illness
BRS	building-related symptom
BTEX	benzene, toluene, ethylbenzene, and xylene
CADR	clean air delivery rate
CB	cigarette butt
CCL	connected-component labeling
CDC	center for disease control
CFD	computational fluid dynamics
CFU	colony-forming unit
CI	confidence interval
COF	cooking oil fume
COPD	chronic obstructive pulmonary disease
CoV	coronavirus
COVID	coronavirus disease
CPC	condensation particle counter
DBD	dielectric barrier discharge
DC	direct current

DEHS	diethylhexyl sebacate
DNA	deoxyribonucleic acid
DNS	direct numerical simulation
DOP	dioctyl phthalate
DRW	discrete random walk
EC	elemental carbon
e-cigarettes	electronic cigarettes
ELA	effective leakage area
EPA	Environmental Protection Agency
ESP	electrostatic precipitator
EU	European Union
HCP	household cleaning product
HEPA	high-efficiency particulate air
HP	hypersensitivity pneumonitis
HVAC	heating, ventilating, and air conditioning
ICRP	International Commission on Radiological Protection
ID	identification
IgE	Immunoglobulin E
IPH	idiopathic pulmonary hemorrhage
LED	light-emitting diode
LES	large-eddy simulation
LPS	lipopolysaccharides
MERS	Middle East Respiratory Syndrome
MERV	minimum efficiency reporting value
MHCD	micro-hollow cathode discharge
MHV	murine hepatitis virus
MMAD	mass median aerodynamic diameter
MPPS	most penetrating particle size
MS	mainstream smoke
N–S	Navier–Stokes
NIST	National Institute of Standards and Technology
OC	organic carbon
ODTS	organic dust toxic syndrome
PAH	polycyclic aromatic hydrocarbon
PAN	polyacrylonitrile
PCO	photocatalytic oxidation
PCP	personal care product
PDA	phase Doppler anemometry
PFU	plaque-forming unit
PG	propylene glycol
PISO	pressure-implicit with splitting of operators

PIV	particle image velocimetry
PM	particulate matter
PNC	particle number concentration
ppb	parts per billion
ppm	parts per million; pages per minute
PRESTO!	pressure staggering option
PRRSV	porcine reproductive and respiratory syndrome virus
PSI-C	particle source in-cell
PTV	particle tracking velocimetry
QAC	quaternary ammonium compound
RANS	Reynolds-averaged Navier–Stokes
RH	relative humidity
RNA	ribonucleic acid
RNS	reactive nitrogen species
RNG	renormalization group
ROM	reactive oxygen metabolite
ROS	reactive oxygen species
rpm	rounds per minute
SARS	severe acute respiratory syndrome
SBS	sick building syndrome
SCA	specific collection area
SEM	scanning electron microscope
SHS	secondhand smoke
SIMPLE	semi-implicit method for pressure-linked equations
SIMPLEC	semi-implicit method for pressure-linked equations-consistent
SIMPLER	semi-implicit method for pressure-linked equations revised
SOA	secondary organic aerosol
SS	side-stream smoke
TB	tracheobronchial
TEM	transmission electron microscope
THS	thirdhand smoke
TLV	threshold limit value
TR	transformer rectifier
UFP	ultrafine particle
ULPA	ultra-low penetration air
USA	United States of America
UV	ultraviolet
UVA	ultraviolet A
UVB	ultraviolet B
UVC	ultraviolet C
UVGI	ultraviolet germicidal irradiation

UV-LIF	ultraviolet-induced fluorescence
VG	vegetable glycerin
VOC	volatile organic compound
VOF	volume of fluid
WHO	World Health Organization

Nomenclature

a	particle radius, Eq. (2.A.3).
a_{nb}	coefficient of a scalar variable for the grid cells neighboring cell P, Eq. (8.20).
a_P	coefficient of a scalar variable at grid cell P, Eq. (8.20).
$a(t)$	particle acceleration at time t, Eq. (3.7).
A	Hamaker constant, Eq. (4.1).
A	cross-sectional area of a fiber, Eq. (11.15); collection area of an electrostatic precipitator, Eq. (11.35).
ACH	air change per hour, Eq. (7.1).
$\boldsymbol{A}_{\mathbf{f}}$	area vector of face f in a grid cell, Eq. (8.16).
A_{p}	projected surface area of a pollen grain, Eq. (7.7); surface area of a droplet, Eq. (8.41).
A_{s}	particle surface area.
B	particle mobility, Eq. (2.15).
B_{m}	Spalding mass number, Eq. (8.42).
$\bar{c}$	particle mean thermal velocity, Eq. (5.10).
c_p	specific heat capacity.
c_{rms}	particle root mean square velocity, Eq. (5.9).
C	particle concentration; non-dimensionless coefficient, Eq. (8.5).
C_0	free stream particle concentration, Eq. (6.3).
$C_{1\varepsilon}$	model constant in the turbulent dissipation rate equation.
$C_{2\varepsilon}$	model constant in the turbulent dissipation rate equation.
$C_{3\varepsilon}$	model constant in the turbulent dissipation rate equation.
C_{bg}	background particle concentration, Eq. (7.3).
C_{c}	slip correction factor, Eq. (2.17).
C_{cell}	particle concentration in the near-wall grid cell, Eq. (8.22).
$C_{\mathrm{d}}(t)$	quantum concentration at time t, Eq. (10.2).
C_{D}	drag coefficient, Eq. (2.4).
$C_{\mathrm{g,p}}$	tracer gas concentrations at the target position, Eq. (10.4).

$C_{g,q}$	airborne quantum concentration at the target position, Eqs. (10.3) and (10.4).
C_{in}	indoor particle concentration, Eq. (7.6).
C_i^{α}	average concentration of pollutant α in zone i, Eq. (8.1).
C_i^{β}	average concentration of pollutant β in zone i, Eq. (8.1).
C_j^{α}	average concentration of pollutant α in zone j, Eq. (8.1).
C_{out}	outdoor particle concentration, Eq. (7.6).
C_r	steady-state particle concentration in room, Eq. (7.3).
C_s	particle concentration in the sampling probe, Eq. (6.3).
C_v	virus concentration in the exhaled droplets, Eq. (10.2).
$C_{v,s}$	vapor concentration at the droplet surface, Eq. (8.38).
$C_{v,\infty}$	vapor concentration in the bulk gas, Eq. (8.38).
$C_{wall,i}$	nominal particle concentration on a wall, Eq. (8.22).
C_{μ}	constant in modeling the turbulent viscosity.
d	particle equivalent diameter; particle diameter; inhalation dose, Eqs. (10.8) and (10.9).
d_{50}	cutoff diameter for 50% particle collection efficiency, Eq. (3.28).
d_a	aerodynamic diameter, Eq. (2.26); the projected area diameter of a pollen grain, Eq. (7.7).
d_b	size of an impact body, Eq. (6.27).
d_c	diameter of a cylinder, Eq. (3.23).
d_e	equivalent volume diameter, Eq. (2.24).
d_f	fiber diameter, Eq. (11.1).
d_g	geometric mean diameter of particles, Eq. (1.3).
d_i	particle size of the ith bin, Eq. (1.2).
d_p	particle diameter.
$\overline{d_p}$	algebraic mean diameter of particles, Eq. (1.2).
$d\boldsymbol{r}$	vector from a grid cell center to a face center, Eq. (8.13).
d_s	Stokes diameter, Eq. (2.26).
D	particle diffusion coefficient, Eq. (5.1).
D_0	duct diameter, Eq. (6.2).
D_f	diffusion transport of a scalar variable across face f, Eq. (8.8).
D_j	jet diameter of an impactor, Eq. (3.24).
D_s	diameter of a sampling probe, Eq. (6.2).
e	restitution coefficient of particles after bounce, Eq. (4.7); charge of an electron, Eq. (11.29).
$\boldsymbol{e_s}$	unit vector connecting two adjacent grid cell centers, Eq. (8.16).
E	overall particle collection efficiency of a filter, Eq. (11.18).
E_D	single-fiber particle collection efficiency due to diffusion, Eq. (11.9).
E_{DR}	single-fiber particle collection efficiency due to interception of the diffusing particles, Eq. (11.10).

$\boldsymbol{E}_e$	electric field strength (vector), Eq. (11.28).
E_e	electric field strength (scalar), Eq. (11.33).
E_G	single-fiber particle collection efficiency due to gravitational settling, Eq. (11.12).
E_I	particle collection efficiency of an impactor, Eq. (3.27); single-fiber particle collection efficiency due to impaction, Eq. (11.5).
E_q	single-fiber particle collection efficiency due to electrostatic image forces, Eq. (11.14).
E_R	single-fiber particle collection efficiency by interception, Eq. (11.2).
ER	emission rate of chemicals or particles, Eq. (7.1).
E_Σ	single-fiber total particle collection efficiency, Eq. (11.20).
f	non-dimensionless coefficient, Eq. (8.4).
$f(d_p)$	frequency function of particle size distribution, Eq. (1.2).
f_v	volatile fraction inside a droplet, Eq. (8.37).
F	force.
$F_{Add,i}$	sum of additional forces acting on a particle, Eq. (8.23).
F_{Adh}	adhesive force, Eq. (4.5).
F_D	drag force, Eq. (2.3).
F_{Diff}	diffusion force, Eq. (5.2).
F_E	electrostatic force, Eq. (4.2).
F_G	gravity, Eq. (2.11).
F_i	non-flow mass contribution rate in zone i, Eq. (8.2).
$F_{i \to j}$	mass rate from zone i to zone j, Eq. (8.1).
$F_{j \to i}$	mass rate from zone j to zone i, Eq. (8.1).
F_n	normal component of the drag force, Eq. (2.6).
F_S	surface tension force, Eq. (4.4).
F_{Th}	thermophoretic force, Eq. (5.25).
F_{van}	van der Waals force, Eq. (4.1).
F_τ	frictional component of the drag force, Eq. (2.7).
g	gravitational acceleration.
G	number of airborne tuberculosis bacilli released per infector per unit time, Eq. (10.7); dimensionless number related to gravitational settling, Eq. (11.11).
G_b	production rate of turbulent kinetic energy due to buoyancy.
G_i^α	release rate of pollutant α in zone i, Eq. (8.1).
G_k	production rate of turbulent kinetic energy due to mean velocity gradients.
h	nozzle size of an impactor, Eq. (3.27); convection heat transfer coefficient, Eq. (8.44).
h_{fg}	latent heat of evaporation of the liquid, Eq. (8.44).
H	thermophoretic coefficient, Eq. (5.28).

i	internal energy.
I	number of source patients, Eq. (10.1).
J	particle diffusion flux, Eq. (5.1).
J_f	mass flow rate of air across face f, Eq. (8.8).
k	Boltzmann constant, Eq. (5.3).
k	turbulent kinetic energy; decay coefficient of pollutants, Eq. (11.36).
k_a	thermal conductivity of the air or gas, Eq. (5.29).
k_c	convection mass transfer coefficient, Eq. (8.38).
k_p	thermal conductivity of particles, Eq. (5.29).
K	proportionality constant in Newton's resistance law, Eq. (2.3); spring coefficient, Eq. (4.6).
K_E	proportionality constant in calculating electrostatic force, Eq. (4.3).
KE	critical kinetic energy.
KE_b	critical kinetic energy required for particle bounce, Eq. (4.7).
Kn	Knudsen number, Eq. (6.30).
Ku	Kuwabara hydrodynamic factor, Eq. (11.2).
L	nozzle length of a rectangular jet impactor, Eq. (3.30); length of a sampling tube, Eq. (6.18).
m	particle mass.
$\dot{m}$	displaced air mass rate by a moving sphere, Eq. (2.1).
m_i	mass of air in zone i, Eq. (8.1).
m_i^α	mass of pollutant α in zone i, Eq. (8.1).
m_p	current mass of a droplet, Eq. (8.37).
$m_{p,0}$	initial mass of a droplet, Eq. (8.37).
M	molecular weight.
$M_{w,v}$	molecular weight of vapor, Eq. (8.41).
n	particle number concentration, Eq. (5.3); number of puffs during the smoking or vaping session, Eq. (7.1); number of copiers, Eq. (7.2); exponent, Eq. (8.5); number of elementary charges, Eq. (11.29).
n_0	initial particle number concentration, Eq. (5.19).
n_i	number of particles in the ith bin of size, Eq. (1.2).
N	total number of particles, Eq. (1.3); particle number concentration, Eq. (11.18).
N_i	total number of droplets or droplet nuclei, Eq. (10.2).
N_{in}	number of particles flowing into the crack, Eq. (7.4).
$N_{penetrated}$	number of particles penetrated through the crack, Eq. (7.4).
N_s	quantity of pathogen-laden aerosols inhaled by the susceptible persons, Eq. (10.1).
$N(t)$	cumulative number of particles deposited per unit area from time 0 to t, Eq. (5.23).
N_v	flux of vapor transfer, Eq. (8.38).

p	pressure, Eq. (2.A.5); pulmonary ventilation rate of a susceptible person, Eq. (10.1).
p_0	ambient pressure, Eq. (2.A.5); osmotic pressure, Eq. (5.3).
p_d	saturated vapor pressure on the droplet surface, Eq. (6.29).
p_s	saturation vapor pressure on a flat liquid surface, Eq. (6.29); breathing rate of the source patient, Eq. (10.4).
$p_{sat}(T_p)$	saturated vapor pressure at temperature T_p, Eq. (8.39).
P	pressure, Eq. (2.22); infection probability, Eqs. (10.1) and (10.7); particle penetration ratio, Eqs. (6.19), (7.4), (7.5), and (11.23).
Pe	Peclet number, Eq. (11.8).
P_D	particle penetration ratio associated with Brownian diffusion, Eq. (7.5).
P_G	particle penetration ratio associated with gravitational settling, Eq. (7.5).
P_i	air pressure at zone i, Eq. (8.4).
P_I	particle penetration ratio associated with inertial impaction, Eq. (7.5).
P_j	pressure at zone j, Eq. (8.4).
Pr	Prandtl number (molecular).
Pr_t	turbulent Prandtl number.
q	amount of charge, Eqs. (4.2) and (11.14); quanta produced by each source patient, Eq. (10.1).
Q	flow rate; room ventilation rate, Eq. (10.7).
Q_0	duct flow rate, Eq. (6.2); nominal flow rate, Eq. (11.39).
Q_{CADR}	clean air delivery rate of an air cleaner, Eq. (11.36).
Q_s	sampling flow rate of a particle sampler, Eq. (6.2).
r	radial position, Eq. (2.A.3).
r_{max}	maximum distance from the droplet surface to its centroid, Eq. (8.35).
R	dimensionless parameter related to single-fiber particle collection efficiency, Eq. (11.1).
Re	Reynolds number.
Re_0	initial Reynolds number of a particle, Eq. (3.20).
Re_p	particle Reynolds number, Eq. (8.24).
R_i^α	removal rate of pollutant α in zone i, Eq. (8.1).
R_u	universal gas constant.
R_ε	additional source term in the turbulent dissipation rate equation.
S	rate of the strain in the turbulent dissipation rate equation; stopping distance, Eq. (3.18).
Sc	Schmidt number.
S_C	particle source term in the drift-flux model, Eq. (8.21); source term of tracer gas concentration.

S_i	source term of energy equation; source term of the ith phase fluid, Eq. (8.29).
S_{ij}	strain rate tensor in the turbulent dissipation rate equation.
S_k	source term of the turbulent kinetic energy equation.
S_P	source term at grid cell P, Eq. (8.20).
Stk	Stokes number, Eq. (3.23).
Stk_{50}	Stokes number for 50% particle collection efficiency, Eq. (3.28).
S_W	particle specific surface area, Eq. (1.1).
S_ε	source term of the turbulent dissipation rate equation.
S_ϕ	source term of scalar variable ϕ, Eq. (8.6).
t	time; thickness of the filter medium, Eq. (11.22).
T	temperature.
T_0	reference temperature.
T_{bp}	boiling point, Eq. (8.36).
T_p	droplet temperature, Eq. (8.36).
T_{vap}	vaporization temperature, Eq. (8.36).
T_∞	local bulk temperature of the gas, Eq. (8.40).
u_i	velocity vector of the ith grid cell, Eq. (8.33).
u_i'	instantaneous fluctuating velocity, Eq. (8.28).
$\overline{u}_i$	time-averaged velocity.
u_j	velocity component in x_j direction, Eq. (8.6).
u_p	velocity vector of a droplet, Eq. (8.33).
$u_{p,i}$	particle velocity component in x_i direction, Eq. (8.23).
U	gas velocity; filtration velocity, Eq. (11.16).
U_0	free stream velocity, Eq. (6.1); face velocity, Eq. (11.15).
U_j	air velocity component in x_j direction, Eq. (8.21).
U_s	gas velocity in the sampling probe, Eq. (6.1).
v	molar volume of the liquid, Eq. (6.29).
V	velocity of a particle.
V_0	initial velocity of a particle, Eq. (3.15).
V_c	critical velocity for particle bounce, Eq. (4.8).
V_{Dep}	particle deposition velocity, Eqs. (5.24) and (8.22).
V_f	final velocity of a particle, Eq. (3.15).
V_i	volume of the ith grid cell, Eq. (8.32).
V_p	particle volume; volume of the identified detached droplet, Eq. (8.31).
V_r	gas velocity in r direction, Eq. (2.A.3).
V_s	particle settling velocity, Eq. (8.22).
$V_{s,j}$	particle settling velocity in x_j direction, Eq. (8.21).
$V(t)$	particle velocity at time t, Eq. (3.6).
V_T	terminal velocity, Eq. (3.1).
$\mathbf{V}_{TE}$	terminal electrostatic velocity (vector), Eq. (11.31).
V_{TE}	terminal electrostatic velocity (scalar), Eq. (11.33).

V_{TF}	terminal velocity under constant external force F, Eq. (3.5).
V_{Th}	thermophoretic velocity, Eq. (5.26).
V_{TS}	terminal settling velocity, Eq. (2.13).
V_{∞}	fluid velocity far from the particle, Eq. (2.A.3).
V_{θ}	gas velocity in θ direction, Eq. (2.A.4).
W	nozzle width of a rectangular jet impactor, Eq. (3.30).
x	separation distance, Eq. (4.1); distance from wall, Eq. (5.19).
$\bar{x}$	root mean square average displacement of particles, Eq. (5.13).
x_{BM}	Brownian displacement.
x_{Grav}	displacement due to gravitational settling.
x_i	coordinate of the ith grid cell, Eq. (8.32).
x_j	coordinate in a Cartesian coordinate system ($j = 1, 2, 3$), Eq. (8.6).
x_{p}	droplet position, Eq. (8.32).
$x(t)$	position of a particle at time t in the X axis, Eq. (3.16).
X	deflection displacement of the micro-cantilever, Eq. (4.6).
X_{v}	local bulk mole fraction of vapor, Eq. (8.40).
$y(t)$	position of a particle at time t in the Y axis, Eq. (3.22).
$Y_{\mathrm{v,s}}$	vapor mass fraction at droplet surface, Eq. (8.43).
$Y_{\mathrm{v},\infty}$	vapor mass fraction in bulk gas, Eq. (8.43).

Greek Symbols

α	a factor to prevent local extrema during reconstruction of gradient, Eq. (8.14); volume fraction of fluid in the multiphase flow, Eq. (8.29); solidity or volume fraction of solid fiber material in the filter, Eq. (11.2).
β	thermal expansion coefficient; correction factor to Stokes's law, Eq. (2.30); deposition fraction of infectious particles in the alveolar region, Eq. (10.7).
γ	surface tension, Eqs. (4.4) and (6.29).
Γ_f	diffusion coefficient at face f, Eq. (8.16).
$\Gamma_{\phi,\mathrm{eff}}$	effective diffusion coefficient for scalar variable ϕ, Eq. (8.6).
δ	wall thickness of a particle sampling head; thickness of the boundary layer, Eq. (6.20).
δ_{g}	geometric standard deviation of particle size distribution, Eq. (1.7).
δy	distance between the grid cell center and the wall, Eq. (8.22).
∂	partial derivative.
Δ	radial displacement of a particle, Eq. (3.26).
Δp	filter pressure drop, Eq. (11.25a).
ΔV	grid cell volume, Eq. (8.8).
∇T	temperature gradient, Eq. (5.25).

$\overline{\nabla\phi}$	average gradient of a scalar variable for two adjacent grid cells, Eq. (8.17).
$\nabla\phi_{c_0}$	gradient of a scalar variable at grid cell c_0, Eq. (8.17).
$\nabla\phi_{c_1}$	gradient of a scalar variable at grid cell c_1, Eq. (8.17).
$\nabla\phi_{\mathrm{upwind}}$	gradient of a scalar variable at the upwind cell, Eq. (8.13).
ε	turbulent dissipation rate.
ε_0	vacuum dielectric constant, Eqs. (4.3) and (11.14).
ε_{f}	relative permittivity of the fiber, Eq. (11.14).
ε_{p}	particle eddy diffusivity, Eq. (8.21); particle emissivity, Eq. (8.44).
η	dynamic viscosity of gas or air; dimensionless parameter in the turbulent dissipation rate equation.
η_0	model constant in the turbulent dissipation rate equation.
η_j^α	filtration efficiency of pollutant α on the flow path from zone j to zone i, Eq. (8.1).
θ	polar angle of the spherical coordinate system, Eq. (2.A.3).
θ_{R}	mean radiation temperature, Eq. (8.44).
$\kappa^{\alpha\beta}$	kinetic reaction coefficient of substances α and β, Eq. (8.1).
λ	gas mean free path, Eq. (2.19).
λ_{p}	particle mean free path, Eq. (5.11).
λ_{UV}	inactivation rate coefficient of UV irradiation, Eq. (10.6).
μ	dimensionless parameter to account for diffusion loss in a tube, Eq. (6.18); dynamic viscosity of air or gas, Eq. (8.23).
μ_{t}	turbulent viscosity.
ξ	particle sampling efficiency, Eq. (6.5); Gaussian random number, Eq. (8.28).
ρ	fluid density, Eq. (8.6).
ρ_0	standard particle density (1,000 kg/m^3), Eq. (2.26).
ρ_{b}	nominal density of the bulk material, Eq. (2.26).
ρ_{g}	gas or air density.
ρ_{p}	particle density.
σ	Stefan–Boltzmann constant, Eq. (8.44).
σ_k	Schmidt number for turbulent kinetic energy.
σ_ε	Schmidt number for turbulent dissipation rate.
τ	relaxation time of a particle, Eq. (3.3).
ϕ	azimuthal angle of the spherical coordinate system, Eq. (2.A.7); scalar variable, Eq. (8.6).
ϕ_f	scalar variable value at face f, Eq. (8.12).
ϕ_{upwind}	scalar variable value at an upwind cell, Eq. (8.12).
χ	dynamic shape factor, Eq. (2.23).
Ψ_{p}	sphericity of a droplet, Eq. (8.35).

About the Companion Website

This book is accompanied by a companion website:

www.wiley.com/go/Zhang/IndoorAerosolDynamics

This website include:

- Electronic sources

Scan this QR code to visit the companion website.

1

Aerosol Fundamentals

1.1 Introduction

Aerosol dynamics is the science of the formation/emission, movement, sedimentation, and interaction of aerosol particles, and their removal from the airborne transmission route. The aerodynamic behaviors of particles are subject to their basic physicochemical properties. In this chapter, the fundamentals of aerosols, including their definition, origin, basic physical and chemical properties, as well as size distribution, are addressed. Additionally, bioaerosols, which are closely related to airborne diseases and allergic symptoms, are briefly introduced.

1.2 Definitions of Aerosols

Aerosols are tiny solid or liquid particles or their hybrid suspended in a gaseous or air medium, and the size of these particles is larger than 0.001 μm but smaller than 100 μm. Particles dispersed in the air are difficult to classify in a scientifically rigorous way. They are usually described in terms of their physical form or source. Table 1.1 provides some common descriptions of aerosols.

1.3 Sources of Aerosols

The sources of atmospheric aerosols encompass the natural and the anthropogenic origins. Among the global emissions of atmospheric aerosols, the contribution of the natural sources is approximately 6.8 times that from the anthropogenic sources, as shown in Table 1.2. However, the anthropogenic aerosols are mainly emitted in the dense urban areas, resulting in higher local concentrations and thus posing greater risks to human health.

Table 1.1 Common descriptions of aerosols.

	Definition	Size range
Aerosol	An assembly of liquid or solid particles or their hybrid suspended in a gaseous medium consisting of particles and the suspending gas	0.001–100 μm
Particle	Small discrete object	0.001–100 μm
Dust	Solid particles formed by crushing or other mechanical breakage of parent materials	Sub-micrometer to more than 100 μm
Fog and mist	Liquid aerosol formed by condensation or atomization	Sub-micrometer to about 100 μm
Fume	Particles from vapor condensation and the possibly subsequent agglomeration	Less than 0.05 μm
Smog	Solid and liquid aerosols from photochemical reactions or vapor condensation	Less than 1 or 2 μm
Smoke	Solid or liquid aerosols from incomplete combustion or condensation of supersaturated vapor	Less than 1 μm

1.3.1 Natural Aerosol Sources

Natural emissions of aerosols are associated with various physical and chemical processes. The main natural sources of atmospheric aerosols are soil/desert dust and sea salt, as shown in Table 1.2. When seawater waves break, tiny droplets of water are released into the air. As these droplets evaporate, salt particles are formed. The gas-to-particle conversion and the photochemical reactions contribute a significant amount of secondary aerosols. More details on the secondary aerosols are addressed in Section 1.3.3. Additional aerosol contributors include botanical debris and volcanic eruptions. The volcanic eruptions release quantities of ash, dust, and sulfate aerosols into the atmosphere. Forest fires also contribute to aerosol emissions, especially severe wildfires.

1.3.2 Anthropogenic Aerosol Sources

Anthropogenic sources of aerosols contribute substantially to air pollution. As shown in Table 1.2, the majority of the anthropogenic aerosols come from the gas-to-particle conversion and the direct emissions by various human activities. Industrial processes, including manufacturing and construction, can generate aerosols in the form of dust, smog, mist, etc., due to grinding, disturbance, combustion, or the release of chemical pollutants. Transportation is also a major source, as fuel-powered vehicles like cars, trucks, and buses release fine

Table 1.2 Relative contributions of natural and anthropogenic sources to the global emissions of atmospheric aerosols.

Source	Contribution to global emissions in mass percentage (%)
Natural	
Soil/desert dust	41.67
Sea salt	36.11
Gas-to-particle conversion[a)]	5.00
Photochemical reactions[b)]	1.67
Botanical debris	1.39
Volcanic eruption	0.83
Forest fires	0.56
Total of the natural sources	87.23
Anthropogenic	
Gas-to-particle conversion[c)]	9.17
Direct emissions (e.g., soot or mechanical friction)	3.33
Photochemical reactions[d)]	0.28
Total of the anthropogenic sources	12.78

a) Including sulfates from SO_2 and H_2S, nitrate from NO_x, ammonium salts from NH_3.
b) Photochemical particle formation primarily from isoprene and monoterpenes vapors from trees.
c) Including sulfates from SO_2 and nitrates from NO_x.
d) Photochemical particle formation primarily from anthropogenic volatile organic compounds.
Source: Hinds and Zhu (2022) / with permission of John Wiley & Sons.

particles and gases into the atmosphere. Additionally, agricultural activities such as tilling and plowing generate dust aerosols. Household activities like heating, cooking, and cleaning can also produce aerosol particles. The burning of solid fuels, such as wood or coal, releases smoke, soot, and combustion byproducts. Physiological activities such as breathing, speaking, coughing, or sneezing can release droplets and aerosols into the surrounding environment, though the amounts are quite limited.

1.3.3 Secondary Aerosols

The atmospheric aerosols of both natural and anthropogenic origins involve secondary aerosols produced by the gas-to-particle conversion and the photochemical reactions. Secondary aerosols are particles formed through chemical reactions from precursor gases such as sulfur dioxide (SO_2), nitrogen oxides (NO_x), volatile organic compounds (VOCs), and ammonia (NH_3). These reactions result in the formation of new particles or the growth of existing particles. The particle size of secondary aerosols is generally in the range of 0.001–2.5 μm (Kavouras and Stephanou 2002), and they are the main source of fine particulate matter ($PM_{2.5}$) in the atmosphere.

1.4 Physical Properties of Aerosol Particles

The physical properties of aerosol particles, including particle size, density, shape, surface area, concentration, and charge, can significantly affect the aerodynamic behavior of the particles. Understanding these fundamental properties and appropriately characterizing them are crucial for indoor aerosol control.

1.4.1 Particle Size

The size of aerosol particles plays an essential role in their behavior, as nearly all dynamic characteristics of aerosols are influenced by particle size. For example, the duration of aerosol particles' suspension in the air and the penetration ratio of inhaled aerosol particles into the lungs are both strongly dependent on the particle size. Aerosol particles exhibit a wide size range, and the sizes of the smallest and largest particles can differ by a factor of up to 100 or even more. This wide range highlights the importance of using appropriate methods to precisely characterize the particle size distribution, as will be discussed in detail in Section 1.7.

Particle size is typically measured in micrometers (μm), and 1 μm is equal to one-millionth of a meter. A particle with a diameter of 1 μm is extremely small, approximately 1/70 the diameter of a human hair. Although some very tiny particles are measured in nanometers (nm), which are even smaller, micrometers are more commonly used when discussing particle size. Aerosol particles are generally classified into three main categories based on size, namely, ultrafine, fine, and coarse particles.

Ultrafine particles (UFPs) are aerosol particles with a diameter less than 0.1 μm (100 nm). Owing to their extremely small size, they exhibit unique dynamic behavior compared to larger particles. For instance, UFPs can stay suspended in the air for long periods and are capable of traveling long distances. Additionally, UFPs pose the greatest threat to human health due to their ability to penetrate epithelial cells and enter the bloodstream. Fine particles are aerosol particles with diameters ranging from 0.1 to 2.5 μm, and they constitute the largest mass ratio of $PM_{2.5}$ (aerosol particles equal to or less than 2.5 μm in diameter). Because of their small size, fine particles can remain airborne for prolonged periods and be transported over long distances by the wind. These particles can be inhaled into

Table 1.3 Size ranges of common indoor aerosol particles.

Category	Emission source	Size range (μm)	References
Combustion	Candle	0.02–1	Afshari et al. (2005)
	Cigarette		
Cooking	Cooking on a gas stove	0.05–0.1	Dennekamp et al. (2001)
	Cooking on an electric stove		
	Frying meat	0.02–1	Afshari et al. (2005)
Household use	Flat iron with steam	0.02–1	Afshari et al. (2005)
	Motor of a vacuum cleaner		
Office equipment	3D printer	0.001–0.1	EPA (2024)
	Laser printer	0.005–0.3	Gu et al. (2020)

the lungs, leading to significant health consequences. Coarse particles, meanwhile, are aerosol particles with a diameter ranging from 2.5 to 10 μm. Due to their larger size, coarse particles tend to settle out of the air under the action of gravity.

Table 1.3 summarizes the size ranges of common indoor aerosol particles generated by different sources. As displayed in the table, the sizes of indoor aerosol particles are generally less than 1 μm. Among them, the particles produced by 3D printing are the smallest, and all are UFPs. The size range of particles produced by laser printing spans from 0.005 to 0.3 μm.

1.4.2 Particle Density

Particle density refers to the mass of particles per unit volume and is typically expressed in kilograms per cubic meter (kg/m^3) or grams per cubic centimeter (g/cm^3). It should be noted that particle density refers specifically to the density of the particles themselves, rather than the overall aerosol density that includes the volume of the suspending air. The density of particles plays a crucial role in their motion in the air. For instance, gravity acts as an external force on particles and is proportional to their density. Particles with a higher density correspond to a stronger inertial force during motion.

Different types of particles exhibit varying densities. For solid or liquid particles, their densities are equivalent to those of their parent materials. However, particles such as smoke or fog may have lower densities than the substances that form them. This is because there may be numerous tiny air pockets inside these particles.

Table 1.4 lists the densities of some common indoor aerosol particles. It is evident that the densities of all particles are generally around 1,000 kg/m^3. Kitchen smoke particles are primarily composed of lipids, combustion byproducts

Table 1.4 Densities of common indoor aerosol particles.

Particle source	Particle name	Density (kg/m^3)	References
Combustion	Candle smoke	910	Pagelsa et al. (2009)
	Cigarette smoke	1,180	Johnson et al. (2014)
Test particles	DEHS particles[a)]	912	Nothhelfer et al. (2025)
	DOP particles[a)]	986	Chen et al. (1990)
Cooking	Cooking fumes	910–1,030	Pikmann et al. (2024)
Material/process emission	3D printing particles	1,105	Matsumura et al. (2023)

a) DEHS and DOP particles refer to aerosol particles generated using diethylhexyl sebacate and dioctyl phthalate, respectively.

(such as carbonaceous particles), moisture, and other organic materials, and the density of these particles depends on the content of each component. In this book, unless otherwise specified, the particle density is assumed to be 1,000 kg/m^3, which is the standard density.

1.4.3 Particle Shape

Liquid aerosol particles tend to be round or spherical because of surface tension. Meanwhile, the shapes of solid aerosol particles can be highly irregular and more complex, for example, with jagged edges or elongated forms, as shown in Figure 1.1. For the sake of convenience in discussing aerosol properties, irregularly shaped particles are often assumed to be spherical, even if they are not. This is mainly because many of the equations and theories describing aerosol dynamic behavior are more straightforward for spherical particles. For non-spherical particles, the corresponding "equivalent diameter" can be used,

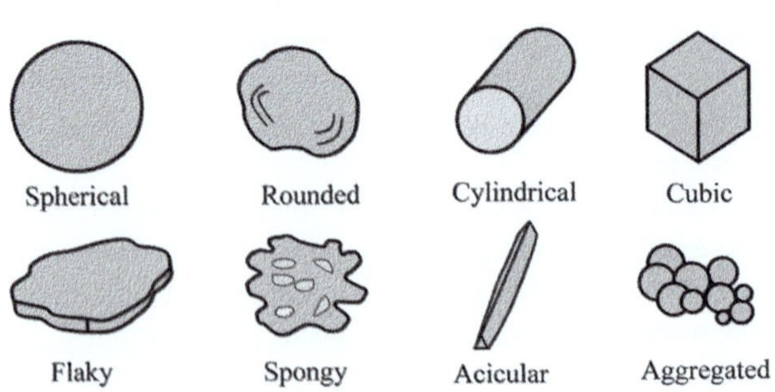

Figure 1.1 Typical shapes of aerosol particles.

Table 1.5 Common equivalent diameters used in aerosol aerodynamics.[a)]

Equivalent diameter	Definition	Mathematical expression
Average diameter	The average diameter of particles measured in n given directions	$d = \frac{1}{n}\sum_{i=1}^{n} d_i$
Surface area diameter	The diameter of a sphere with the same surface area as the studied particle	$d = \sqrt{\frac{A_s}{\pi}}$
Volume diameter	The diameter of a sphere with the same volume as the studied particle	$d = \sqrt[3]{\frac{6V_p}{\pi}}$
Mass diameter	The diameter of a sphere with the same mass and density as the studied particle	$d = \sqrt[3]{\frac{6m}{\pi\rho_p}}$
Aerodynamic diameter	The diameter of a sphere with a density of 1,000 kg/m^3, having the same terminal settling velocity as the studied particle	$d = \sqrt{\frac{18V_{TS}\eta}{1,000g}}$
Stokes diameter	The diameter of a sphere having the same density and settling velocity as the studied particle	$d = \sqrt{\frac{18V_{TS}\eta}{\rho_p g}}$

a) In this table, A_s is the surface of a particle, V_p is the particle volume, m is the particle mass, ρ_p denotes particle density, V_{TS} is the terminal settling velocity of a particle, η is the dynamic viscosity of air, and g is the gravitational acceleration.

which is the diameter of a spherical particle that has the same physical properties, like volume or surface area, as the irregularly shaped particle. Table 1.5 defines some common equivalent diameters used in aerosol aerodynamic studies. The aerodynamic and Stokes diameters mentioned in Table 1.5 are described in detail in Chapter 2.

The use of equivalent diameter aids in the analysis of the dynamic properties of particles without greatly impairing accuracy. In most cases, ignoring the exact shape of a particle does not significantly affect the outcome (Hinds and Zhu 2022). However, for particles with very unusual shapes, like long, thin fibers or highly branched structures found in some types of smoke, the impact of shape becomes more evident. In such situations, these particles can be characterized by their fractal dimension.

1.4.4 Particle Surface Area

Surface area is defined as the total area of the particle's outer surface. In general, the surface area of fine particles with equal mass is greater than that of coarse particles. For instance, a cube with a side length of 1 cm has a surface area of 6×1^2 cm^2, while a smaller cube with a side length of $1/n$ cm has a surface area of $6 \times (1/n)^2$ cm^2. In a cube with a side length of 1 cm, there are n^3 such smaller cubes, resulting in a total surface area of $6n$ cm^2, which indicates that the surface area has increased by a factor of n. The surface area of particles influences not only their settling velocity but also the health effects of aerosols that deposit in the alveolar region because the contact area of the aerosols with the tissue cells is a critical parameter affecting particle toxicity (Ruzer and Harley 2012).

In addition to the particle surface area, the concept of the specific surface area is widely used in aerosol studies, and it is defined as the total surface area per unit mass of particles. Considering a spherical particle, the relationship between the specific surface area S_w (m^2/kg) and particle diameter is given by:

$$S_\mathrm{w} = \frac{\pi d_\mathrm{p}^2}{\frac{1}{6}\pi d_\mathrm{p}^3 \rho_\mathrm{p}} = \frac{6}{d_\mathrm{p}\rho_\mathrm{p}} \tag{1.1}$$

where ρ_p is the particle density (kg/m^3) and d_p is the particle diameter (m).

In Eq. (1.1), it can be seen that the specific surface area of particles is inversely proportional to their size, with smaller particles having larger specific surface areas. For particles with identical density, the specific surface area of a particle with a size of 0.1 μm is 100 times that of a particle with a size of 10 μm. Additionally, the specific surface area is influenced by factors such as particle porosity, pore size distribution, shape, and surface roughness. As the specific surface area of particles increases, their surface energy also rises, which enhances the surface reactivity. Furthermore, this parameter is crucial for understanding the coagulation, adhesion, and adsorption behavior of aerosol particles.

1.4.5 Particle Concentration

Aerosol particle concentration indicates the amount of particulate matter present in the air. The particle concentration can be expressed in terms of mass concentration or number concentration. Mass concentration refers to the total mass of particles present in a specific volume of air and is typically expressed in units such as grams per cubic meter (g/m^3), milligrams per cubic meter (mg/m^3), or micrograms per cubic meter (μg/m^3). The particle mass concentration serves as a key indicator of air pollution levels and potential health risks. For instance, the World Health Organization (WHO), the US Environmental Protection Agency (EPA), and the Chinese national standard (GB/T 18883-2022) have established exposure limits of $PM_{2.5}$ and PM_{10} based on the particle mass concentration, as shown in Table 1.6.

The particle number concentration indicates the number of particles present in a given volume of air, typically expressed in units like particles per cubic centimeter

Table 1.6 $PM_{2.5}$ and PM_{10} concentration limits.

Organization/entity	$PM_{2.5}$ (μg/m³)	PM_{10} (μg/m³)	Comment
WHO[a)]	15	45	24-hour average
EPA	35	150	
Chinese national standard	50	100	

a) Only the air quality guideline (AQG) levels by the WHO are presented.

Table 1.7 Comparison of particle concentration when expressed in different units.

Item	Mass concentration (μg/m³)	Parts per million (volume/volume) (ppm)	Parts per million (mass/mass) (ppm)
Indoor $PM_{2.5}$ concentration limit specified by the WHO	15	1.5×10^{-5}	0.01
Indoor PM_{10} concentration limit specified by the WHO	45	4.5×10^{-5}	0.04

(particles/cm³) or particles per cubic meter (particles/m³). This metric is especially useful for quantifying specific types of particles, such as biological aerosols or fibers, where the particle count is more significant than the total particle mass.

The volume ratio or mass ratio in parts per million (ppm), i.e., the ratio of the particle volume or mass to the suspending gas medium's volume or mass, is not employed for aerosol concentration because the use of this unit generally yields very small values, especially for the volume ratio. Table 1.7 compares the quantities expressed in various units. It is evident that when volume ratios are employed, the particle concentration values are six orders of magnitude lower than the mass concentration of the particles.

1.4.6 Electrical Properties of Aerosol Particles

The electrical properties of aerosol particles are mainly reflected in the magnitude and polarity of their charges. Almost all particles, whether generated naturally or anthropogenically, carry some charges. For most particles, the magnitudes of positive and negative charges are approximately equal, leading to the overall electrical neutrality of particles suspended in the air.

The charging of particles is primarily attributed to collisions between particles, natural radiation, attachment of external particles or electrons, etc. The magnitude of the particle charge varies with the particle size, mass, humidity, temperature, etc. In general, larger particles can carry more charges than smaller particles due to the greater surface area of the former, which provides more opportunities to acquire and retain charges from surrounding ions. Higher humidity, for example, can lead to the dissipation of charges through the condensation of water vapor.

The charges carried by particles significantly affect their dynamic behavior. Charged particles are subjected to electrostatic forces when exposed to an electric field, which alters their trajectories. This principle is widely utilized in electrostatic precipitators. Additionally, the polarity of the charges on aerosol particles plays a crucial role in their agglomeration and coagulation behavior. Particles with opposite charges attract each other, promoting the agglomeration and formation of larger particles, which settle more quickly in the air. In contrast, particles with the same charges repel each other, inhibiting agglomeration and prolonging their suspension time in the air. Furthermore, the electrostatic attraction between charged particles and approaching surfaces influences deposition behavior. When charged particles pass through a surface with an opposite charge, they are strongly attracted, leading to an increased deposition velocity and deposition rate. For instance, the charging of particles with diameters of 0.5, 1.0, and 2.0 μm was claimed to increase their deposition velocity by up to 72% in indoor space (Din et al. 2020).

1.5 Chemical Properties

Aerosol particles may be composed of various chemical substances, including organic compounds, inorganic salts, metals, and biological materials. Aerosols can be acidic, neutral, or alkaline, depending on their chemical compositions. Furthermore, aerosols can be divided into soluble aerosols and insoluble aerosols, depending on whether they can dissolve in water or other solutions. Soluble aerosols, such as salts, can dissolve in the moisture present in the air, leading to probable chemical reactions or changes in the aerosols' size and dynamic behavior. Insoluble aerosols, such as certain types of dust, are generally unaffected by humidity, which well maintains their original shape and size. In indoor environments, the formation of certain aerosols is even attributed to chemical reactions. For instance, secondary organic aerosols (SOA) may arise from chemical reactions between ozone and VOCs emitted by cleaning agents, paints, and building materials. These reactions contribute to an increase in particle concentrations indoors.

1.6 Bioaerosol

When we discuss aerosols, our minds often go immediately to dust or smoke. However, there are also large numbers of bioaerosols in the air. The definition and sources of bioaerosols are briefly introduced in the following sections.

1.6.1 Definition

Bioaerosols are diverse particles derived from living organisms, including pollens, fragments of fungal mycelium or spores, bacteria and spores, viruses, protozoa, excrement or fragments of insects, and hair or dander from mammals, along with other residues or byproducts from these organisms. Bioaerosols are widely distributed in both indoor and outdoor air. Among particles larger than 0.2 μm, approximately 5–50% originate from biological particles (Ruzer and Harley 2012). The commonly used units for characterizing the concentration of bioaerosols are particles/m^3 and CFU/m^3, where CFU/m^3 represents the number of viable bacteria or fungi in the air that can form colonies after being cultured.

The size range of bioaerosols is very wide. Viruses are the smallest live bioaerosols, with sizes ranging from 0.02 to 0.3 μm. The size range of bacteria is typically between 0.25 and 10 μm, the size range of fungi is between 1 and 60 μm, while the diameters of pollen range from a few micrometers to around 100 μm. The size ranges and sources of common bioaerosols are presented in Table 1.8.

Although most bioaerosols in the air are harmless, some bioaerosol particles may be infectious or allergenic or carry toxic components or metabolites. Living organisms may reproduce, but this is not a prerequisite for allergenic or toxic effects. Dead cells or cell remnants may also have an impact on human health.

Table 1.8 A summary of common bioaerosols in terms of size ranges and sources.

Bioaerosol	Particle size range (μm)	Source	References
Virus	0.02–0.3	Infected individuals, body fluids, sewage, etc.	Louten (2016)
Bacterium	0.25–10	Respiratory tracts, skin, soil, feces, wastewater, etc.	Aryal (2022)
Fungus	1–60	Decaying plants, soil, wastewater, and damp environments	Golan and Pringle (2017)
Pollen	5–100	Flowers, trees, and grasses	Ryan (1999)
Protozoa	1–50	Soil, water bodies, and contaminated environments	Baron (1996)
Dust mite debris	0.5–50	Household dust, bedding, and carpets	Ryan (1999)
Hair	70–90	Humans and animals	Loussouarn et al. (2016)
Dander	0.4–100	Human and animal skin shedding	Ryan (1999)

1.6.2 Sources of Bioaerosols

A portion of bioaerosols originate from outdoor sources such as soil, surface water, and plants. Microorganisms in soil can be aerosolized due to mechanical disturbances like farming activities, soil excavation, landfill operations, and dust storms. Bacteria are the most abundant organisms in surface soil, with concentrations exceeding 10^{10} cells/g (Maier et al. 2008). Many natural and artificial water bodies, such as sewage oxidation ponds and cooling water systems, contain large quantities of microorganisms, including Gram-negative bacteria and actinomycetes. These microorganisms can form bioaerosols through bubble bursts when subjected to disturbances. Some plants release pollen in order to reproduce. The fungal spores or bacteria on the surfaces of living and dead plants can also be aerosolized due to high-speed airflow sweeping.

The processing of organic materials in many occupational environments can become a high-intensity source of bioaerosols; these materials include plants, organic fertilizers, wastewater, and contaminated liquids. In both industrial and non-industrial environments, microorganisms growing in a building's HVAC (heating, ventilating, and air conditioning) system or the structure of the building itself can also be specific bioaerosol sources. In general, high humidity is a prerequisite for microbial growth. Building interiors with water leakage or condensation can become sources of fungal and other biological aerosols.

In non-industrial indoor environments such as office and residential buildings, the main source of bioaerosols, especially for bacteria and viruses, is humans. High concentrations of bacteria in indoor air may not necessarily pose a health hazard, but they are a sign of poor ventilation of the interior environment. Culturable bacteria and fungi (above 5,000 and 500 CFU/m^3, respectively) have been recommended as indicators of the presence of abnormal indoor bioaerosol sources or insufficient ventilation (Reponen et al. 1992).

1.7 Size Distribution of Aerosol Particles

Aerosols can be divided into monodisperse aerosols and polydisperse aerosols. Among the monodisperse aerosols, the size of the particles can be described in terms of the particle diameter alone. However, monodisperse aerosols can only be generated in the laboratory. Most aerosols are polydisperse, with particle sizes spanning several orders of magnitude. Given the wide variation in particle size and the strong influence of size on the aerodynamic properties of aerosols, it is essential to use statistical methods to characterize these size distributions.

1.7.1 Overview of Particle Sizes

Particle size is commonly characterized in terms of mean, mode, median, and geometric mean size. The mean, or arithmetic average, is obtained by summing all particle sizes and then dividing by the total number of particles. The algebraic mean diameter can be calculated as:

$$\bar{d}_\text{p} = \frac{\sum n_i d_i}{\sum n_i} = \int_0^\infty d_\text{p} f(d_\text{p}) \text{d}d_\text{p} \tag{1.2}$$

where n_i is the number of particles in the ith group, d_i is the algebraic mean of the upper and lower limits of the particle size range or the square root of the product of the upper and lower limits of the particle size range, and $f(d_\text{p})$ is the frequency function.

The mode is the particle size that appears most frequently. The median represents the midpoint in the distribution, where half of the particles are smaller and half are larger than the median size. The geometric mean d_g is calculated as the Nth root of the product of N particles:

$$d_\text{g} = (d_1 d_2 d_3 \cdots d_N)^{1/N} \tag{1.3}$$

For size data grouped into M intervals, the geometric mean involves the characteristic diameters and the number of particles in each interval:

$$d_\text{g} = \left(d_1^{n_1} d_2^{n_2} d_3^{n_3} \cdots d_M^{n_M}\right)^{1/N} \tag{1.4}$$

where n_1, n_2, n_3, …, n_M are the number of particles in each interval, and d_1, d_2, d_3, …, d_M are the midpoints or characteristic diameters in corresponding intervals.

The geometric mean diameter can also be expressed in terms of $\ln(d)$ by converting Eq. (1.4) to natural logarithms as:

$$\ln d_\text{g} = \frac{\sum n_i \ln d_i}{N} \tag{1.5}$$

$$d_\text{g} = \exp\left(\frac{\sum n_i \ln d_i}{N}\right) \tag{1.6}$$

1.7.2 Particle Size Distribution Functions

Although the size of aerosol particles can be expressed using the aforementioned characteristic quantities, it is more commonly expressed in functional forms in practical applications. In general, the distributions of particle size are random, but they can be approximated into certain patterns and thus represented by specific distribution functions. Among these, the most widely used is the log-normal distribution function. The log-normal distribution is derived from the normal distribution, which is introduced in the following sections.

1.7.2.1 Normal Distribution

The normal distribution, also known as the Gaussian distribution, is a probability distribution used to describe the sizes of particles when they cluster around a central value. In a normal distribution, most particles have sizes close to the mean, with a few particles having sizes that deviate significantly from it. The number distribution function of the normal distribution is given by:

$$\frac{dN}{d(d_p)} = \frac{N}{\sqrt{2\pi}\delta_g} \exp\left[-\frac{(d_p - \bar{d}_p)^2}{2\delta_g{}^2}\right] \tag{1.7}$$

where N is the total number of collected particles, d_p(μm) is the particle diameter, and $\bar{d}_p$ denotes the algebraic mean diameter. Meanwhile, δ_g represents the geometric standard deviation and can be expressed as:

$$\delta_g = \left(\frac{\sum n_i(d_p - \bar{d}_p)^2}{N-1}\right)^{1/2} \tag{1.8}$$

The value of δ_g is generally greater than 0, and it is used to describe the degree of divergence of aerosol particle sizes. A larger value of δ_g indicates a greater degree of dispersion among aerosol particles. If $\delta_g = 0$, the aerosol is classified as a monodisperse system, where all particles possess the same diameter.

The normal distribution function is rarely employed to describe aerosol particle size distributions, as most aerosols display a skewed distribution with a long tail at larger sizes. Notably, the normal distribution is symmetrical and is mainly applicable to monodisperse test aerosols, certain pollens and spores, specifically prepared polystyrene latex spheres, and so on.

1.7.2.2 Lognormal Distribution

As mentioned above, a few particle distributions conform to the normal distribution since particle distribution curves are generally skewed. This skewness can be eliminated by taking the logarithmic operation of the particle size, resulting in a lognormal distribution. The lognormal distribution has become one of the most prevalent statistical distributions in aerosol science. The lognormal distribution of particle sizes can be formulated as:

$$\frac{dN}{d(\log(d_p))} = \frac{N}{\sqrt{2\pi}\log(\delta_g)} \exp\left[-\frac{(\log(d_p) - \log(\bar{d}_p))^2}{2(\log(\delta_g))^2}\right] \tag{1.9}$$

where log is the base 10 or constant e logarithm. The terms $\log(\bar{d}_p)$ and $\log(\delta_g)$ can be formulated as:

$$\log(\bar{d}_p) = \frac{\sum n_i \log(d_i)}{N} \tag{1.10}$$

$$\log(\delta_g) = \left(\frac{\sum n_i(\log(d_i) - \log(\bar{d}_p))^2}{N-1}\right)^{1/2} \tag{1.11}$$

where n_i is the number of particles in size group i, having a midpoint of size d_i, and where $N = \sum n_i$. As compared with the normal distribution, the lognormal distribution replaces $\overline{d}_p$ with $\log(\overline{d}_p)$ and δ_g with $\log(\delta_g)$.

Example 1.1

The number of laboratory-generated oil mist particles collected in 1 ml of air by an electrostatic low-voltage impactor (ELPI+, Dekati, Finland) is shown in Table 1.9. The ELPI can measure particles between 6 nm and 10 μm, with 14 size intervals. The size of a particle in a specific size interval is equal to or greater than the lower limit and smaller than the upper limit of that interval. Please plot the lognormal distribution curve of the particle sizes.

Solution

To draw a lognormal distribution curve, three key parameters need to be determined, namely, $\log(\overline{d}_p)$, $\log(\delta_g)$, and N in Eq. (1.9). According to Eq. (1.10), $\log(\overline{d}_p)$ can be obtained (base 10 logarithm), with a value of −1.06. Then, according to Eq. (1.11), $\log(\delta_g)$ can be calculated, with a value of 0.54. The value of N is 187,882 from Table 1.9. The logarithmic normal distribution of oil mist particle sizes is illustrated in Figure 1.2.

Table 1.9 Number of oil mist particles collected in 1 ml of air in the lab.

Size range (μm)	Particle count	Midpoint of size range (μm)
0.006–0.016	102,302	0.011
0.016–0.027	31,704	0.0215
0.027–0.054	13,781	0.0405
0.054–0.093	10,528	0.0735
0.093–0.154	6,821	0.1235
0.154–0.26	6,275	0.207
0.26–0.379	5,390	0.3195
0.379–0.609	5,000	0.494
0.609–0.942	4,307	0.7755
0.942–1.59	1,024	1.266
1.59–2.38	500	1.985
2.38–3.97	124	3.175
3.97–6.64	88	5.305
6.64–10	38	8.32
0.006–10 (Total)	187,882	N/A

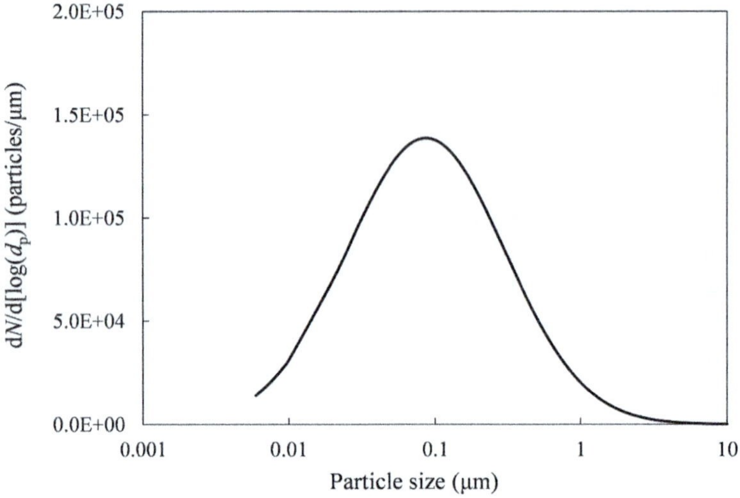

Figure 1.2 Logarithmic normal distribution of oil mist particle sizes.

1.8 Summary

This chapter mainly introduces the definition and sources of aerosols, the physical properties of aerosol particles (including particle size, shape, density, surface area, electrical charges, and aerosol concentration), and chemical properties, as well as the definition and sources of bioaerosols, and methods for charactering the size distribution of polydisperse aerosol particles. The preceding content is the basis for studying indoor aerosol dynamics and will be used often in subsequent chapters.

The particle size of aerosol particles is typically measured in micrometers. The particle size of indoor aerosols is commonly around 1 μm, and their density is generally more or less 1,000 kg/m^3. Due to the irregular shapes of particles, especially solid particles, equivalent particle size is often used as a characteristic parameter. The surface area of particles is a key factor in evaluating inhalation exposure, and fine particles of equal mass have a larger surface area than do coarse particles. Additionally, the specific surface area is frequently used to characterize particle surface activity, which affects particle aggregation and coagulation significantly. Aerosol concentration can be expressed in terms of mass concentration and number concentration. The former is mainly used to evaluate indoor air pollution, while the latter is mostly used to quantify biological aerosol particles or fiber particles. As for the size distributions of aerosol particles, they are often polydisperse and are typically described by a lognormal distribution function.

Problems

1.1 Calculate the aerodynamic diameter of an irregular particle with a terminal settling velocity of 0.001 m/s. Assume the air dynamic viscosity is 1.8×10^{-5} Pa · s.
[Answer: 5.8 μm]

1.2 The size distribution of aerosol particles is as follows. Please calculate the geometric mean diameter.

Size range (μm)	Particle count
0.1–0.3	50
0.3–0.5	30
0.5–1.0	20

[Answer: 0.3 μm]

1.3 An irregularly shaped particle has a volume of 2×10^{-18} m^3. Please calculate its volume-equivalent diameter.
[Answer: 1.6 μm]

1.4 A spherical aerosol particle has a diameter of 0.5 μm and a density of 1,200 kg/m^3. Please calculate its specific surface area.
[Answer: 10,000 m^2/kg]

1.5 Explain the possible mechanism to generate secondary organic aerosols (SOA) in indoor environments and discuss their impact on the indoor air quality.

1.6 Explain why the lognormal distribution is more commonly used than the normal distribution to describe the particle size distribution, and discuss the physical meaning of the geometric standard deviation in the lognormal distribution.

References

Afshari, A., Matson, U., and Ekberg, L.E. (2005). Characterization of indoor sources of fine and ultrafine particles: a study conducted in a full-scale chamber. *Indoor Air* 15: 141–150.

Aryal, S. (2022). Different size, shape and arrangement of bacterial cells. https://microbiologyinfo.com/different-size-shape-and-arrangement-of-bacterial-cells (accessed 5 September 2025).

Baron, S. (1996). *Medical Microbiology*, 4e. University of Texas Medical Branch at Galveston.

Chen, B.T., Cheng, Y.S., and Yeh, H.C. (1990). A study of density effect and droplet deformation in the TSI aerodynamic particle sizer. *Aerosol Science and Technology* 12: 278–285.

Dennekamp, M., Howarth, S., Dick, C. et al. (2001). Ultrafine particles and nitrogen oxides generated by gas and electric cooking. *Occupational and Environmental Medicine* 58: 511–516.

Din, T., McGrath, J.A., and Byrne, M.A. (2020). A test chamber investigation of the effect of charging on aerosol deposition on indoor surfaces. *Aerosol and Air Quality Research* 20: 2669–2680.

Golan, J. and Pringle, A. (2017). Long distance dispersal of fungi. *Microbiology Spectrum* 5: 1–24.

Gu, J., Karrasch, S., and Salthammer, T. (2020). Review of the characteristics and possible health effects of particles emitted from laser printing devices. *Indoor Air* 30: 396–421.

Hinds, W.C. (2012). *Aerosol Technology: Properties, Behavior, and Measurement of Airborne Particles*, 2e. Wiley.

Hinds, W.C. and Zhu, Y. (2022). *Aerosol Technology: Properties, Behavior, and Measurement of Airborne Particles*, 3e. Wiley.

Johnson, T.J., Olfert, J.S., Cabot, R. et al. (2014). Steady-state measurement of the effective particle density of cigarette smoke. *Journal of Aerosol Science* 75: 9–16.

Kavouras, I.G. and Stephanou, E.G. (2002). Particle size distribution of organic primary and secondary aerosol constituents in urban, background marine, and forest atmosphere. *Journal of Geophysical Research* 107: 366–378.

Loussouarn, G., Lozano, I., Panhard, S. et al. (2016). Diversity in human hair growth, diameter, colour and shape. An in vivo study on young adults from 24 different ethnic groups observed in the five continents. *European Journal of Dermatology* 26: 144–154.

Louten, J. (2016). Virus structure and classification. *Essential Human Virology* 6: 19–29.

Maier, R.M., Pepper, I.L., and Gerba, C.P. (2008). *Environmental Microbiology*, 2e. Academic Press.

Matsumura, S., Kawamura, S., Kandpal, L., and Vangla, P. (2023). 3D printed porous particle and its geotechnical properties. *Acta Geotechnica* 18: 5735–5753.

Nothhelfer, M., Todea, A.M., Bathen, D. et al. (2025). Performance evaluation of five different low-cost particulate matter sensors for monodisperse test aerosols. *Aerosol and Air Quality Research* 25: 1–17.

Pagelsa, J., Wierzbickaa, A., Nilsson, E. et al. (2009). Chemical composition and mass emission factors of candle smoke particles. *Journal of Aerosol Science* 40: 193–208.

Pikmann, J., Drewnick, F., Fachinger, F., and Borrmann, S. (2024). Particulate emissions from cooking: emission factors, emission dynamics, and mass spectrometric analysis for different cooking methods. *Atmospheric Chemistry and Physics* 24: 12295–12321.

Reponen, T., Nevalainen, A., Jantunen, M. et al. (1992). Normal range criteria for indoor air bacteria and fungal spores in a subarctic climate. *Indoor Air* 2: 26–31.

Ruzer, L.S. and Harley, N.H. (2012). *Aerosols Handbook: Measurement, Dosimetry, and Health Effects*, 2e. CPC Press.

Ryan, T. (1999). HEPA's the buzzword for vacuums. https://archives.starbulletin.com/1999/04/02/features/story2.html?utm_source=chatgpt.com (accessed 5 September 2025).

US Environmental Protection Agency (2024). 3D printing research at EPA. https://www.epa.gov/chemical-research/3d-printing-research-epa (accessed 5 September 2025).

2

Particle Motion Basics

2.1 Introduction

The uniform linear motion of aerosol particles represents the simplest and most fundamental mode of particle movement in aerosol dynamics. This type of uniform motion generally arises from the action of two forces: a constant external force, such as gravity or an electrical force, and the drag force exerted by the surrounding gas on the particle. In this chapter, the methods for calculating the drag force, along with some related details, are introduced.

2.2 Newton's Resistance Law

The calculation of particle drag force originates from Newton's resistance law, which was initially developed to determine the drag force on a cannonball moving through the air. This law is also applicable to other spherical objects, but the Reynolds number (Re) of the spherical object should be greater than 1,000. The derivation of Newton's resistance law for a sphere is described as follows.

When a sphere drops through still air, it must push aside the air in its path, as shown in Figure 2.1. For a sphere with diameter d, the displaced air volume rate is equal to the projected area of the sphere multiplied by its velocity V. The displaced air mass rate can thus be calculated as:

$$\dot{m} = \rho_g \frac{\pi}{4} d^2 V \tag{2.1}$$

where ρ_g is the air density.

The rate of change of the displaced air's momentum is:

$$\dot{m}V = \rho_g \frac{\pi}{4} d^2 V^2 \tag{2.2}$$

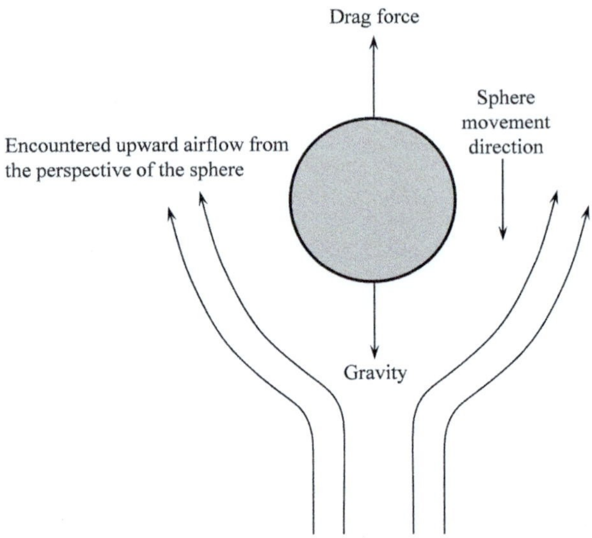

Figure 2.1 Schematic diagram of the linear falling motion of a sphere in still air and the encountered upward airflow from the perspective of the sphere.

The rate of change of momentum is equal to the force exerted on the sphere by the gas, which is called the gas resistance force or the drag force F_D:

$$F_D = K\rho_g \frac{\pi}{4} d^2 V^2 \tag{2.3}$$

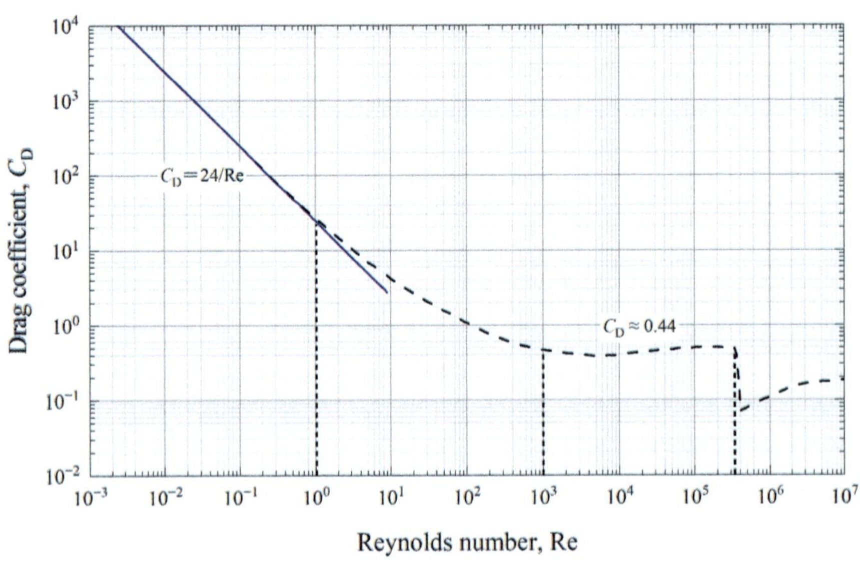

Figure 2.2 Drag coefficient versus Reynolds number for a spherical object.

where K is a constant of proportionality. This equation is only applicable when the Reynolds number of the sphere is greater than 1,000. Replacing K with the drag coefficient C_D, we obtain the more general form of Newton's resistance force as:

$$F_D = C_D \rho_g \frac{\pi}{8} d^2 V^2 \tag{2.4}$$

In this equation, C_D is 0.44 when Re ranges from 1,000 to 2×10^5. Please note that the calculation of the Reynolds number is based on the characteristic length of the spherical diameter. For Re < 1,000, C_D is no longer a constant, as shown in Figure 2.2. More specifically, C_D is equal to $\frac{24}{\text{Re}}$ when Re < 1. For Re between 1 and 1,000, C_D can be calculated as (Hinds and Zhu 2022):

$$C_D = \frac{24}{\text{Re}}(1 + 0.15\text{Re}^{0.687}) \tag{2.5}$$

2.3 Stokes's Law

The derivation of Eq. (2.3) is based on the fact that the viscous effect during motion can be neglected as compared with the inertial effect. However, due to the generally low velocities and small sizes of aerosol particles, viscous forces play a much larger role than inertial forces. Under such conditions, Stokes formulated a drag expression in 1851. Stokes's law is particularly relevant to understanding the dynamic behavior of aerosol particles, especially the particles' settling in the air.

Stokes's law can be derived from the Navier–Stokes equations, which govern Newtonian fluid motion due to the pressure gradient, viscous forces, and body forces. These equations are complex, nonlinear partial differential equations and are thus impossible to solve directly. To simplify the solution, Stokes made several key assumptions:

1) Incompressible fluid: the gas density is constant.
2) Negligible inertial forces: inertial forces are negligible compared to the viscous forces.
3) No nearby obstacles: there are no walls or other particles close to the moving particle.
4) Constant particle motion: the particle moves at a steady speed.
5) Rigid spherical particle: the particle is perfectly spherical and rigid.
6) No slip of the fluid at the particle surface: the gas's relative velocity at the particle surface is zero.

Based on the aforementioned assumptions, the Navier–Stokes equations can be simplified into linear equations that are easier to solve. In Stokes's derivation, the main focus is on the forces acting on the spherical particle as it moves through the fluid. The fluid exerts both normal (perpendicular) and tangential (parallel) forces on the particle's surface. Interested readers can find details of the derivation in the Appendix of this chapter. The normal force acts in the direction opposite to the particle motion and can be expressed as:

$$F_n = \pi \eta V d_p \tag{2.6}$$

where η is the dynamic viscosity of the fluid, V is the velocity of the particle, and d_p is the particle diameter.

The frictional force is parallel to the surface of the particle, and the overall frictional force is also along the opposite direction of particle motion, which can be written as:

$$F_\tau = 2\pi \eta V d_p \tag{2.7}$$

The total drag force on the particle is the sum of the two aforementioned forces, expressed as:

$$F_D = 3\pi \eta V d_p \tag{2.8}$$

Equation (2.8) is Stokes's law, which is valid only when the Reynolds number of the particle is less than 1. It is instructive to compare the drag force given by Stokes's law and by Newton's law:

$$F_D = 3\pi \eta V d_p = C_D \rho_g \frac{\pi}{8} d_p^2 V^2, \quad \text{for Re} < 1 \tag{2.9}$$

It can be seen that Stokes's law accounts for viscosity but excludes inertia-related variables like ρ_g, whereas Newton's law includes ρ_g. Solving Eq. (2.9) yields the following expression for the drag coefficient C_D:

$$C_D = \frac{24\eta}{\rho_g V d_p} = \frac{24}{\text{Re}} \tag{2.10}$$

This equation corresponds to the linear section on the leftmost side of Figure 2.2.

As mentioned earlier, several assumptions are used to derive Stokes's law. Let us examine whether these assumptions limit the applicability of Stokes's law. The assumption of an incompressible fluid does not mean that air itself is incompressible, but rather that the air around the particle does not compress significantly as the particle moves through it. Since the relative velocity between air and particles is much smaller than the speed of sound, this assumption is almost always true for aerosol particles.

When a particle is located within 10 particle diameters from a wall, the drag force acting on the particle may have to be corrected due to interaction with the wall (Hinds and Zhu 2022); further details are described in Section 2.9. For nonrigid spheres, such as water droplets, the correction to Stokes's law is often negligible. For instance, the settling velocity of a droplet is only 0.6% faster than the velocity predicted by Stokes's law when one accounts for the internal fluid circulation within the droplet due to surface friction (Baron and Willeke 2001). Some specific cases, such as nonzero velocity at the particle's interphase surface and the impact of nonspherical particle shapes, are discussed in Sections 2.6 and 2.7, respectively.

2.4 Settling Velocity

Stokes's law plays a crucial role in determining the settling velocity of aerosol particles in still air due to gravity. When a particle is released, it quickly reaches a constant speed known as the terminal settling velocity, at which the upward drag force exerted by the air is balanced by the downward gravitational force. This relationship is described by the following equations:

$$F_D = F_G = mg \tag{2.11}$$

$$3\pi\eta V d_p = (\rho_p - \rho_g) g \frac{\pi}{6} d_p^3 \tag{2.12}$$

where g, ρ_p, and ρ_g are the acceleration of gravity, the density of the particle, and the density of the gas, respectively. In fact, ρ_g can be neglected because ρ_p is commonly much larger than the gas density. Solving Eq. (2.12) yields the terminal setting velocity V_{TS} as:

$$V_{TS} = \frac{\rho_p d_p^2 g}{18\eta}, \quad \text{for } d_p > 1\ \mu\text{m and Re} < 1.0 \tag{2.13}$$

The terminal settling velocity is proportional to the square of the particle diameter and inversely proportional to the dynamic viscosity of air. Note that Eq. (2.13) is applicable only to particles larger than 1.0 μm, unless the slip correction factor, discussed in Section 2.6, is considered. Under standard conditions, the simplified format of Eq. (2.13) for spherical particles with a density of 1,000 kg/m^3 is (Hinds and Zhu 2022):

$$V_{TS} \cong 3 \times 10^{-5} d_p^2\ (\text{m/s}), \quad \text{for } 1 < d_p < 100\ \mu\text{m} \tag{2.14}$$

where d_p is expressed in μm.

2.5 Particle Mobility

According to Stokes's law, the drag force is directly proportional to velocity. From Eq. (2.8), the particle mobility, B, a measure of the relative ease of producing steady motion for an aerosol particle, can be defined as:

$$B = \frac{V}{F_D} = \frac{1}{3\pi\eta d_p}, \quad \text{for } d_p > 1\ \mu\text{m and Re} < 1.0 \tag{2.15}$$

The unit for particle mobility is m/(N·s). Eq. (2.15) is applicable only to particles with diameters larger than 1 μm. For particles smaller than 1 μm, B must be multiplied by the slip correction factor, which is defined and discussed in Section 2.6. The terminal settling velocity of the aerosol particle is simply the product of the gravitational or drag force with the mobility, expressed as:

$$V_{\mathrm{TS}} = F_{\mathrm{D}}B = F_{\mathrm{G}}B \tag{2.16}$$

Example 2.1
Calculate the terminal settling velocity, drag force, and mobility of a rigid spherical particle with a diameter of 2.5 μm in still air. The particle density is 1,000 kg/m^3.

Solution
From Eq. (2.13), the terminal settling velocity of the particle can be determined as:

$$V_{\mathrm{TS}} = \frac{\rho_{\mathrm{p}} d_{\mathrm{p}}^2 g}{18\eta} = \frac{1,000 \times (2.5 \times 10^{-6})^2 \times 9.81}{18 \times 1.81 \times 10^{-5}} = 1.9 \times 10^{-4}\ \mathrm{m/s}$$

Next, the particle's Reynolds number is calculated to determine whether or not it is in the Stokes region:

$$\mathrm{Re} = \frac{\rho_{\mathrm{g}} d_{\mathrm{p}} V_{\mathrm{TS}}}{\eta} = \frac{1.205 \times (2.5 \times 10^{-6}) \times (1.9 \times 10^{-4})}{1.81 \times 10^{-5}} = 3.16 \times 10^{-5}$$

Since Re is less than 1, the particle motion is in the Stokes region. The drag force on the particle is given by:

$$F_{\mathrm{D}} = 3\pi\eta V d_{\mathrm{p}} = 3\pi(1.81 \times 10^{-5})(1.9 \times 10^{-4})(2.5 \times 10^{-6}) = 8.1 \times 10^{-14}\ \mathrm{N}$$

The particle mobility is:

$$B = \frac{V}{F_{\mathrm{D}}} = \frac{1.9 \times 10^{-4}}{8.1 \times 10^{-14}} = 2.3 \times 10^{9}\ \mathrm{m/N \cdot s}$$

2.6 Slip Correction Factor

Stokes's law in the preceding discussion assumes that the gas molecules at the particle's surface are stationary relatively to the particle. This assumption is not valid for very small particles, especially those with sizes close to the gas's molecular mean free path. For small particles, a phenomenon known as "slip" occurs, where the relative gas velocity at the particle's surface is not zero. The slip effect causes these small particles to settle more quickly than predicted by Stokes's law, as the actual drag force is lower than the estimation provided by the law. To address this issue, Cunningham introduced a correction factor in 1910. The Cunningham correction factor, C_c, is always greater than one, and it compensates for the slip effect by reducing the drag force calculated using Stokes's law as:

$$F_D = \frac{3\pi\eta V d_p}{C_c} \tag{2.17}$$

where

$$C_c = 1 + \frac{2.52\lambda}{d_p} \tag{2.18}$$

Here, λ is the mean free path of gas molecules, which can be calculated as (Chapman and Cowling 1990):

$$\lambda = \frac{\eta}{\rho_g}\sqrt{\frac{\pi M}{2R_u T}} \tag{2.19}$$

where M is the molecular weight, R_u is the universal gas constant, and T is the absolute temperature (K).

Incorporating the Cunningham correction factor into Eq. (2.17) extends the applicability of Stokes's law to particles with diameters as small as 0.1 μm. To account for particles less than 0.1 μm in diameter, the following equation can be used for all particle sizes at standard pressure conditions (Allen and Raabe 1982, 1985):

$$C_c = 1 + \frac{\lambda}{d_p}\left[2.34 + 1.05\exp\left(-0.39\frac{d_p}{\lambda}\right)\right] \tag{2.20}$$

The slip-corrected form of the terminal settling velocity can be cast as:

$$V_{TS} = \frac{\rho_p d_p^2 g C_c}{18\eta}, \quad \text{for Re} < 1.0 \tag{2.21}$$

Equation (2.21) is valid for particles of all sizes when Re is less than 1.0, and C_c is given by Eq. (2.20).

Table 2.1 and Figure 2.3 present the slip correction factors for different particle sizes under the standard conditions. The slip correction factor is 1.17 for a particle of 1.0 μm, which implies that the particle settles 17% faster than that predicted by Eq. (2.13). For particles smaller than 1 μm, the slip effect becomes increasingly significant, as shown in Figure 2.3. In fact, to ensure accuracy, the slip correction factor should also be applied for particles with diameters less than 5 μm.

The slip correction factor is impacted not only by particle size but also by pressure, as pressure affects the mean free path of the molecules. A more general form of Eq. (2.20) for air at different pressures is (Hinds and Zhu 2022):

$$C_c = 1 + \frac{1}{Pd_p}[15.60 + 7.00\exp(-0.059Pd_p)] \tag{2.22}$$

where P is the absolute pressure, kPa; and d_p is the particle diameter, μm.

Table 2.1 Slip correction factor for particles of different sizes under standard conditions (20 °C and 101 kPa).

d_p (μm)	C_c	d_p (μm)	C_c
0.01	22.680	1.0	1.168
0.1	2.911	2.0	1.084
0.2	1.890	3.0	1.056
0.3	1.574	4.0	1.042
0.4	1.424	5.0	1.034
0.5	1.337	6.0	1.028
0.6	1.280	7.0	1.024
0.7	1.240	8.0	1.021
0.8	1.210	9.0	1.019
0.9	1.186	10.0	1.017

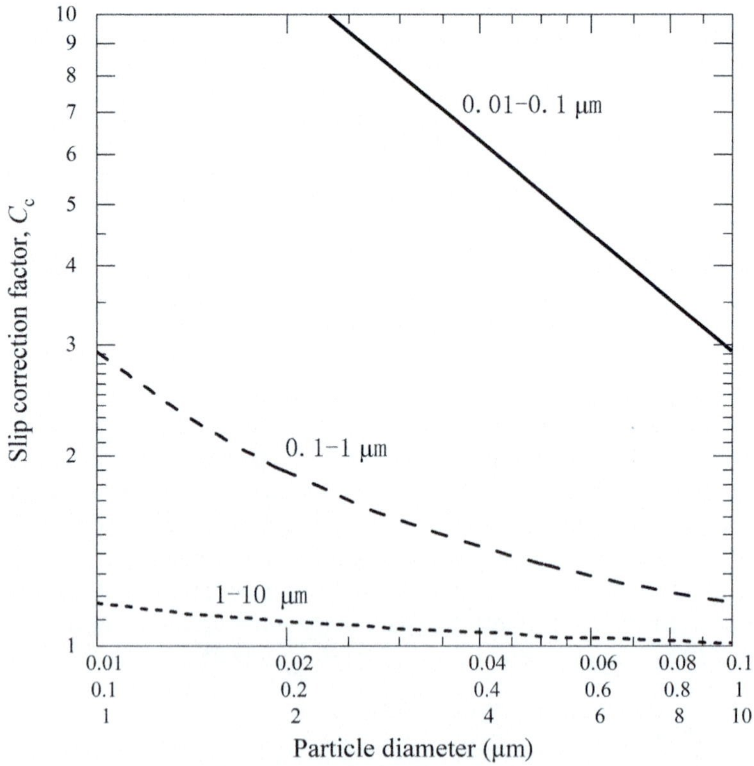

Figure 2.3 Slip correction factor at standard conditions. *Source:* Hinds and Zhu (2022) / with permission of John Wiley & Sons.

Table 2.2 Effect of pressure on terminal settling velocity of standard-density (1,000 kg/m^3) spherical particles at 293.15 K.

Particle diameter (μm)	V_{TS} at different pressures (m/s)		
	$P = 0.1$ atm	$P = 1.0$ atm	$P = 10$ atm
0.001	6.9×10^{-8}	6.9×10^{-9}	6.9×10^{-10}
0.01	6.9×10^{-7}	7.0×10^{-8}	8.7×10^{-9}
0.1	7.0×10^{-6}	8.8×10^{-7}	3.5×10^{-7}
1	8.8×10^{-5}	3.5×10^{-5}	3.1×10^{-5}
10	0.0035	0.0031	0.0029
100	0.29	0.25	0.17

Source: Hinds and Zhu (2022) / with permission of John Wiley & Sons.

Equation (2.22) reveals the influence of pressure on the slip correction and offers a straightforward method for calculating slip correction factors under varying pressure conditions. Furthermore, the product of pressure and particle diameter in Eq. (2.22) can be considered as a new parameter. This implies that the slip correction factor remains constant as long as the product of pressure and particle diameter remains unchanged.

Although pressure is not directly included in Stokes's law, it influences the settling velocity of small particles by affecting the mean free path, which in turn impacts the slip correction factor. Table 2.2 displays the effect of pressure on the terminal settling velocity of standard-density (1,000 kg/m^3) spherical particles at 293.15 K. As can be seen in the table, higher pressures result in lower settling velocities, especially for particles smaller than 0.1 μm.

Example 2.2

Calculate the terminal settling velocity of a 0.5 μm silica sphere at 20 °C and an ambient pressure of 150 kPa. The density of silica is 2,600 kg/m^3.

Solution

According to Eq. (2.22), the slip correction factor of the particle can be calculated as:

$$C_c = 1 + \frac{1}{Pd_p}[15.60 + 7.00\exp(-0.059Pd_p)]$$

$$= 1 + \frac{1}{150 \times 0.5}[15.60 + 7.00\exp(-0.059 \times 150 \times 0.5)] = 1.209$$

Then the terminal settling velocity is:

$$V_{TS} = \frac{\rho_p d_p^2 g C_c}{18\eta} = \frac{2,600 \times (0.5 \times 10^{-6})^2 \times 9.81 \times 1.209}{18 \times 1.81 \times 10^{-5}} = 2.4 \times 10^{-5}\ \text{m/s}$$

2.7 Nonspherical Shape Correction

The equations mentioned earlier for the calculation of the drag force and terminal settling velocity are based on the assumption that particles are spherical. In reality, however, many particles deviate from this shape. Some exhibit alternative geometries, such as cubes (e.g., sea salt particles), cylinders (e.g., bacteria and fibers), single crystals, or aggregates of spheres. Others, like clumped particles or crushed materials, exhibit irregular shapes. These shape variations can significantly influence both the drag force and terminal settling velocity.

To account for the effect of particle shape on drag force, we may adopt the dynamic shape factor, χ. This factor is defined as the ratio of the actual drag force on a nonspherical particle to the resistance of a spherical particle with the identical volume and velocity as the nonspherical particle. The dynamic shape factor is defined as:

$$\chi = \frac{F_{\mathrm{D}}}{3\pi\eta V d_{\mathrm{e}}} \tag{2.23}$$

where d_{e} is the equivalent volume diameter, which is defined as the diameter of a spherical particle with the same volume as the irregularly shaped particle.

When the dynamic shape factor is considered, Stokes's law for irregularly shaped particles can be written as:

$$F_{\mathrm{D}} = 3\pi\eta V d_{\mathrm{e}}\chi \tag{2.24}$$

Furthermore, the terminal settling velocity can be cast as:

$$V_{\mathrm{TS}} = \frac{\rho_{\mathrm{p}} d_{\mathrm{e}}^2 g}{18\eta\chi} \tag{2.25}$$

Table 2.3 Dynamic shape factors for particles with different shapes.

	Dynamic shape factor, χ			
		Axial ratio		
Shape		**2**	**5**	**10**
Sphere	1.00	N/A	N/A	N/A
Cube	1.08	N/A	N/A	N/A
Cylinder	N/A	1.09	1.23	1.43
Straight chain	N/A	1.10	1.35	1.68
Compact three spheres	1.15	N/A	N/A	N/A
Compact four spheres	1.17	N/A	N/A	N/A

Source: Hinds and Zhu (2022) / with permission of John Wiley & Sons.

Table 2.3 lists the dynamic shape factors of particles in different shapes. These values are obtained by measuring the terminal settling velocities of particles of different shapes in liquids. Except for some streamlined shapes, the dynamic shape factor is generally greater than 1.0. This indicates that nonspherical particles tend to settle more slowly than spherical particles of equivalent volume.

The above addresses the correction of the drag force and terminal settling velocity for nonspherical particles. The slip correction for nonspherical particles, meanwhile, is very complex, and available studies indicate that the slip correction factor calculated for an equivalent volume sphere is sufficient for most irregularly shaped particles (Dahneke 1973).

2.8 Aerodynamic Diameter

The aerodynamic diameter, denoted as d_a, is a commonly used equivalent diameter in aerosol dynamics. It is defined as the diameter of a sphere with a density of 1,000 kg/m^3 that has the same terminal settling velocity as the studied particle. Another related, but less frequently used, equivalent diameter is the Stokes diameter, d_s, which represents the diameter of a sphere with the same density and terminal settling velocity as the particle. Equation (2.26) can be used to describe these diameters, without considering slip correction:

$$V_{TS} = \frac{\rho_p d_e^2 g}{18\eta\chi} = \frac{\rho_0 d_a^2 g}{18\eta} = \frac{\rho_b d_s^2 g}{18\eta} \tag{2.26}$$

where ρ_0 is the standard particle density (1,000 kg/m^3), and ρ_b denotes the nominal density of the bulk material of the particle rather than the actual particle density. In fact, the aerodynamic diameter of particles can be regarded as the diameter of water droplets with the same terminal settling velocity. For example, a particle with an aerodynamic diameter of 1 μm behaves aerodynamically like a 1 μm water droplet, irrespective of its shape, density, or physical dimensions.

Figure 2.4 compares the equivalent volume diameter, aerodynamic diameter, and Stokes diameter of an irregularly shaped particle. They all have the same

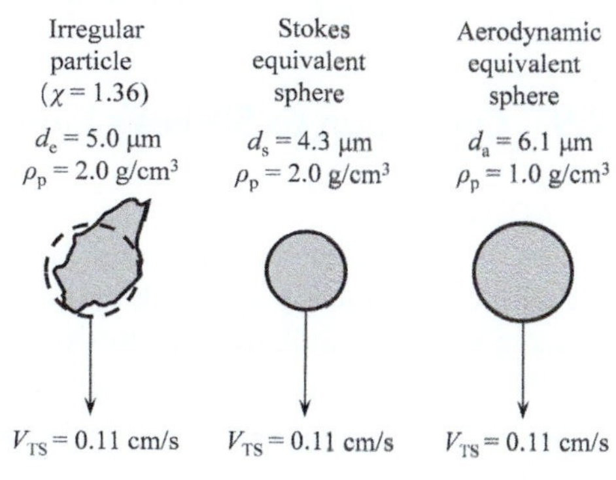

Figure 2.4 An irregularly shaped particle and its equivalent spheres.

terminal settling velocity in spite of distinct shapes or densities. Both the aerodynamic and Stokes diameters are defined according to their aerodynamic behavior rather than their geometric properties. The aerodynamic diameter is crucial for characterizing processes such as filtration and respiratory deposition. In many cases, if the aerodynamic diameter of a particle is known, it is not necessary to know its actual size, geometric shape, density, etc.

Rearranging Eq. (2.26) yields:

$$d_a = d_e \left(\frac{\rho_p}{\rho_0 \chi} \right)^{1/2} = d_s \left(\frac{\rho_b}{\rho_0} \right)^{1/2} \tag{2.27}$$

For spherical particles, Eq. (2.27) can be cast as:

$$d_a = d_p \left(\frac{\rho_p}{\rho_0} \right)^{1/2} \tag{2.28}$$

From Eq. (2.28), it is apparent that if the density of a spherical particle is greater than 1,000 kg/m^3, d_a is always larger than d_p. Note that Eqs. (2.26)–(2.28) are only valid when the particle's Reynolds number is less than 1.0. Moreover, for particles smaller than 1 μm, the slip correction factor should be accounted for, which yields:

$$d_a = d_p \left(\frac{C_c(d_p)}{C_c(d_a)} \right)^{1/2} \left(\frac{\rho_p}{\rho_0} \right)^{1/2} \tag{2.29}$$

where $C_c(d_p)$ and $C_c(d_a)$ are the corresponding slip correction factors for d_p and d_a, respectively.

Example 2.3

Calculate the aerodynamic diameter of a particle with an equivalent diameter of 15 μm and a density of 2,600 kg/m^3. Assume that the shape factor of the particle is 1.88.

Solution

The aerodynamic diameter of the particle can be calculated by Eq. (2.27):

$$d_a = d_e \left(\frac{\rho_p}{\rho_0 \chi} \right)^{1/2} = 15 \left(\frac{2,600}{1,000 \times 1.88} \right)^{1/2} = 17.6\ \mu\text{m}$$

2.9 Correction to Stokes's Law near Boundary Walls

Stokes's law described in Section 2.3 is derived under the assumption that particles move in an unbounded medium. However, in indoor environments, aerosol particles can frequently approach and impact boundary walls. When a particle is near a boundary wall, the velocity profiles around a spherical particle may not exactly follow those presented in Eqs. (2.A.3) and (2.A.4) due to the boundary

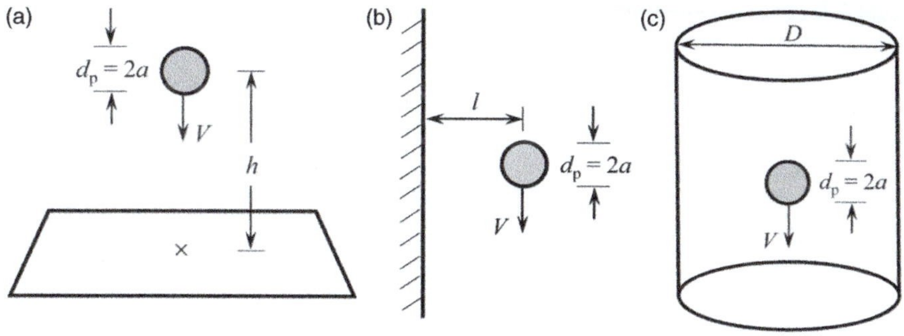

Figure 2.5 Three typical types of particle motion near rigid boundary walls: (a) perpendicular motion approaching a horizontal infinite plane, (b) parallel motion along a vertical infinite plane, and (c) parallel motion along an infinitely long cylindrical tube.

effect. Consequently, the drag force acting on the particle requires a correction, which may be cast as:

$$F_D = 3\pi\eta V d_p \beta \tag{2.30}$$

where β is a correction factor. Analytical expressions of β are provided for the following three typical cases with confined particle motion, namely: (i) perpendicular motion approaching a rigid wall, as shown in Figure 2.5a, (ii) parallel motion along a rigid wall, as shown in Figure 2.5b, and (iii) parallel motion along an infinitely long cylindrical tube, as illustrated in Figure 2.5c.

When a particle with a radius of a perpendicularly approaches a horizontal infinite plane, as shown in Figure 2.5a, the correction factor β can be expressed as (Brenner 1961):

$$\beta = \frac{4}{3}\sinh\alpha \sum_{n=1}^{\infty} \frac{n(n+1)}{(2n-1)(2n+3)} \left[\frac{2\sinh(2n+1)\alpha + (2n+1)\sinh 2\alpha}{4\sinh^2(n+1/2)\alpha - (2n+1)^2\sinh^2\alpha} - 1 \right] \tag{2.31}$$

where n denotes the order of the polynomial expansion, α is a parameter related to the particle radius a and the distance h between the particle center and the horizontal plane. And α is defined as:

$$\alpha = \ln\left(\frac{h}{a} + \sqrt{\left(\frac{h}{a}\right)^2 - 1} \right) \tag{2.32}$$

The corresponding values of β at different ratios of h/a are summarized in Table 2.4. It can be seen that β increases as a particle approaches the rigid wall. For simplicity, when $h/a > 10$, β can be approximated as (Brenner 1961):

$$\beta \approx 1 + \frac{9}{8}\frac{a}{h} \tag{2.33}$$

Table 2.4 Drag correction factor β for a particle settling above a rigid surface.

α	h/a	a/h	β
0	1	1	∞
0.5	1.13	0.89	9.25
1.0	1.54	0.65	3.04
1.5	2.35	0.43	1.84
2.0	3.76	0.27	1.41
2.5	6.13	0.16	1.22
3.0	10.07	0.10	1.13
∞	∞	0	1

Source: Brenner (1961) / with permission of Elsevier.

When a particle with a radius of a moves in parallel with an infinite plane, as illustrated in Figure 2.5b, β is (Goldman et al. 1967):

$$\beta = \frac{1}{1 - 0.5625(a/l) + 0.125(a/l)^3 - 0.1758(a/l)^4 - 0.0625(a/l)^5} \tag{2.34}$$

where l denotes the distance between the particle and the vertical plane.

For a particle of diameter d_p (d_p=2a) moving along the axis of an infinitely long cylindrical tube, as shown in Figure 2.5c, and when $d_p/D \leqslant 0.6$, β can be expressed as (Haberman and Sayre 1958):

$$\beta = \frac{1 - 0.75857(d_p/D)^5}{1 - 2.1050(d_p/D) + 2.0865(d_p/D)^3 - 1.7068(d_p/D)^5 + 0.72603(d_p/D)^6} \tag{2.35}$$

2.10 Summary

This chapter addresses Stokes's law for calculating the drag force of particles, along with particle settling velocity, particle mobility, correction factors for the submicron and nonspherical particles and for the boundary effect, and the aerodynamic diameter commonly used in aerosol dynamics studies.

Stokes's law reveals that the drag force on a particle is proportional to the particle's velocity, diameter, and the carrier gas's dynamic viscosity, while being independent of particle density. It is vital to note that Stokes's law applies only to conditions under which the particle's Reynolds number is less than 1, and may also require correction when particle motion is close to boundary walls.

When a particle is subjected solely to the drag force and gravity, it quickly reaches a steady, terminal settling velocity. The terminal settling velocity increases with particle density and diameter. Particle mobility, defined as the ratio of the terminal settling velocity to the drag force, is used to evaluate the ease with which steady motion is produced for an aerosol particle.

The slip effect can be meaningful for small particles, where molecular slip causes the drag force to be lower than that predicted by Stokes's law. For particles, particularly those smaller than 1 μm, it is necessary to incorporate a slip correction factor when calculating the drag force and terminal settling velocity.

The derivations of Stokes's law and particle settling velocity in this chapter are based on spherical particles. For nonspherical particles, dynamic shape factors must be employed to correct the drag force. As the dynamic shape factor increases, the drag force acting on the particle also increases, resulting in a decrease in its terminal settling velocity.

The aerodynamic diameter is defined as the diameter of a sphere with a density of 1,000 kg/m^3 that has the same terminal settling velocity as the studied particle. Particles with the same aerodynamic diameter have identical terminal settling velocities and similar particle movement behavior.

Problems

2.1 Calculate the terminal settling velocity, drag force, and mobility of a rigid spherical particle with a diameter of 1 μm in still air under standard conditions. The particle density is 1,000 kg/m^3.
[Answer: 3.0×10^{-5} m/s; 5.1×10^{-15} N; 5.9×10^{9} m/(N·s)]

2.2 A cubic particle (dynamic shape factor $\chi = 1.08$) has an equivalent volume diameter $d_e = 2$ μm. Calculate its aerodynamic diameter, terminal settling velocity, and the drag force at terminal settling velocity. The density of the particle is 1,200 kg/m^3.
[Answer: 2.1 μm; 1.3×10^{-4} m/s; 4.7×10^{-14} N]

2.3 A particle settles at 20 °C and an ambient pressure of 50 kPa. Calculate the slip correction factor and terminal settling velocity. The particle has a diameter of 0.8 μm and a density of 1,200 kg/m^3.
[Answer: 1.4; 3.2×10^{-5} m/s]

2.4 Calculate the terminal settling velocity of fibers with a diameter of 1 μm and a length of 10 μm. The fiber density is 2,500 kg/m^3, and slip correction can be omitted.
[Answer: 3.29×10^{-4} m/s]

2.5 A spherical particle has a terminal settling velocity of 1 cm/s. The density of the particle is 1,800 kg/m^3. Please calculate the particle diameter.
[Answer: 13.6 μm]

2.6 Convert the equivalent volume diameter of a nonspherical particle (d_e = 10 μm, ρ_p = 1,800 kg/m^3, dynamic shape factor χ = 1.25) to its aerodynamic diameter.
[Answer: 12 μm]

2.7 Compare Newton's resistance law and Stokes's law from the perspectives of the dominant forces and the applicability of the laws varying with the Reynolds number.

2.8 Why is the aerodynamic diameter more commonly used than the geometric diameter in aerosol science? Furthermore, compare the aerodynamic diameter and Stokes diameter in terms of their definitions and applications.

2.9 The Cunningham slip correction factor C_c becomes significant for particles with diameters less than 1 μm. Explain the cause of the "slip effect" and describe how the slip correction factor modifies Stokes's law. Additionally, discuss the impacts of particle size, gas pressure, and air temperature on the value of C_c.

2.10 The dynamic shape factor quantifies the deviation of a nonspherical particle's drag force compared to that of a spherical particle. Explain how it influences the terminal settling velocity and aerodynamic diameter of irregularly shaped particles. Why do most nonspherical particles settle more slowly than their spherical counterparts with equivalent volume?

A. Appendix: Derivation of Stokes's Law

The motion of Newtonian fluid, such as air and water, is governed by the Navier–Stokes (N–S) equation. For incompressible flows, when the body forces are ignored, the N–S equation can be written as:

$$\rho\left(\frac{\partial \boldsymbol{V}}{\partial t} + \boldsymbol{V}\cdot\nabla\boldsymbol{V}\right) = -\nabla p + \eta\nabla^2(\boldsymbol{V}) \tag{2.A.1}$$

where $\boldsymbol{V}$ is the velocity vector of the fluid at any given point in space, ρ is the fluid density, p is the pressure, and η is the fluid's dynamic viscosity.

For steady flow, $\frac{\partial V}{\partial t}$ is zero. Additionally, in the case of low-Reynolds-number flow, where viscous forces dominate over inertial forces, the inertial term $\boldsymbol{V} \cdot \nabla \boldsymbol{V}$ can be neglected. This results in a simplified form of the N–S equation as:

$$\nabla p = \eta \nabla^2(\boldsymbol{V}) \tag{2.A.2}$$

This equation can describe the flow behavior of a fluid around a small, spherical particle. For convenience, the particle is assumed to be stationary, and the fluid flows with a velocity V_∞, as shown in Figure 2.A.1. V_∞ is the fluid velocity far from the particle. The fluid flow direction is assumed to be along the Z-axis.

To obtain the velocity distribution around the particle, we solve the simplified N–S equation in spherical coordinates. The resulting velocity components are:

$$V_r = V_\infty \left[1 - \frac{3}{2}\left(\frac{a}{r}\right) + \frac{1}{2}\left(\frac{a}{r}\right)^3\right] \cos\theta \tag{2.A.3}$$

$$V_\theta = -V_\infty \left[1 - \frac{3}{4}\left(\frac{a}{r}\right) - \frac{1}{4}\left(\frac{a}{4}\right)^3\right] \sin\theta \tag{2.A.4}$$

where a represents the radius of the spherical particle, and r, θ, and ϕ are based on the spherical coordinates. The solution is symmetric with respect to ϕ, meaning that the velocity distribution is independent of ϕ.

This velocity distribution results in two forces on the particle: a normal force due to pressure p and a tangential force due to shear stress τ. The pressure and

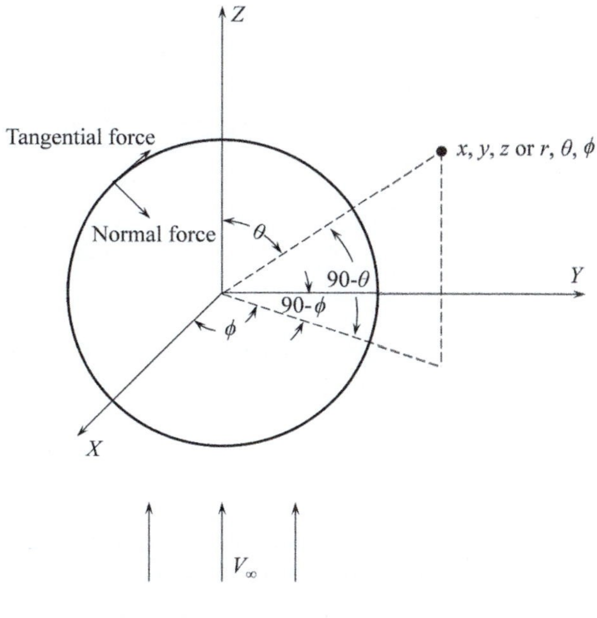

Figure 2.A.1 Forces acting on a stationary spherical particle in a flowing fluid. *Source:* Hinds and Zhu (2022) / with permission of John Wiley & Sons.

shear stress at any point near the particle are given by:

$$p = p_0 - \frac{3\eta V_\infty}{2a}\left(\frac{a}{r}\right)^2 \cos\theta \quad (2.A.5)$$

$$\tau = \frac{3\eta V_\infty}{2a}\left(\frac{a}{r}\right)^4 \sin\theta \quad (2.A.6)$$

where p_0 is the ambient pressure, measured at a location far from the particle ($r \gg a$).

To determine the total drag force exerted by the fluid on the particle, we integrate both the pressure force (normal) and the frictional force (tangential) over the surface of the sphere, as shown in Figure 2.A.2. For the pressure force in the Z-direction:

$$F_\mathrm{n} = \int_S -p\cos\theta \mathrm{d}S = \int_0^{2\pi}\int_0^{\pi}\left(-p_0 + \frac{3\eta V_\infty}{2a}\cos^2\theta\right) a^2 \sin\theta \mathrm{d}\theta \mathrm{d}\phi \quad (2.A.7)$$

By integrating Eq. (2.A.7), we obtain the normal (form) component of the Stokes drag:

$$F_\mathrm{n} = 2\pi\eta V_\infty a \quad (2.A.8)$$

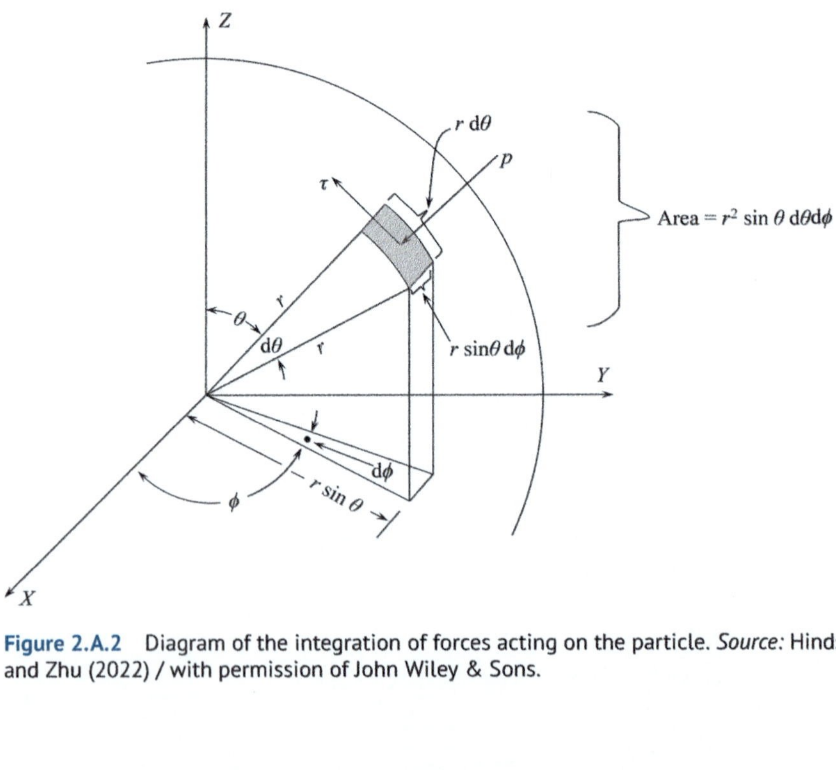

Figure 2.A.2 Diagram of the integration of forces acting on the particle. *Source:* Hinds and Zhu (2022) / with permission of John Wiley & Sons.

Similarly, the tangential (frictional) force is:

$$F_\tau = \int_S \tau \sin\theta \mathrm{d}S = \int_0^{2\pi} \int_0^{\pi} \left(\frac{3}{2} \frac{\eta V_\infty}{a} \sin^3\theta \right) a^2 \mathrm{d}\theta \mathrm{d}\phi \tag{2.A.9}$$

$$F_\tau = 4\pi\eta V_\infty a \tag{2.A.10}$$

When these two components are combined, the total drag force acting on the particle, commonly known as Stokes's law, is obtained as:

$$F_\mathrm{D} = 3\pi\eta V_\infty d_\mathrm{p} \tag{2.A.11}$$

References

Allen, M.D. and Raabe, O.G. (1982). Re-evaluation of Millikan's oil drop data for the motion of small particles in air. *Journal of Aerosol Science* 13: 537–547.

Allen, M.D. and Raabe, O.G. (1985). Slip correction measurements of spherical solid aerosol particles in an improved Millikan apparatus. *Aerosol Science and Technology* 4: 269–286.

Baron, P.A. and Willeke, K. (2001). *Aerosol Measurement: Principles, Techniques, and Applications*. Wiley.

Brenner, H. (1961). The slow motion of a sphere through a viscous fluid towards a plane surface. *Chemical Engineering Science* 16: 242–251.

Chapman, S. and Cowling, T.G. (1990). *The Mathematical Theory of Non-uniform Gases*, 3e. Cambridge University Press.

Dahneke, B.E. (1973). Slip correction factors for non-spherical bodies – III: the form of the general law. *Journal of Aerosol Science* 4: 163–170.

Goldman, A.J., Cox, R.G., and Brenner, H. (1967). Slow viscous motion of a sphere parallel to a plane wall – I motion through a quiescent fluid. *Chemical Engineering Science* 22: 637–651.

Haberman, W.L. and Sayre, R.M. (1958). Motion of rigid and fluid spheres in stationary and moving liquids inside cylindrical tubes. David Taylor Model Basin, Report No. 1143.

Hinds, W.C. and Zhu, Y. (2022). *Aerosol Technology: Properties, Behavior, and Measurement of Airborne Particles*, 3e. Wiley.

3

Straight-Line Acceleration and Curvilinear Particle Motion

3.1 Introduction

In Chapter 2, the fundamental steady, straight-line motion of particles is addressed. In the present chapter, our focus shifts to the exploration of the more complicated forms of particle movement, namely, straight-line acceleration due to constant or varying forces, and curvilinear motion. For a better understanding of these two modes of motion, the concepts of relaxation time, stopping distance, and Stokes number are introduced. Additionally, a specific and particularly important form of particle curvilinear motion, as encountered in inertial impactors, is examined.

3.2 Relaxation Time

Before discussing the straight-line acceleration and curvilinear motion of aerosol particles, let us introduce a new specific term: relaxation time. The terminal velocity of a particle in the Stokes region depends directly on the net external force F acting on the particle. Typical examples of external forces include gravity, centrifugal forces, and electrostatic forces. Notably, the drag force is not classified as an external force. The terminal velocity of the aerosol particle is the product of the external force and the mobility, expressed as:

$$V_{\mathrm{T}} = BF \tag{3.1}$$

where V_{T} is the terminal velocity, and B is the particle mobility.

When the external force involves only gravity, F_{G}, Eq. (3.1) can be written as:

$$V_{\mathrm{TS}} = BF_{\mathrm{G}} = Bmg \tag{3.2}$$

The product of particle mass and mobility, denoted as mB, is known as the "relaxation time," which is represented by the symbol τ and has a unit of time. A particle's mass is related to its diameter and density, while particle mobility can be expressed by Eq. (2.15) in Chapter 2. Therefore, a more detailed expression of the relaxation time is given as:

$$\tau = mB = \rho_{\mathrm{p}} \frac{\pi}{6} d_{\mathrm{p}}^3 \left(\frac{C_{\mathrm{c}}}{3\pi\eta d_{\mathrm{p}} \chi} \right) = \frac{\rho_{\mathrm{p}} d_{\mathrm{p}}^2 C_{\mathrm{c}}}{18\eta\chi} \tag{3.3}$$

The relaxation time characterizes the time required for a particle to adapt or "relax" its velocity to a new condition of forces. In fact, the relaxation time has the same meaning as the concept of inertia commonly referenced in physics. The relaxation time of particles is determined solely by their mass and mobility, and is independent of forces acting on the particles. Although the relaxation time is an inherent property of particles, we should note that Eq. (3.3) includes air viscosity and the slip correction factor, both of which are influenced by the surrounding air temperature and pressure. Moreover, the application of the relaxation time is confined to particle motion within the Stokes region, where the particle's Reynolds number (Re) is less than 1.

Table 3.1 provides a comparison of relaxation time across a range of particle sizes. Notably, relaxation time increases rapidly with particle size, as shown in Eq. (3.3).

With the relaxation time, we can simplify the calculation of the particle terminal settling velocity. Substituting τ into Eq. (2.13) in Chapter 2, or substituting Eq. (3.3) to Eq. (3.2), yields:

$$V_{\mathrm{TS}} = \tau g \tag{3.4}$$

Thus, the particle terminal settling velocity is the product of relaxation time and the gravitational acceleration when the only external force acting on the particles

Table 3.1 Relaxation time for standard density particles (1,000 kg/m^3) under standard conditions.

Particle diameter (μm)	Relaxation time (s)
0.01	7.0×10^{-9}
0.1	9.0×10^{-8}
1.0	3.5×10^{-6}
10.0	3.1×10^{-4}
100	3.1×10^{-2}

Source: Hinds and Zhu (2022) / with permission of John Wiley & Sons.

is gravity. If the external force F acts on a particle with mass m, the more general form of Eq. (3.4) is:

$$V_{\mathrm{TF}} = \tau \frac{F}{m} \tag{3.5}$$

3.3 Straight-Line Particle Acceleration

The equations derived in Chapter 2 for the terminal settling velocity disregard the process of particle acceleration, focusing solely on the equilibrium state in which all the forces acting on the particle are balanced. Now, let us direct our attention to the acceleration of a particle released with an initial velocity of zero in still air. During this acceleration process, we are interested in the dynamic motion over a certain time period.

At any moment during the process of particle acceleration, Newton's second law can be used:

$$\sum F = \frac{\mathrm{d}[mV(t)]}{\mathrm{d}t} \tag{3.6}$$

where $V(t)$ represents the instantaneous velocity of a particle at time t. Since the mass of particles is generally constant, Eq. (3.6) can be rewritten as:

$$\sum F = m\frac{\mathrm{d}V(t)}{\mathrm{d}t} = ma(t) \tag{3.7}$$

where $a(t)$ is the instantaneous acceleration and is subject to gravity and the drag force. The drag force is dependent on the particle velocity at any given moment. Note that the acceleration mentioned here is solely due to the forces acting on the particle, and the additional air acceleration is omitted. During the particle acceleration process, the drag force varies with the velocity, as indicated by Eq. (2.8). Assuming that the gravitational force is positive and no slip correction is needed, Eq. (3.7) can be further written as:

$$F_{\mathrm{G}} - F_{\mathrm{D}} = mg - 3\pi\eta V(t)d_{\mathrm{p}} = m\frac{\mathrm{d}V(t)}{\mathrm{d}t} \tag{3.8}$$

Multiplying both sides of Eq. (3.8) by the particle mobility B yields:

$$mBg - 3\pi\eta V(t)Bd_{\mathrm{p}} = mB\frac{\mathrm{d}V(t)}{\mathrm{d}t} \tag{3.9}$$

Considering that $\tau = mB$ and $B = (3\pi\eta d_{\mathrm{p}})^{-1}$, Eq. (3.9) can be formulated as:

$$\tau g - V(t) = \tau\frac{\mathrm{d}V(t)}{\mathrm{d}t} \tag{3.10}$$

Substituting τg with V_{TS}, rearranging, and then integrating on both sides, we obtain:

$$\int_0^t \frac{\mathrm{d}t}{\tau} = \int_0^{V(t)} \frac{\mathrm{d}V(t)}{V_{\mathrm{TS}} - V(t)} \tag{3.11}$$

The integration on both sides of Eq. (3.11) yields:

$$\frac{t}{\tau} = -\ln[V_{TS} - V(t)] + \ln V_{TS} \tag{3.12}$$

Multiplying both sides of Eq. (3.12) by −1 and exponentiating both sides, we can cast the equation further as:

$$e^{-t/\tau} = \frac{V_{TS} - V(t)}{V_{TS}} \tag{3.13}$$

Then $V(t)$ can be formulated as:

$$V(t) = V_{TS}(1 - e^{-t/\tau}) \tag{3.14}$$

Equation (3.14) describes the variation of the particle velocity, $V(t)$, with time t, after the particle's release in still air in a gravitational field. $V(t)$ is 0 at $t = 0$ and is equal to V_{TS} for $t \gg \tau$. As illustrated in Figure 3.1, the particle achieves 63% of its terminal settling velocity after a time interval of τ, and 95% of its terminal settling velocity after $t = 3\tau$. In fact, from a precise mathematical perspective, $V(t)$ will never reach V_{TS}.

Table 3.2 displays the time required for standard particles to reach their terminal settling velocity under standard conditions. Larger particle diameters correspond

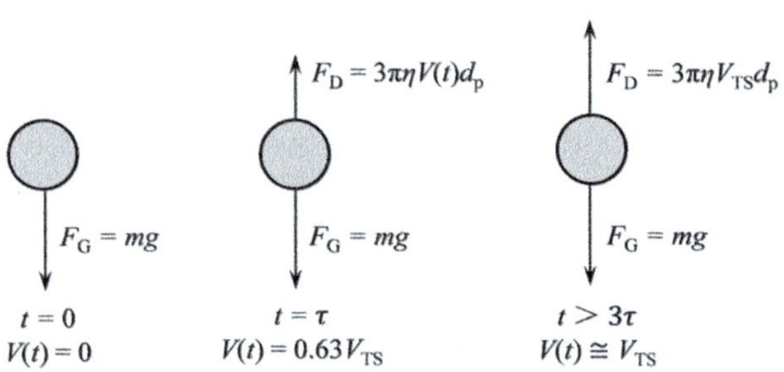

Figure 3.1 Acceleration of a particle in still air. *Source:* Hinds and Zhu (2022) / with permission of John Wiley & Sons.

Table 3.2 Time required for particles to reach their terminal settling velocity (spherical particles with standard density [1,000 kg/m^3] at 293.15 K).

Particle diameter (μm)	Time to reach terminal settling velocity ($t = 3\tau$, ms)
0.01	2.0×10^{-5}
0.1	2.7×10^{-4}
1.0	1.1×10^{-2}
10.0	0.94
100	92

Source: Hinds and Zhu (2022) / with permission of John Wiley & Sons.

to longer times to achieve terminal velocity. Furthermore, it can be seen that even for particles with a diameter of 100 μm, the time required to reach terminal velocity is significantly less than 1 second. Therefore, in practice, it is reasonable to disregard the very short time required for particles to reach their terminal velocity, considering the relatively large timescale of the corresponding research.

Equation (3.14) represents a simplified scenario derived from the general equation of motion. When the velocity of a particle at $t = 0$ is V_0, and its terminal velocity along a specific axis is V_f, the general equation of particle motion can be expressed as:

$$V(t) = V_f - (V_f - V_0)e^{-t/\tau} \tag{3.15}$$

If a particle is released in still air and accelerates toward its terminal settling velocity, the initial velocity V_0 is zero and the final velocity V_f is equal to V_{TS}. More commonly, when a static particle is released into a horizontal airflow with a velocity U, its terminal velocity is also U if the gravitational force is neglected. Substituting these quantities into Eq. (3.15) yields Eq. (3.14), only with V_{TS} replaced by U.

Example 3.1

A spherical particle with a diameter of 10 μm and a density of 2,000 kg/m^3 is released into still air under standard conditions. The air has a dynamic viscosity of 1.81×10^{-5} Pa · s. Assuming that there is no slip correction and the initial velocity of the particle is 0, calculate:

1) The relaxation time (τ) of the particle.
2) The velocity of the particle at $t = 2\tau$.
3) The time for the particle to reach 90% of its terminal velocity.

Solution

According to Eq. (3.3), the relaxation time of the particle can be determined as:

$$\tau = \frac{\rho_p d_p^2}{18\eta} = \frac{2{,}000 \times (10 \times 10^{-6})^2}{18 \times 1.81 \times 10^{-5}} = 6.14 \times 10^{-4} \text{ s}$$

i.e.,

$$\tau = 0.614 \text{ ms}$$

The velocity of the particle at any time t is given by:

$$V(t) = V_{TS}(1 - e^{-t/\tau})$$

The terminal velocity V_{TS} is calculated by means of Eq. (2.13) as:

$$V_{TS} = \frac{\rho_p d_p^2 g}{18\eta} = \frac{2{,}000 \times (10 \times 10^{-6})^2 \times 9.81}{18 \times 1.81 \times 10^{-5}} = 6.02 \times 10^{-3} \text{ m/s}$$

At $t = 2\tau$,

$$V(2\tau) = V_{TS}(1 - e^{-2}) = 6.02 \times 10^{-3} \times (1 - 0.135) = 5.21 \times 10^{-3} \text{ m/s}$$

i.e.,

$$V(2\tau) = 5.21 \text{ mm/s}$$

The time required for the particle to reach 90% of its terminal velocity is:

$$V(t) = 0.9\ V_{TS} = V_{TS}(1 - e^{-t/\tau})$$

Rearranging it, we obtain:

$$0.9 = 1 - e^{-t/\tau}$$

Then:

$$t = -\tau \ln(0.1)$$

i.e.,

$$t = 6.14 \times 10^{-4} \times 2.303 = 1.41 \times 10^{-3} \text{ s} = 1.41 \text{ ms}$$

3.4 Stopping Distance

Let us further analyze the acceleration process for a particle under constant external forces. When we substitute $V(t)$ with dx/dt in Eq. (3.15) and integrate, we obtain the displacement along the X-axis as a function of time, denoted as $x(t)$:

$$\int_0^{x(t)} dx = \int_0^t V_f dt - \int_0^t (V_f - V_0)e^{-t/\tau} dt \tag{3.16}$$

$$x(t) = V_f t - (V_f - V_0)\tau(1 - e^{-t/\tau}) \tag{3.17}$$

Equation (3.17) provides the position, $x(t)$, of an accelerating particle at any given time, t. This equation is particularly useful for determining the instantaneous position of a particle. A particular case not previously addressed is the calculation of the maximum distance that a particle moving with an initial velocity, V_0, can travel in still air without external forces. In this scenario, $V_f = 0$ and $t \gg \tau$, and Eq. (3.17) leads to the following:

$$S = V_0 \tau \tag{3.18}$$

where S represents the total distance traveled by a particle. As expressed by Eq. (3.3), τ is characterized as the product of m and B. Therefore, the stopping distance can be described as the product of a particle's mobility and its initial momentum:

$$S = BmV_0 \tag{3.19}$$

In theory, it would take infinite time for a particle to travel its entire stopping distance. Notably, a particle travels 95% of that distance in a duration of 3τ. The stopping distance represents the definite range that a particle would travel in still air if the external force abruptly ceased to act on the particle. A particular application of the stopping distance arises when a particle is moving with airflow through a 90° elbow. Under such conditions, the stopping distance denotes the distance that the particle continues to travel in its original direction.

The earlier discussion of stopping distance is based on the assumption that the particle motion is in the Stokes region. However, in practical engineering scenarios, particle motion often falls outside the Stokes regime. For instance, particles generated by grinding wheels typically have both larger diameters and higher initial velocities. To solve this problem, the following empirical equation can be used to calculate the stopping distance of a particle with an initial Reynolds number below 1,500 (Mercer 1973):

$$S = \frac{\rho_p d_p}{\rho_g}\left[\mathrm{Re}_0^{1/3} - \sqrt{6}\arctan\left(\frac{\mathrm{Re}_0^{1/3}}{\sqrt{6}}\right)\right], \quad \text{for } 1 \leq \mathrm{Re}_0 < 1,500 \tag{3.20}$$

where Re_0 denotes the initial Reynolds number of the particle.

Table 3.3 provides stopping distances for standard-density particles ranging from 0.01 to 100 μm in diameter. Note that the results for particles of 10 and 100 μm are calculated based on Eq. (3.20), while the rest are based on Eq. (3.18). Despite the high initial velocity of 10 m/s, these particles exhibit short stopping distances, and they slow down rapidly. This rapid deceleration illustrates the limited travel distance of aerosol particles, even for larger ones. In fact, particle motion is mainly affected by the surrounding airflow rather than by the trajectories predicted by Eq. (3.18) or (3.20).

Table 3.3 Stopping distance and the time to travel 95% of the stopping distance for standard spherical particles with an initial velocity of 10 m/s.

Particle diameter (μm)	Stopping distance (mm)	Time to reach 95% of stopping distance (ms)
0.01	7.0×10^{-5}	2.0×10^{-5}
0.1	9.0×10^{-4}	2.7×10^{-4}
1.0	0.035	1.1×10^{-2}
10.0	2.3	0.85
100	127	65

Source: Hinds and Zhu (2022)/with permission of John Wiley & Sons.

Example 3.2

The initial velocity of spherical particles ejected by a rotating grinding wheel is 10 m/s, and the particle density is 3,000 kg/m^3. Calculate the stopping distances for particles with diameters of 1 and 20 μm, respectively, in still air (under standard conditions). The slip correction can be ignored.

Solution

For particles with a diameter of 1 μm, the initial Reynolds number is:

$$\mathrm{Re}_0 = \frac{\rho_g d_p V_0}{\eta} = \frac{1.205 \times 10^{-6} \times 10}{1.81 \times 10^{-5}} = 0.67$$

According to Eq. (3.15):

$$S = \tau V_0 = \frac{\rho_p d_p^2}{18\eta} V_0 = \frac{3{,}000 \times (10^{-6})^2 \times 10}{18 \times 1.81 \times 10^{-5}}$$
$$= 9.21 \times 10^{-5}\ \mathrm{m} = 0.0921\ \mathrm{mm}$$

Meanwhile, for particles with a diameter of 20 μm, the initial Reynolds number is:

$$\mathrm{Re}_0 = \frac{\rho_g d_p V_0}{\eta} = \frac{1.205 \times 20 \times 10^{-6} \times 10}{1.81 \times 10^{-5}} = 13.3$$

Because the initial Reynolds number of these particles is larger than 1, Eq. (3.20) is used to calculate their stopping stance:

$$S = \frac{\rho_p d_p}{\rho_g}\left[\mathrm{Re}_0^{1/3} - \sqrt{6}\arctan\left(\frac{\mathrm{Re}_0^{1/3}}{\sqrt{6}}\right)\right]$$
$$= \frac{3{,}000 \times 20 \times 10^{-6}}{1.205}\left[13.3^{1/3} - \sqrt{6}\arctan\left(\frac{13.3^{1/3}}{\sqrt{6}}\right)\right]$$
$$= 0.024\ \mathrm{m} = 24\ \mathrm{mm}$$

3.5 Curvilinear Motion

Particles in a fluid rarely travel in straight lines. Instead, their trajectories are often curved due to a combination of force actions, such as gravity, drag force, inertial force, buoyancy, electrical force, and magnetic force. One common cause of curved motion is the imbalance between the inertial force and the drag force. For instance, when particles travel through a bend in a duct system, their inertia prevents them from following the flow streamlines perfectly. Larger or denser particles, which have higher inertia, deviate more significantly, creating curved trajectories that are more pronounced.

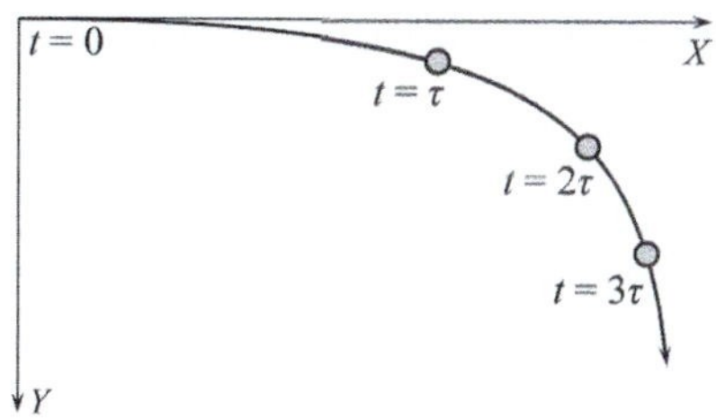

Figure 3.2 Trajectory of a particle injected into the still air horizontally.

Another example is the interaction between gravity and the drag force. A simple case of such motion is shown in Figure 3.2, where a particle is injected into still air horizontally with a velocity V_0. The motion of the particle can be considered as a combination of horizontal motion in the X-direction with only the drag force, and vertical motion in the Y-direction under the gravitational and drag forces. The motion equations in the two directions are:

$$x(t) = V_0\tau(1 - e^{-t/\tau}) \tag{3.21}$$

$$y(t) = V_{TS}t - V_{TS}\tau(1 - e^{-t/\tau}) \tag{3.22}$$

Solving x and y at different times allows us to plot the path of the particle, as shown in Figure 3.2.

3.6 Stokes Number

The curvilinear motion of particles can be characterized by a dimensionless parameter known as the Stokes number (Stk). It represents the ratio of the stopping distance of a particle to the characteristic dimension of an obstacle. For instance, in the case of the flow perpendicular to a cylinder with a diameter of d_c, as shown in Figure 3.3, the Stokes number can be expressed as:

$$\text{Stk} = \frac{S}{d_c} = \frac{\tau V_0}{d_c}, \quad \text{for Re}_0 < 1.0 \tag{3.23}$$

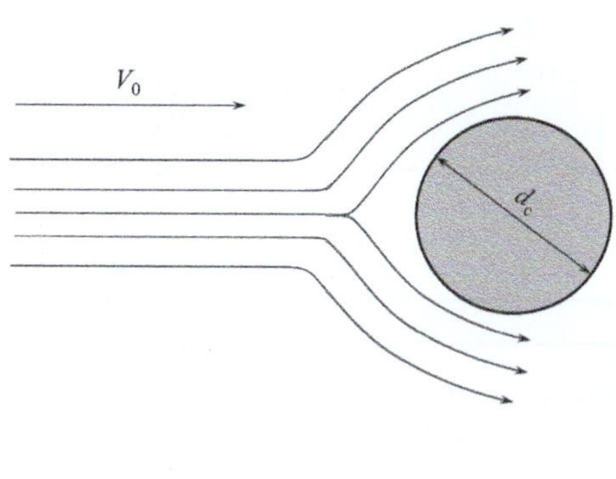

Figure 3.3 The streamlines of airflow passing around a cylinder with a diameter of d_c.

where V_0 is the velocity of incoming air while it is still far from the cylinder, and Re_0 is equal to $\rho_g d_p V_0/\eta$. If Re_0 is larger than 1.0, Eq. (3.20) can be used to calculate the stopping distance outputted to Eq. (3.23). The Stokes number also represents the ratio of the particle relaxation time to the transit time for passing around the obstacle. If Stk $\gg$1, particles maintain a straight trajectory even when the gas streams change their flow direction. In contrast, if Stk $\ll$1, particles follow the gas streamlines perfectly. As the Stokes number increases, particles become more resistant to changing their direction with changes in the direction of the gas streamlines. This parameter is employed to characterize the inertial impaction, as discussed in the next section.

3.7 Inertial Impaction

Impaction represents a specific type of curvilinear motion that is applied widely in the collection and measurement of aerosol particles. Given its significance, impaction should be analyzed theoretically and experimentally. A typical application of the impaction principle is the inertial impactor, as shown in Figure 3.4. An impaction plate deflects the airflow, causing an abrupt 90° bend of the streamlines. Particles with inertia exceeding a specific threshold fail to follow the streamlines and collide with the flat plate. For the sake of simplicity, we assume that the particles stick to the plate once they hit it. In contrast, small particles follow the streamlines and pass through the impactor. Consequently, the inertial impactor collects particles larger than a certain aerodynamic diameter, while it allows particles smaller than this diameter to pass through.

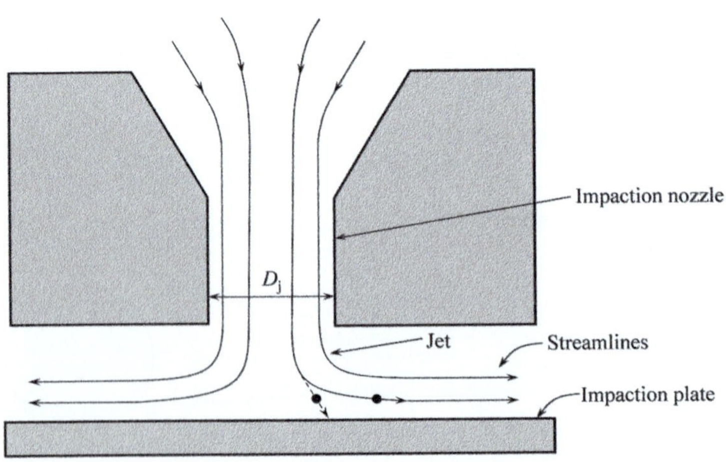

Figure 3.4 Schematic diagram of an impactor. *Source:* Hinds and Zhu (2022) / with permission of John Wiley & Sons.

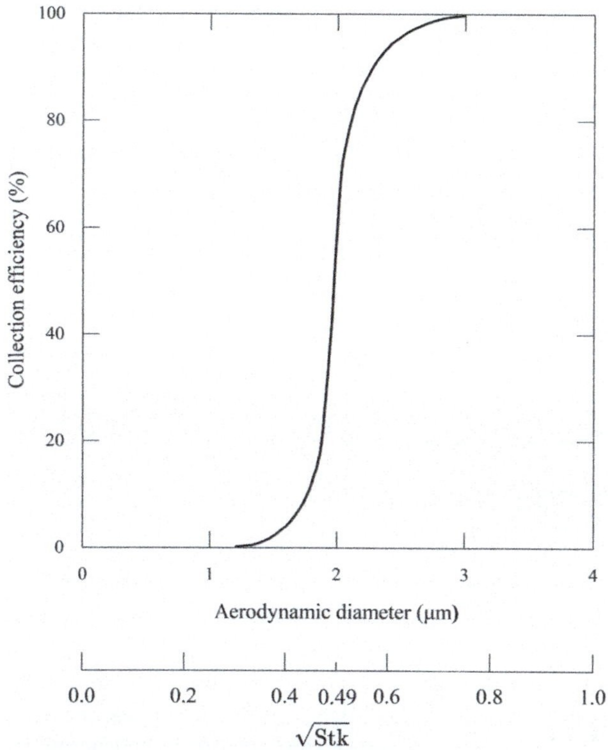

Figure 3.5 Particle collection efficiency of a typical impactor. *Source:* Hinds and Zhu (2022) / with permission of John Wiley & Sons.

The theory of impactors aims to characterize the mathematical relationship between the collection efficiency of the impactor and the particle size, as shown in Figure 3.5. The key parameter that characterizes the particle collection efficiency is the Stokes number, also known as the impaction parameter. Figure 3.5 also illustrates the relationship between the collection efficiency and the square root of the Stokes number, which is proportional to particle size. For an impactor, the Stokes number is defined as the ratio of the particle stopping distance (τU) to the jet radius ($D_j/2$):

$$\text{Stk} = \frac{\tau U}{D_j/2} = \frac{\rho_p d_p^2 U C_c}{9 \eta D_j} \tag{3.24}$$

The characteristic dimension in Eq. (3.24) is different from that in Eq. (3.23), as the dimension in Eq. (3.24) is defined as the radius of the nozzle ($D_j/2$) rather than a characteristic dimension of an obstacle. In the case of an impactor equipped with a rectangular nozzle, the half-width of the jet should be substituted for the jet radius in Eq. (3.24).

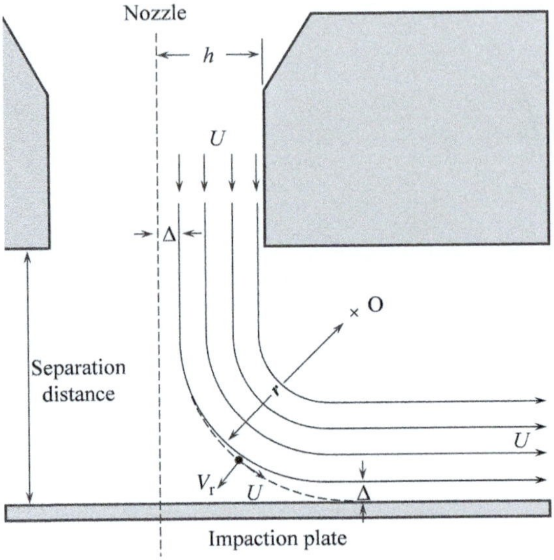

Figure 3.6 Simplified impactor model. *Source:* Hinds and Zhu (2022) / with permission of John Wiley & Sons.

The analysis presented earlier demonstrates the process of impaction and the significance of the relevant parameters. The collision process is analyzed mathematically in more detail below. As shown in Figure 3.6, a rectangular jet impactor is used to analyze the relationship between the particle collection efficiency and Stokes number. Due to symmetry, only half of the impactor is considered here. When a particle exits the nozzle along a streamline, the centrifugal force acting on the particle propels it toward the impaction plate with a radial velocity V_r, which can be expressed as:

$$V_r = \tau a_r = \frac{\tau U^2}{r} \tag{3.25}$$

where r is the radius of curvature of the streamline. It is assumed that the air velocity U remains constant in the curved-streamline region. The radial displacement Δ of the particle from its original streamline can be determined by multiplying its radial velocity by the time needed to traverse the curved section (a quarter circle) of the streamline:

$$\Delta = V_r t = \frac{\tau U^2}{r}\left(\frac{2\pi r}{4U}\right) = \frac{\pi}{2}\tau U \tag{3.26}$$

As illustrated in Figure 3.6, any particle passing through the nozzle within a distance Δ from the centerline of the jet will ultimately collide with the plate. Particles whose distance from the vertical axis is larger than Δ will not reach the impaction surface. Thus, the impaction efficiency E_I, namely, the fraction of particles collected, is equal to the ratio of Δ and h, as follows:

$$E_I = \frac{\Delta}{h} = \frac{\pi\tau U}{2h} = \frac{\pi}{2}(\mathrm{Stk}) \tag{3.27}$$

Table 3.4 Stokes numbers for 50% particle collection efficiency by an impactor.

Type of impactor	Stk_{50}	$(Stk_{50})^{0.5}$
Circular jet	0.24	0.49
Rectangular jet	0.59	0.77

Source: Hinds and Zhu (2022) / John Wiley & Sons.

Although this simplified analysis does not fully characterize impaction efficiency, it demonstrates that the Stokes number, as defined by Eq. (3.24), is the key parameter for the characterization of the inertial impaction.

Inertial impactors are capable of completely collecting particles larger than a specific aerodynamic diameter while allowing particles smaller than this diameter to pass through. This specific diameter is defined as the cutoff size, cutoff diameter, or d_{50}. The design of an impactor is crucial for achieving size-based particle classification. For most impactors, it is not necessary to obtain a complete collection efficiency curve as a function of particle size. A well-designed impactor can be characterized by a single parameter, namely, the Stk_{50}, which is the Stokes number that yields 50% collection efficiency. Stk_{50} represents the location of the ideal collection efficiency curve that best fits the actual efficiency curve, as illustrated in Figure 3.7. In practical applications, analysis of impactor efficiency can be simplified by assuming that the mass of particles larger than the cutoff size but not collected (the upper shaded area in Figure 3.7) is equal to the mass of particles smaller than the cutoff size but collected (the lower shaded area in Figure 3.7). This assumption allows the particle collection efficiency of the impactor to be characterized by a single value (Stk).

Table 3.4 presents Stk_{50} for two types of impactors that meet specific Reynolds number and geometric design criteria. As long as an impactor meets these design standards, its Stk_{50} remains constant, independent of nozzle diameter or velocity. For rectangular nozzle impactors, the Stk_{50} in Table 3.4 is 0.59, while the Stk_{50} calculated by the simplified Eq. (3.27) is 0.32. The recommended design criteria for impactors are as follows. First, the Reynolds number of the airflow through the nozzle throat should be maintained between 500 and 3,000. For a specific cutoff size, the Reynolds number can be adjusted by adopting multiple nozzles in parallel. Second, the separation distance between the nozzle and the impaction plate should be based on the nozzle diameter. For circular nozzles, the ratio of separation distance to nozzle diameter should be between 1 and 5, while for rectangular nozzles, the ratio should be between 1.5 and 5. Generally, lower ratios are preferred.

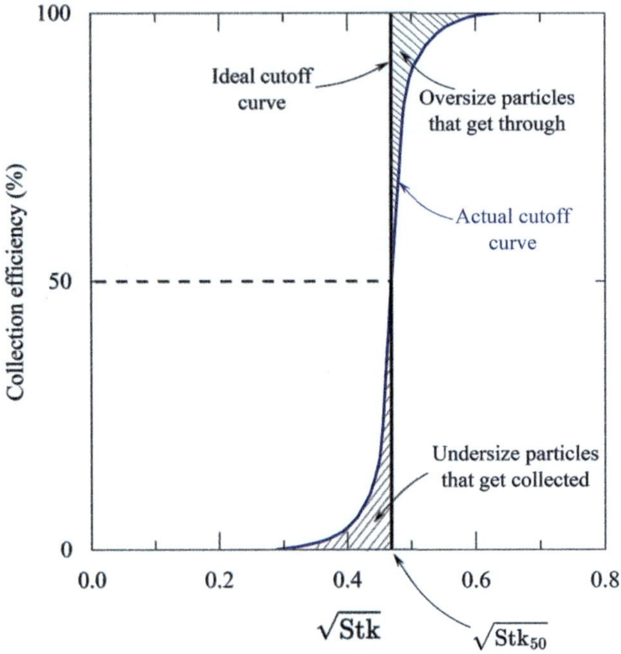

Figure 3.7 Actual and ideal impactor cutoff curves. *Source:* Hinds and Zhu (2022) / with permission of John Wiley & Sons.

When Eq. (3.24) is rearranged, the cutoff diameter or d_{50} can be cast as:

$$d_{50}\sqrt{C_c} = \left[\frac{9\eta D_j(\mathrm{Stk}_{50})}{\rho_p U}\right]^{1/2} \tag{3.28}$$

where Stk_{50} can be obtained from Table 3.4. When the velocity U in Eq. (3.28) is replaced with the jet flow rate, the equation can be expressed in the following form for a circular jet impactor:

$$d_{50}\sqrt{C_c} = \left[\frac{9\pi\eta D_j^{\,3}(\mathrm{Stk}_{50})}{4\rho_p Q}\right]^{1/2} \tag{3.29}$$

and for a rectangular jet impactor with length L and width W:

$$d_{50}\sqrt{C_c} = \left[\frac{9\eta L W^2(\mathrm{Stk}_{50})}{\rho_p Q}\right]^{1/2} \tag{3.30}$$

For conventional impactors, d_{50} can be calculated from $d_{50}\sqrt{C_c}$ by means of the following empirical equation (Hinds and Zhu 2022):

$$d_{50} = d_{50}\sqrt{C_c} - 0.078, \text{ with } d_{50} \text{ in } \mu\text{m} \tag{3.31}$$

Example 3.3

Calculate the cutoff diameters d_{50} of both circular and rectangular jet impactors operating at 1.67×10^{-5} m^3/s. The circular jet impactor has a diameter of 1 mm, and the rectangular jet impactor has dimensions of 2 mm in length and 1 mm in width. Both impactors are assumed to satisfy the recommended design criteria.

Solution

For the round jet impactor, Stk_{50} is 0.24 according to Table 3.4. Substituting the related data into Eq. (3.29) yields:

$$d_{50}\sqrt{C_c} = \left[\frac{9\pi\times1.81\times10^{-5}\times(10^{-3})^3\times0.24}{4\times1{,}000\times1.67\times10^{-5}}\right]^{1/2}$$
$$= 1.36\times10^{-6}\text{m} = 1.36\ \mu\text{m}$$

From Eq. (3.31):

$$d_{50} = d_{50}\sqrt{C_c} - 0.078 = 1.36 - 0.078 = 1.28\ \mu\text{m}$$

For the rectangular jet impactor, Stk_{50} is 0.59 from Table 3.4. Substituting the related data into Eq. (3.30) yields:

$$d_{50}\sqrt{C_c} = \left[\frac{9\times1.81\times10^{-5}\times(2\times10^{-3})\times(10^{-3})^2\times0.59}{1{,}000\times1.67\times10^{-5}}\right]^{1/2}$$
$$= 3.39\times10^{-6}\ \text{m} = 3.39\ \mu\text{m}$$

Then d_{50} can be calculated as:

$$d_{50} = d_{50}\sqrt{C_c} - 0.078 = 3.39 - 0.078 = 3.3\ \mu\text{m}$$

3.8 Summary

This chapter addresses the straight-line acceleration and curvilinear motion of particles, along with key concepts such as relaxation time, particle stopping distance, and the Stokes number. The main points are as follows:

The particle relaxation time is the product of a particle's mass and its mobility, and characterizes the time required for the particle to adjust to a new state when subjected to a variation in the external forces. The particle relaxation time is an inherent property of particles, solely dependent on their mass and mobility. Furthermore, it should be noted that Eq. (3.3) can only be used to calculate the particle relaxation time when the particle's Reynolds number is less than 1.

The formula for calculating the transient velocity of a particle during linear acceleration can be derived from Newton's second law. When a particle is subjected only to the gravitational and drag forces, the time required for it to reach its terminal settling velocity is approximately three times its relaxation time. Even for particles with a diameter of 100 μm, this time is less than 0.1 seconds. Given that the timescales involved in studying particle motion are typically much longer, the time required for particles to reach their terminal velocity can often be ignored.

The particle stopping distance represents the maximum distance a particle can travel in still air when the external force acting on it suddenly disappears. From a precise mathematical perspective, the particle would take infinitely long to travel the entire stopping distance. However, in practice, the particle can travel 95% of the stopping distance within three times the relaxation time.

The curvilinear motion of particles can be characterized by a dimensionless number, namely, the Stokes number, which is the ratio of the stopping distance of the particles to the characteristic size of obstacles. The closer the Stokes number is to zero, the better a particle's ability to follow the airflow. Inertial impactors are designed in accordance with the theory of curvilinear particle motion. The particle collection efficiency of an impactor can be represented by Stk_{50}, which denotes the Stokes number at which the particle collection efficiency is 50%.

Problems

3.1 A spherical particle with a diameter of 5 μm and a density of 1,500 kg/m^3 is released in still air under standard conditions. Assume no slip correction. Calculate the particle's relaxation time and terminal settling velocity.
[Answer: 1.15×10^{-4} seconds; 1.1×10^{-3} m/s]

3.2 Consider a spherical particle with a diameter of 8 μm and a density of 2,200 kg/m^3 moving in a rotating airflow system. If the centrifugal acceleration is 1,000 m/s^2 under standard conditions, calculate the particle's relaxation time and its terminal velocity under the centrifugal force.
[Answer: 4.3×10^{-4} seconds; 0.43 m/s]

3.3 A 20 μm particle (density: 2,500 kg/m^3) is released in still air. Calculate the time required for the particle to reach 95% of its terminal velocity and its velocity at $t = 2\tau$.
[Answer: 9.3×10^{-3} seconds; 2.6×10^{-2} m/s]

3.4 A grinding wheel ejects a spherical particle (diameter: 5 μm, density: 3,500 kg/m^3) into still air with an initial velocity of 2 m/s. Assume standard air conditions and no slip correction. Calculate the particle's initial Reynolds number and stopping distance.
[Answer: 0.67; 5.4×10^{-4} m]

3.5 A particle (diameter: 25 μm, density: 4,000 kg/m^3) is ejected into still air with an initial velocity of 20 m/s. Calculate its initial Reynolds number and stopping distance.
[Answer: 33.3; 8.0×10^{-2} m]

3.6 A spherical particle with a diameter of 2 μm and a density of 1,000 kg/m^3 is carried by an airstream moving at 10 m/s. The flow encounters a cylindrical obstacle with a diameter of 5 mm. Calculate the Stokes number.
[Answer: 2.2×10^{-2} seconds]

3.7 A rectangular jet impactor (nozzle width: 1.5 mm, length: 3 mm) operates at a flow rate of 2.0×10^{-5} m^3/s. Calculate the cutoff diameter d_{50}.
[Answer: 5.6 μm]

3.8 Explain how the relaxation time τ governs the transient behavior of particles during acceleration or deceleration. Discuss why this parameter is critical for predicting particle motion in both straight-line and curvilinear scenarios, and how the variations in air temperature or pressure affect the value of τ.

3.9 The equations for relaxation time and terminal velocity are derived under the assumptions of Stokes regime (Re < 1). Discuss the challenges associated with modeling particle motion when Re > 1, such as in high-velocity airflow or for larger particles.

References

Hinds, W.C. and Zhu, Y. (2022). *Aerosol Technology: Properties, Behavior, and Measurement of Airborne Particles*, 3e. Wiley.
Mercer, T.T. (1973). *Aerosol Technology in Hazard Evaluation*. Academic Press.

4

Adhesion and Detachment of Particles

4.1 Introduction

Particle adhesion refers to the attachment of particles to solid surfaces or to one another. Adhesion of particles plays a critical role in a number of areas. For instance, agglomerates can be formed when particles adhere to each other, which in turn affects the particle size distribution and the aerodynamic properties of the particulate systems. The efficiency of particle removal by means of filtration and precipitation depends significantly on the capability of particles to remain attached to the surfaces of a particle collector. This chapter introduces topics related to particle adhesion, including the adhesive forces, particle detachment, particle resuspension, and particle bounce.

4.2 Adhesive Forces

Common adhesive forces include the van der Waals force, electrostatic force, and the force arising from the surface tension of adsorbed liquid films between the particles and a surface or among particles. These forces are dependent on numerous factors, such as the particle material, shape, and size, and the roughness of the contacting surface. Additionally, the relative humidity of the surrounding air, the temperature, and the contact duration can significantly impact the strength of adhesion. The following sections provide a detailed explanation of several common adhesive forces.

4.2.1 van der Waals Force

The van der Waals force is the most significant adhesive force, as a long-range attractive interaction between molecules, which differs from short-range forces

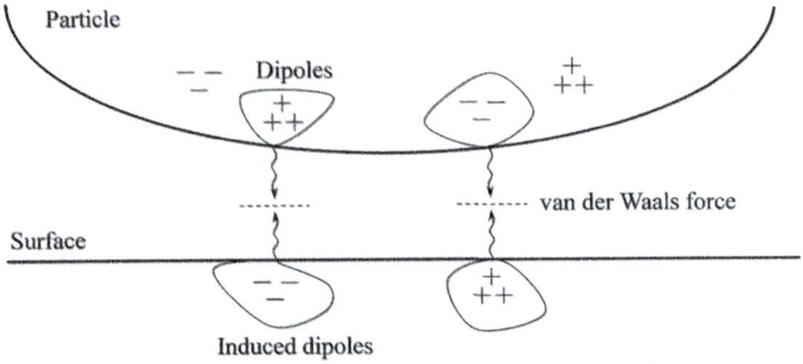

Figure 4.1 Schematic diagram of van der Waals adhesive force.

such as chemical bonds. It arises from the random motion of electrons within materials, creating temporary charge imbalances known as dipoles. These dipoles, in turn, induce complementary dipoles in neighboring materials, resulting in attractive forces, as illustrated in Figure 4.1. It is imperative to note that the van der Waals force rapidly decreases as the distance between the particles and surface increases, such that the action of this force is confined to a range of only a few molecular diameters away from the surface (Ranade 1987).

In fact, when a particle comes into contact with a surface, the interaction typically occurs at only a few asperities due to the irregularities that are present on most material surfaces, as shown in Figure 4.2. The majority of the surface remains separated from the particle by an average distance, denoted as x, which depends on the scale of the surface roughness. For smooth surfaces, the separation distance is approximately 0.4 nm (Hinds and Zhu 2022), which is close to the molecular

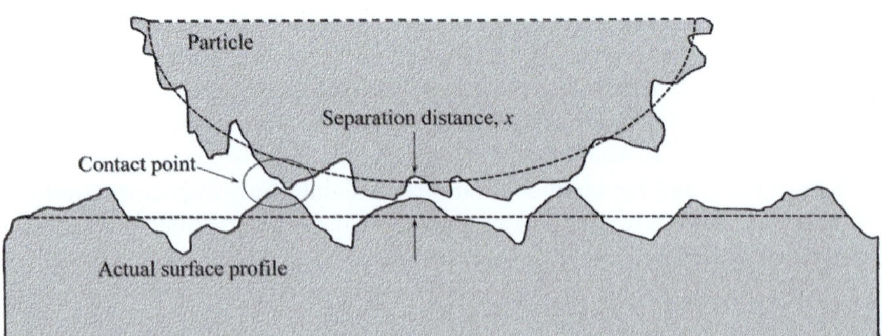

Figure 4.2 Schematic diagram of particle–surface contact geometry. *Source:* Hinds and Zhu (2022) / with permission of John Wiley & Sons.

geometric scale. The van der Waals force is the sum of the attractive forces between all complementary dipoles and is expressed as:

$$F_{\text{van}} = \frac{Ad_{\text{p}}}{12x^2} \tag{4.1}$$

where A is the Hamaker constant and varies with the material, typically ranging from 6×10^{-20} to 150×10^{-20} J for common materials. Note that Eq. (4.1) applies only to ideal, rigid materials that exhibit negligible flattening in the contact area.

After the initial contact between a particle and a surface, the van der Waals force and other attractive forces gradually alter the surface asperities, reducing the separation distance and increasing the contact area. This process continues until a balance is reached between the attractive forces and the forces that resist deformation, and the process may take several hours. The hardness of the surface in contact with the particles determines the final contact area, thereby influencing the strength of the adhesive force. For soft materials like plastics, flattening can increase the adhesive forces by more than 100 times (Tsai et al. 1991).

4.2.2 Electrostatic Force

Particles with a diameter of 0.1 μm or larger typically carry a small net charge q, which induces an equivalent but opposite charge on the adjacent surface. This phenomenon results in an attractive electrostatic force, given by:

$$F_{\text{E}} = \frac{K_{\text{E}}q^2}{x_q^2} \tag{4.2}$$

where K_{E} is a proportionality constant, and x_q is the separation distance between the opposite charges. Here K_{E} is defined as:

$$K_{\text{E}} = \frac{1}{4\pi\varepsilon_0} \tag{4.3}$$

where ε_0 denotes the vacuum dielectric constant, and $\varepsilon_0 = 8.85 \times 10^{-12}\ \text{C}^2/(\text{N}\cdot\text{m}^2)$.

Electrostatic forces are particularly important for insulating materials, as they can retain electric charge for extended periods, especially when the air humidity is low. The charge carried by particles larger than 0.1 μm is approximately proportional to the square root of the particle diameter (Hinds and Zhu 2022). Consequently, the electrostatic adhesive force is also proportional to the particle diameter.

4.2.3 Surface Tension Force

When the air relative humidity exceeds 60%, capillary condensation can occur between particles and the surfaces with which they come into contact

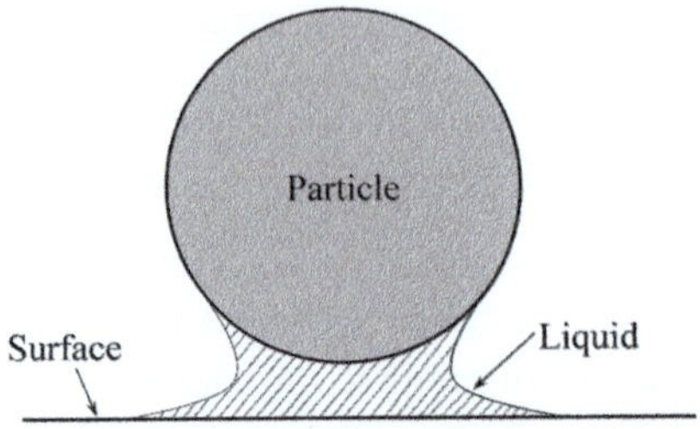

Figure 4.3 Adhesive force due to the surface tension of the liquid film between a particle and a surface.

(Ranade 1987). As shown in Figure 4.3, the surface tension of the liquid drawn into the capillary space in the contact zone generates an attractive force between the particle and the surface. For ideal, smooth surfaces with relative humidity over 90% in the surroundings, this force can be expressed as:

$$F_S = 2\pi\gamma d_p \tag{4.4}$$

where γ represents the surface tension of the liquid. Note that this formula is only applicable to smooth surfaces. Indeed, for surfaces at lower relative humidities, the surface tension force is governed by the curvature of the asperities in the contact zones, rather than by the particle diameter. Since this curvature can vary significantly among particles, a wide range of surface tension forces may arise, even for particles of the same size.

4.2.4 Overall Adhesive Forces and Some Comparisons

The adhesive forces mentioned earlier can be measured by determining the force required to detach a particle from a surface. This measurement can be accomplished by direct methods, such as using atomic force microscopy (AFM), or indirect techniques, such as applying vibrations or air currents to detach the particles. An empirical formula, derived from direct measurements of glass and quartz particles (>20 μm) at 25 °C, has been proposed (Corn 1961) as:

$$F_{Adh} = 0.063 d_p [1 + 0.009(\%RH)] \tag{4.5}$$

where F_{Adh} is the adhesive force, N; d_p is the particle diameter, m; and %RH represents the relative humidity in percentage.

Example 4.1

Calculate the air velocity required to detach a 2 μm spherical particle from a 100 μm filter fiber under conditions of 90% relative humidity. Assume that the drag force acting on the particle is governed by Stokes's law and that the presence of the fiber does not influence the air velocity.

Solution

To detach the particle from the fiber, the drag force must be equal to the adhesive force. The drag force can be obtained by Stokes's law as:

$$F_D = 3\pi\eta V d_p$$

The adhesive force can be calculated by Eq. (4.5):

$$F_{Adh} = 0.063 d_p[1 + 0.009(\%RH)]$$

Then,

$$F_D = F_{Adh}$$

which yields:

$$3\pi\eta V d_p = 0.063 d_p[1 + 0.009(\%RH)]$$

Then,

$$V = \frac{0.063[1 + 0.009(\%RH)]}{3\pi\eta} = \frac{0.063(1 + 0.009 \times 90)}{3\pi(1.81 \times 10^{-5})} = 669 \text{ m/s}$$

The magnitudes of the three adhesive forces discussed earlier are all directly proportional to the first power of the particle diameter. Table 4.1 compares different adhesive forces acting on a 2.5 μm particle at rest on a rigid surface, with respect to the gravitational force. It can be seen that the adhesive forces acting on the particle are much larger than the gravitational force, which further underscores the importance of particle adhesion in aerosol dynamics. Moreover, at a relative humidity of 90%, the overall adhesive force acting on the particle is much greater than the van der Waals force and the electrostatic force. Previous studies have indicated that, except for particles with very large amounts of charges, the van der Waals and surface tension forces generally exceed the electrostatic forces (Hinds and Zhu 2022).

Table 4.1 The adhesive forces and gravitational force acting on a 2.5 μm particle with standard density (1,000 kg/m^3).

Type of force	Value (nN)	References	Note
Overall adhesive force	10–125	Feng and Hays (2003)	Dependent upon the particle and surface properties
van der Waals force	285	N/A	Calculated by Eq. (4.5) at 90% RH
Electrostatic force	1.8–63	Feng and Hays (2003)	Dependent upon the electric charge distribution on the particle surface
Gravitational force	8×10^{-5}	N/A	N/A

4.3 Detachment of Particles

The detachment of particles from their adhered surfaces or their parent material is an important topic of indoor aerosol dynamics, as it may significantly contribute to indoor aerosol emission. This section examines the methods used to measure particle detachment forces and compares the gravitational, adhesive, and detachment forces for particles of various sizes.

4.3.1 Methods for Measurement of Particle Detachment

4.3.1.1 Centrifugal Separation

Several experimental techniques are employed to measure the force required to detach particles from a surface. One such method is centrifugal separation. In this approach, particles are generally placed on a rotating disc, and the rotation speed is gradually increased. As the speed rises, the centrifugal force acting on the particles increases until it eventually overcomes the adhesive forces. The centrifugal force at which particles begin to detach provides a measure of the adhesive forces.

Figure 4.4 presents a schematic diagram of a centrifuge used to measure the binding force of *Aspergillus niger* (*A. niger*) spores with their growing colony (Zhang et al. 2024a, 2024b). The diameter of an *A. niger* spore typically ranges from 2 to 5 μm. A centrifugal tube is positioned in the tube holder of the centrifuge, at an angle of 34° from the vertical. A culture plate inoculated with *A. niger* colonies is aligned parallel to the rotational axis, and the distance between the culture plate and the axis is *R*. On the opposite side of the culture plate, a parallel recipient plate is used to collect the detached spores. The recipient plate is coated with Tween 20 glue to ensure that the spores adhere to the surface after contact. The gap between the culture plate and the recipient plate is 4 mm. In this way, the binding forces between the *A. niger* spores and the colonies are measured.

It shall be noted that the binding force between the *A. niger* spores and the growing colony includes not only the adhesive force among spores as described in Section 4.2 but also the biological binding force provided by the septum between two adjacent spores. The septum contains three lamellae, and the center lamella is fragile and soluble in water. Detachment of a spore shall break the septum first.

4.3.1.2 Air Blowing for Particle Separation

The second method implements the separation of particles by means of blowing airflow. This method estimates the critical drag force required to detach aerosol particles by measuring the critical airspeed at which the particles are separated. The drag force acting on the aerosol particles can be calculated using Stokes's law.

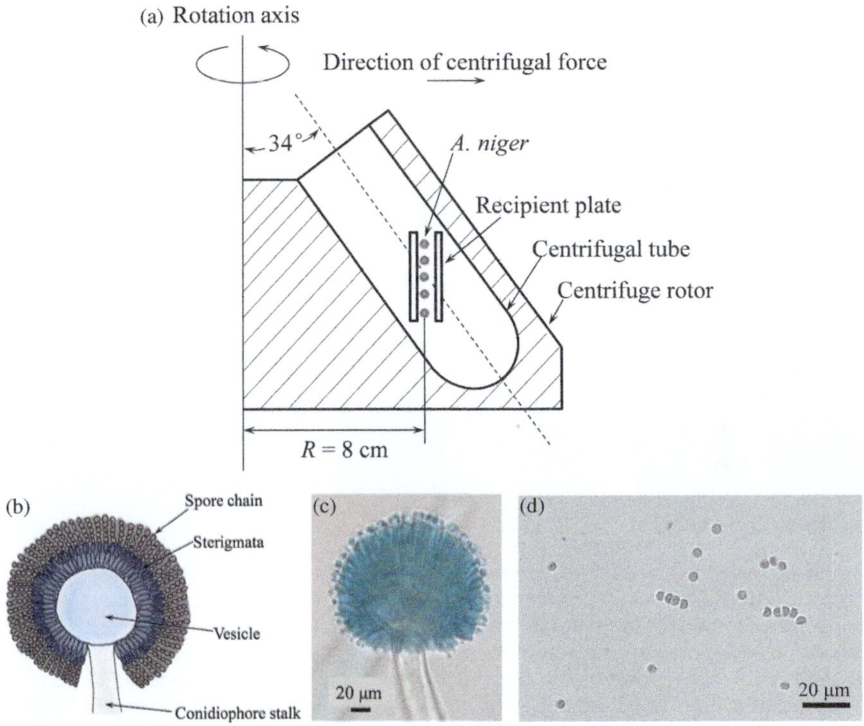

Figure 4.4 A centrifuge used to measure the forces to detach *Aspergillus niger* spores: (a) schematic diagram of a centrifuge to separate the spores from the growing colonies. *Source:* Zhang et al. (2024b)/with permission of Elsevier; (b) schematics of an *A. niger* conidial head. *Source:* Zhang et al. (2024a) / with permission of Elsevier; (c) microscopic picture of an *A. niger* conidial head after dyeing, with the equivalent diameter of approximately 155 μm. *Source:* Zhang et al. (2024a) / with permission of Elsevier; (d) microscopic picture of the collected *A. niger* spores on the recipient plate after the centrifugal separation with the rotating speed of 2,400 rpm.

Figure 4.5 provides a schematic diagram of the wind tunnel setup used to measure the separation of fungal spores from their growing colonies (Zhang et al. 2024b). The wind tunnel consists of a fan, a divergent section, a test section, a convergent section, and a filter. The dimensions of the test section are 36 cm (length) × 20 cm (width) × 20 cm (height). During measurement, the *A. niger* colony is placed at the center of the test section, and a glass slide coated with Tween 20 glue is positioned downstream of the colony to collect the detached spores. The distance between the colony and the glass slide is 3.5 cm. The wind speed in the test section is gradually increased by regulating the fan input frequency until the detachment of the spores. Then the threshold speed is recorded for calculating the drag force to detach the spores.

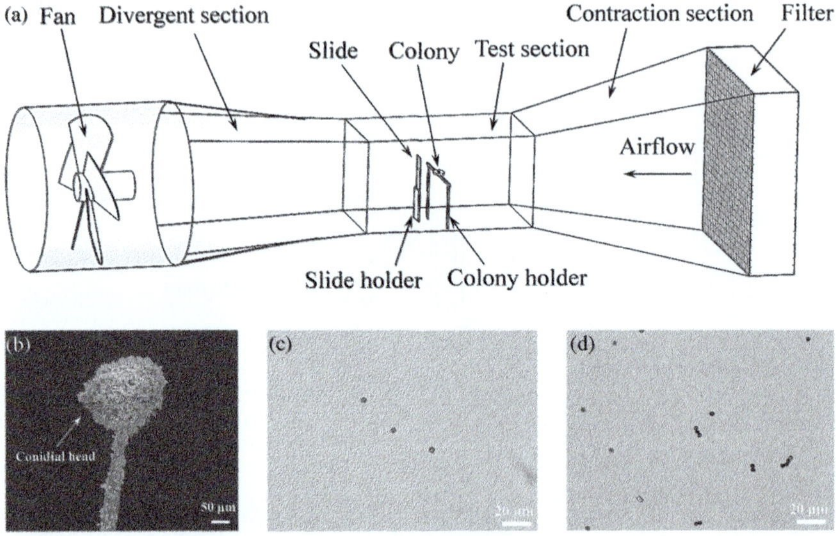

Figure 4.5 A wind tunnel used for measuring the separation of *Aspergillus niger* spores from the growing colonies: (a) schematic diagram of the wind tunnel test facility. *Source:* Zhang et al. (2024b)/with permission of Elsevier; (b) picture of an *A. niger* conidial head by a scanning electron microscope (SEM) before the air-blowing test; (c) collected *A. niger* spores by the slide after the air-blowing test with a speed of 3.5 m/s; (d) collected *A. niger* spores by the slide after the air-blowing test with a speed of 5 m/s. *Source:* Li et al. (2018) / with permission of John Wiley & Sons.

4.3.1.3 Atomic Force Microscopy Separation

The third method employs a sophisticated AFM to separate particles. AFM allows for the direct measurement of the adhesive force between individual particles and the surface. A controlled force is applied through the AFM tip, and the force required to detach the particle is quantified.

Figure 4.6 shows a schematic diagram of the AFM measurement system, which consists of a piezoelectric scanner, micro-cantilever, laser diode, position detector, feedback system, controller, and computer. During measurement, the micro-cantilever remains fixed, while the piezoelectric scanner holding the sample moves in the X, Y, and Z directions. When the micro-cantilever tip contacts the sample, a small deflection occurs, and its displacement is monitored by the position detector. The detector calculates the displacement or bending of the micro-cantilever tip by detecting changes in the position of the laser spot emitted by the laser diode. The distance between the micro-cantilever tip and the sample is varied, and thus a force–displacement curve is obtained. The force between the micro-cantilever and the sample is calculated using Hooke's law:

$$F = KX \tag{4.6}$$

where F is the elastic force by the micro-cantilever, N; K is the spring constant, N/m; and X is the deflection displacement of the micro-cantilever, m.

4.3.1.4 Vibration Separation

The last method for measuring the particle detachment force is by means of a vibration technique in which the surface that holds the particles is subjected to controlled vibrations. The frequency and amplitude of the vibrations are gradually

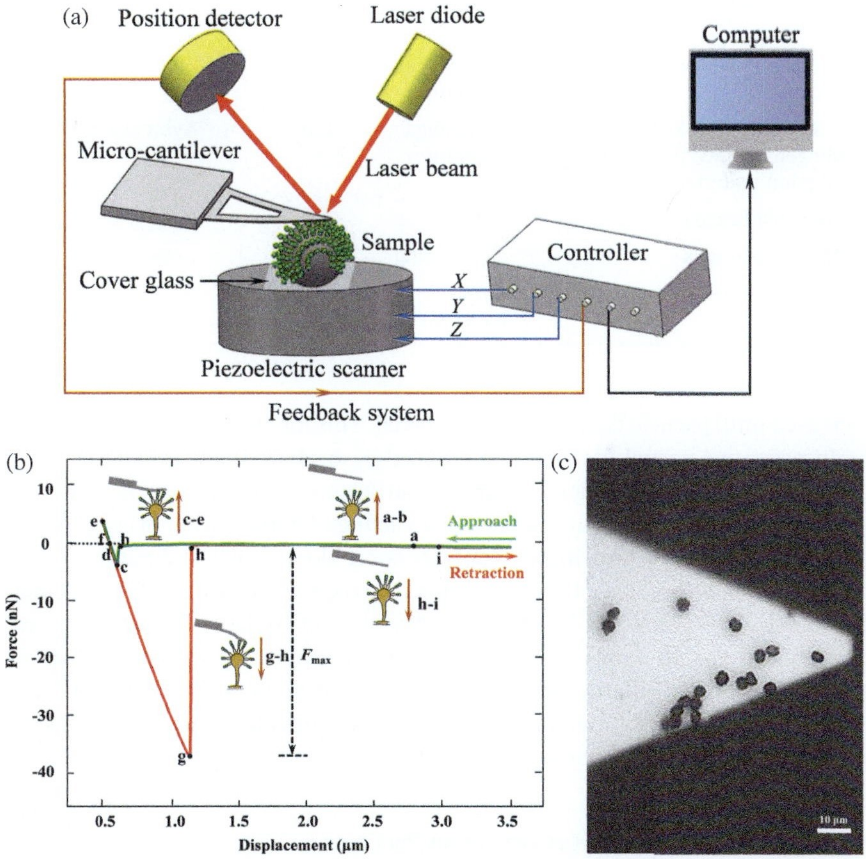

Figure 4.6 An atomic force microscope (AFM) used to measure the force required to detach *A. niger* spores from the growing colonies: (a) schematic diagram of the AFM to separate the *A. niger* spores. *Source:* Adapted from Guo et al. (2014); (b) attractive or repulsive forces versus displacement of the cantilever when contacting with and separating the spores. *Source:* Li et al. (2018) / with permission of John Wiley & Sons; (c) microscopic picture of the adhered spores on the cantilever. *Source:* Li et al. (2018) / with permission of John Wiley & Sons.

Table 4.2 The force required for the detachment of some aerosol particles.

Particle type	Measurement method	Force (nN)	References
10-day-old *Aspergillus niger* spores	AFM	1.98	Li et al. (2018)
10-day-old *Aspergillus niger* spores	Air blowing	2.53	Li et al. (2018)
4-day-old *Aspergillus niger* spores	AFM	3.27	Li et al. (2018)
66-hour-old *Aspergillus niger* spores	Centrifugal separation	0.65	Zhang et al. (2024b)
Spores of the corn leaf spot pathogen (formed on the dried field plant material)	Centrifugal separation	140	Aylor (1975)
Brown rust fungus	Air blowing	27	Geagea et al. (1997)
Yellow rust fungus	Air blowing	40	Geagea et al. (1997)
Tin spheres (20–30 μm) on a silicon wafer	Surface vibration	35–190	Hein et al. (2002)

increased until particle detachment is observed. This method is particularly useful for investigating particle detachment induced by mechanical disturbances.

The force required to detach aerosol particles is typically on the order of nano-Newtons (nN). Furthermore, the detachment force varies in accordance with the particle type, growing age (only for some bioaerosols), and measurement method, as shown in Table 4.2. For example, the required forces by AFM to detach the spores are 3.27 nN and 1.98 nN for 4-day-old and 10-day-old *A. niger* spores, respectively. For 66-hour-old *A. niger* spores, the centrifugal method requires a force of 0.65 nN. Different test methods provided highly different results. For brown and yellow rust fungi, the sweeping airflow provides a drag force of 27 nN and 40 nN to detach the spores, respectively.

4.3.2 Comparison of Forces Acting on Particles

Adhesive forces generally increase linearly with particle diameter, while detachment forces exhibit a cubic increase with particle diameter for both the vibrational and centrifugal separation, and a quadratic increase for air sweeping. This relationship implies that as particle size decreases, detachment of particles from surfaces becomes increasingly difficult. The phenomenon aligns with common intuition, namely, that larger particles, such as grains of sand, can be easily

Table 4.3 Comparison of adhesive, gravitational, and drag forces when air sweeps across spherical particles with standard density (1,000 kg/m^3).

Diameter (μm)	Adhesive force[a] (nN)	Gravity (nN)	Drag force at 10 m/s (nN)
0.1	10	5×10^{-9}	0.2
1.0	10^2	5×10^{-6}	2
10	10^3	5×10^{-3}	30
100	10^4	5	600

a) In this table, the adhesive force is calculated by Eq. (4.5) with a relative humidity of 50%.
Source: Hinds and Zhu (2022) / with permission of John Wiley & Sons.

removed by shaking or airflow blowing, whereas smaller particles, like soot, are not easy to remove or displace. Notably, the adhesive force acting on particles smaller than 10 μm is significantly greater than other forces, as shown in Table 4.3. Although discrete individual particles smaller than 10 μm are generally difficult to detach from a surface, a thick layer of such particles can be removed easily.

4.4 Particle Resuspension

Particle resuspension refers to the process by which particles that have settled or adhered to a surface are lifted back into the air. Resuspension contributes significantly to the quantities of airborne particles present in indoor environments. Resuspension of particles may be due to air sweeping, mechanical disturbances, or particle collisions. The following sections address the factors influencing particle resuspension, boundary-layer resuspension, and some typical particle resuspension rates.

4.4.1 Major Factors Affecting Particle Resuspension

Unlike the static adhesion of particles, wherein particles remain firmly attached to a surface, resuspension involves the detachment and subsequent movement of particles away from the surface. Before becoming airborne, particles may roll or slide along the surface, and significantly less force is required to sustain the particle motion than for the particle detachment from the stationary status. In fact, the force required for resuspension is only approximately 1% of that required for detachment from the stationary status (Hinds and Zhu 2022). The main factors influencing particle resuspension are addressed in the following.

4.4.1.1 Aerodynamic Action

Aerodynamic action plays a critical role in particle resuspension. When air flows over a surface, it may generate lift forces acting on particles attached to the surface. If the lift forces exceed the adhesive forces between the particles and the surface, the particles are detached and probably carried into the sweeping air, resulting in particle resuspension. Note that the particle resuspension may also be analyzed from the perspective of energy accumulation, when there is a transfer of turbulent energy to a particle under turbulent airflow (Ziskind 2006). Additionally, airflow disturbances may drive particles to roll, slide, or sway, further weakening their adhesion and promoting detachment and resuspension of the particles into the air.

Figure 4.7 illustrates the evolution of a dust cloud after a jet impingement on dust deposited on a stainless-steel plate (Mei et al. 2016). It can be observed that

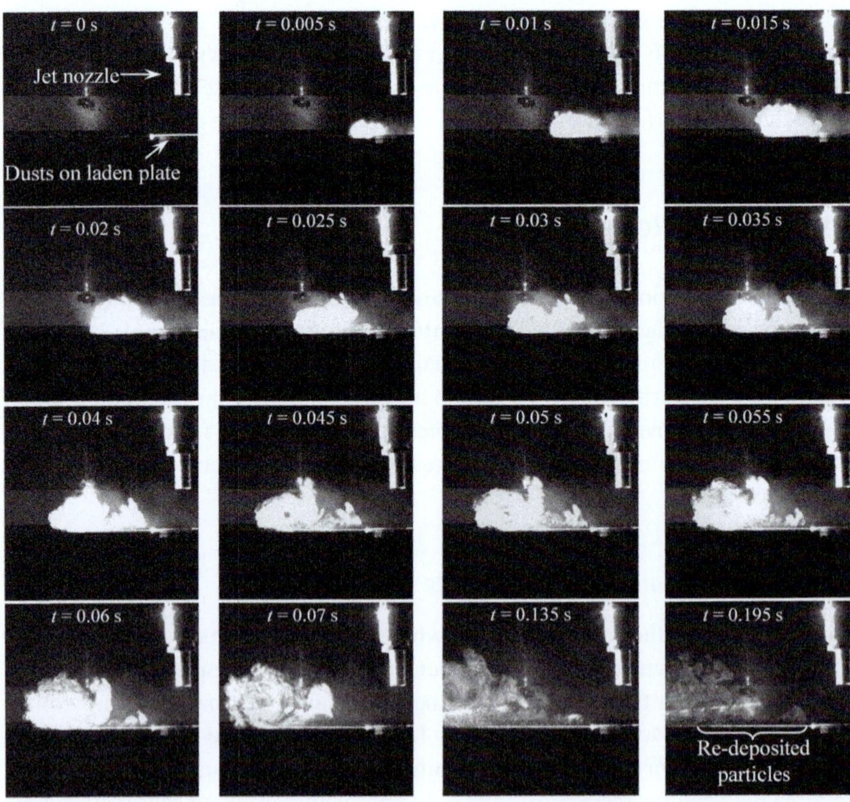

Figure 4.7 Jet-induced particle resuspension from a rough stainless-steel plate and the formed dust cloud. *Source:* Mei et al. (2016) / with permission of Taylor & Francis.

during the initial stage of jet impingement, as shown at $t = 0.01$ seconds, particles are quickly detached from the surface of the deposition plate. As the jet impingement continues, the spatial region affected by the suspended particles gradually expands. At $t = 0.195$ seconds, a portion of the re-deposited dust is visible outside the initial particle-laden substrate.

4.4.1.2 Mechanical Disturbances

Mechanical disturbances, such as vibrations of surfaces, surface cleaning, or moving objects on surfaces, have direct impacts on particles initially deposited on surfaces, which may lead to resuspension of particles away from surfaces. Figure 4.8 illustrates the dynamic process of particle resuspension as a brush sweeps across a cardboard surface with deposited particles on it. When the brush contacts the particle-laden cardboard, its tip moves forward along the surface, causing particles to accumulate, as shown at $t = 0.18$ seconds. At $t = 0.36$ seconds, the bristles leave the surface and the particles are lifted up, forming a particle cloud that continues to develop and diffuse. The left side of the particle collection plate is initially black, indicating the absence of particles on the surface. After sweeping by the brush ($t = 0.60$ seconds), the surface turns white, indicating the re-deposition of resuspended particles.

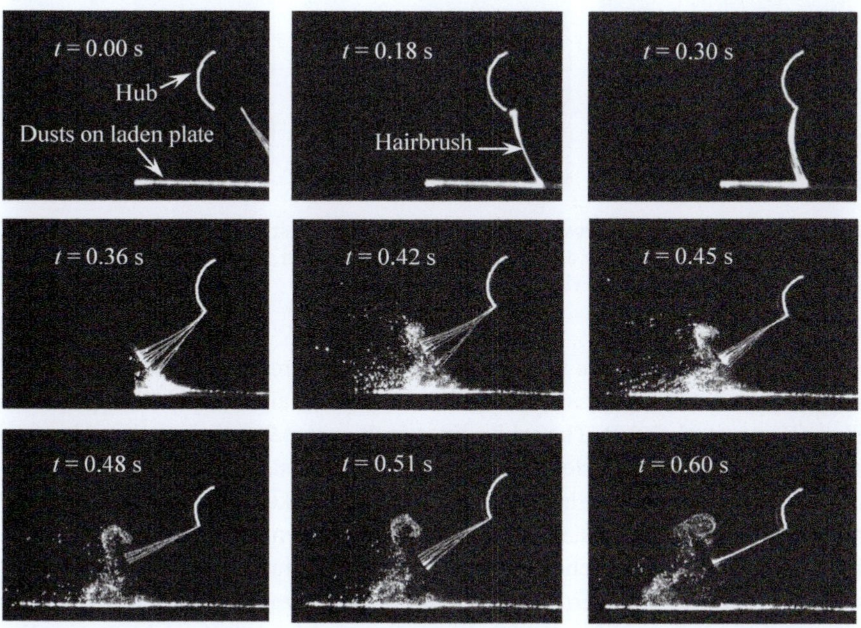

Figure 4.8 Particle resuspension from cardboard surfaces when swept by a hairbrush.

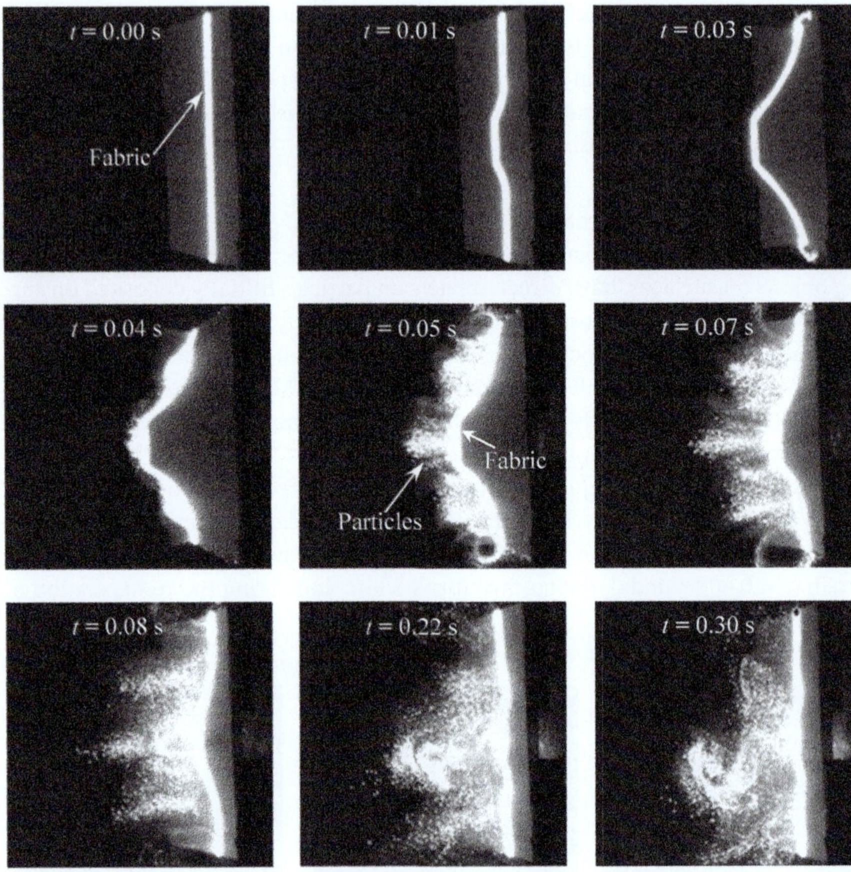

Figure 4.9 Separation and resuspension of particles deposited on a fabric surface due to a mechanical strike by a horizontal rod from the right. *Source:* Wang et al. (2019) / with permission of Taylor & Francis.

Figure 4.9 illustrates the particle detachment and resuspension from a fabric surface due to a mechanical strike, where the material of the fabric is polyester fiber (Wang et al. 2019). The white lines in Figure 4.9 represent the side view of the fabric, with a particle layer attached to its left side. At $t = 0.01$ seconds, a horizontal rod on the right-hand side (not present) strikes the fabric, which moves to the left, but no particles are detached at this moment. As the fabric continues to move leftward, the attached particles begin to separate from the fabric surface at $t = 0.04$ seconds. Furthermore, the fabric's displacement reaches its maximum value at this moment, before the fabric starts moving back to the right. During the period from $t = 0.05$ seconds to $t = 0.08$ seconds, the separated particles decelerate and are sucked back to the fabric surface due to the sucking airflow to the right. At $t = 0.08$ seconds, a cloud of detached particles is observed near the center of the fabric. Subsequently, as shown at $t = 0.22$ seconds and $t = 0.30$ seconds, the particles fall due to gravity, and a more distinct particle cloud can be seen.

4.4.1.3 Electrostatic Action

Electrostatic action can significantly influence particle adhesion and resuspension. Both particles and surfaces typically contain electric charges, and the electrostatic attraction or repulsion between them can either strengthen or weaken adhesive forces. Changes in environmental conditions, such as variations in humidity, can also impact these interactions, thereby affecting the resuspension process.

4.4.1.4 Surface Properties

The characteristics of the surface on which particles are deposited play a crucial role in resuspension. The particle resuspension rate is subject to factors such as surface roughness, texture, and material composition (Ranade 1987). For example, the presence of coatings or grease on the surface can alter adhesive forces, further affecting the dynamics of resuspension (Hinds and Zhu 2022).

4.4.1.5 Particle Properties

The properties of particles, such as size, shape, density, and hygroscopicity, also influence particle resuspension (Henry et al. 2023). Smaller particles tend to adhere to surfaces more strongly due to their higher surface area-to-mass ratios, while larger particles are more susceptible to the aerodynamic lift forces. Figure 4.10 presents some representative bulk air velocities required to re-entrain different sizes of glass beads from glass surfaces (Corn and Stein 1965). As expected, larger particles and higher air velocities increase the re-entrainment ratio. Additionally, the chemical composition of particles might affect the particle–surface interactions, which in turn influence resuspension, especially in the presence of moisture or other reactants (Hinds and Zhu 2022).

4.4.2 Resuspension in Boundary Layers

In a typical turbulent airstream, a thin layer of laminar airflow forms near the surface and is known as the boundary layer. Particles within this boundary layer are partially shielded from re-entrainment due to their full embedding in it. However, occasional bursts of turbulent eddies can penetrate the boundary layer and detach and resuspend these particles. Thus, resuspension is impacted not only by airflow speed but also by the duration of turbulence exposure, as prolonged or

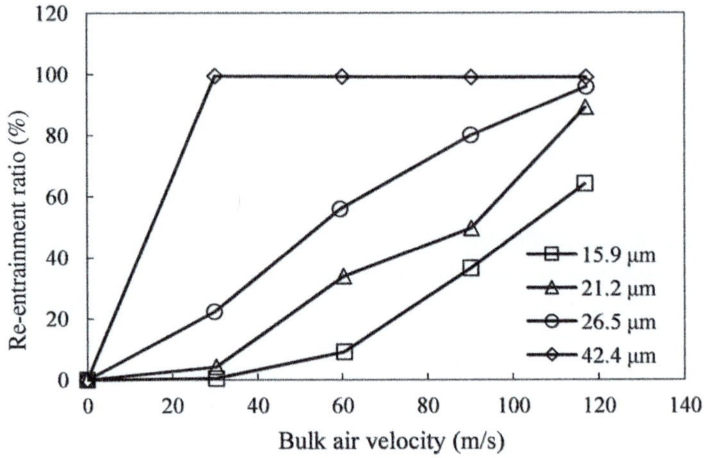

Figure 4.10 Variation in particle resuspension ratio with particle size and air sweeping velocity. *Source:* Adapted from Corn and Stein (1965).

intermittent turbulence increases the probability of particle detachment (Braaten et al. 1990).

4.4.3 Particle Resuspension Rates

Table 4.4 summarizes the particle resuspension rates due to some common types of disturbance. Among these types, jet impingement results in the highest resuspension rate. In cases of jet impingement, the particle resuspension rate from rough stainless-steel surfaces is much higher than that from smooth surfaces. This is mainly because increased surface roughness provides more sites for particle detachment, thereby enhancing resuspension. With the exception of jet impingement, the resuspension rates induced by various types of disturbance are significantly lower than the particle emission by smoking, for which emission rates of $PM_{2.5}$ and PM_5 can both reach 1.7 mg/min (Brauer et al. 2000).

Example 4.2

Determine the mainstream air velocity in parallel with a surface that is required to detach a spherical particle from the surface. The diameter and density of the particle are 10 μm and 1,000 kg/m^3, respectively. Assume that the surface has a laminar boundary layer with a thickness of 0.1 mm and a linear velocity gradient. For simplicity, assume that the particle becomes detached when the drag force, calculated at the particle centerline velocity, is equal to 1% of the adhesive force. The relative humidity is 50%, and Stokes's law is applicable for drag force calculations.

Table 4.4 Comparison of resuspending particle rates under different types of disturbance.[a)]

Disturbance type	$PM_{2.5}$ (mg/min)	PM_5 (mg/min)	References
Jet impingement (smooth stainless steel)	0.9–24.19	5.58–45.88	Mei et al. (2016)
Jet impingement (rough stainless steel)	4.1–110.64	20.96–188.68	
Vacuum cleaning (FF)	0.47	0.73	Ferro et al. (2004)
Vacuum cleaning (BR)	0.17	0.36	
One person walking (FF)	0.12	0.50	
Two people walking (FF)	0.45	1.44	
One person walking (BM)	0.06	0.28	
Folding blanket (BR)	0.03	0.10	
Making bed/folding clothes (BR)	0.15	0.44	
Dancing on a wooden floor (BM)	0.03	0.11	
Dancing on carpet (BM)	0.19	0.95	
Dry scraping furniture (FF)	0.32	0.94	

a) In this table, "FF" denotes measurement locations in living and dining rooms on the first floor, "BR" indicates a first-floor bedroom, "BM" represents the basement, and "Dry scraping furniture" refers to the process of using a dry cloth to scrape particles from furniture surfaces.

Solution

According to Eq. (4.5), the adhesive force between the particle and surface can be calculated as:

$$F_{\mathrm{Adh}} = 0.063d_{\mathrm{p}}[1 + 0.009(\%\mathrm{RH})] = 0.063 \times 10 \times 10^{-6}(1 + 0.009 \times 50)$$
$$= 9.135 \times 10^{-7}\mathrm{N}$$

The drag force required to detach the particle is:

$$F_{\mathrm{D}} = 0.01F_{\mathrm{Adh}} = 0.01 \times 9.135 \times 10^{-7} = 9.135 \times 10^{-9}\ \mathrm{N}$$

The drag force can also be solved by Stokes's law as:

$$F_{\mathrm{D}} = 3\pi\eta V d_{\mathrm{p}}$$

Rearranging to solve for velocity V (velocity at the particle centerline), we obtain:

$$V = \frac{F_{\mathrm{D}}}{3\pi\eta d_{\mathrm{p}}} = \frac{9.135 \times 10^{-9}}{3\pi \times (1.81 \times 10^{-5}) \times (10 \times 10^{-6})} = 5.36\ \mathrm{m/s}$$

The velocity within the laminar boundary layer varies linearly; therefore:

$$\frac{dV}{dy} = \frac{V_\infty}{\delta}$$

The velocity at the particle's centerline (at height $y = d_p/2$) is calculated as:

$$V = \frac{V_\infty}{\delta}\frac{d_p}{2}$$

where V_∞ is the mainstream velocity, and δ is the thickness of the boundary layer.

Rearranging for V_∞:

$$V_\infty = \frac{2V\delta}{d_p} = \frac{2 \times 5.36 \times (1 \times 10^{-4})}{10 \times 10^{-6}} = 107 \text{ m/s}$$

4.5 Particle Bounce

When a solid particle at low speed impacts a surface, its kinetic energy is very probably lost due to the deformation of both the particle and the surface. However, as the impact velocity increases, a portion of the kinetic energy is dissipated during deformation, while the other portion can be elastically converted into rebound kinetic energy. If the rebound energy exceeds the adhesive energy, which is the energy required to overcome the adhesive force, the particles will rebound from the surface. It should be noted that the rebound velocity does not always increase with the impact velocity. This is because at higher impact velocities, both the particle and the surface undergo more significant deformation. As a result, a larger portion of the kinetic energy is consumed in the deformation process, rather than being converted into rebound energy.

In addition to the particle approach velocity, several factors, including particle size, hardness, surface roughness, and surface coatings, can influence the particle bounce. Larger and harder particles are more likely to bounce because they do not deform easily and thus retain more energy for rebounding. Coating with oil or grease increases the adhesion energy and promotes energy dissipation, thus significantly reducing the likelihood of bounce. Coatings are commonly used in filters and impactors to prevent particle loss (Boskovic et al. 2007).

To predict when and how particle bounce will occur, two methods can be used. One method establishes a threshold for the bouncing kinetic energy, while the other defines a critical velocity V_c, above either of which bounce is likely to occur. The kinetic energy required for particle bounce can be formulated as (Dahneke 1971):

$$KE_b = \frac{d_p A(1 - e^2)}{2xe^2} \tag{4.7}$$

where x represents the separation distance between the particle and surface, as defined in Section 4.2.1, A denotes the Hamaker constant, and e represents the coefficient of restitution, which is equivalent to the ratio of the rebound velocity to the approach velocity. According to Wall et al. (1990), typical values for e range from 0.73 to 0.81. A and e are dependent solely on the material composition of the particle and surface.

The critical velocity to judge the particle bounce can be given as (Hinds and Zhu 2022):

$$V_c = \frac{\beta}{d_a} \tag{4.8}$$

where β is a constant that relies on the materials used and the geometric shape of the collision interface, and d_a is the aerodynamic diameter of the particles. Commonly, $\beta = 2 \times 10^{-6}$ m^2/s represents the maximum velocity below which bounce will not occur on uncoated metal impaction plates (Cheng and Yeh 1979).

Example 4.3
What impactor jet velocity is required for sampling of solid particles up to 10 μm in size without causing them to bounce on an uncoated impaction plate?

Solution
Using Eq. (4.8), the critical jet velocity can be calculated as:

$$V_c = \frac{\beta}{d_a} = \frac{2 \times 10^{-6}}{10 \times 10^{-6}} = 0.2 \text{ m/s}$$

4.6 Summary

This chapter introduces the adhesion, detachment, resuspension, and bounce of aerosol particles to/from surfaces. The main points are as follows:

Particle adhesive forces commonly include the van der Waals force, electrostatic force, and surface tension force. The magnitudes of these forces are proportional to the first power of the particle diameter. With the exception of highly charged particles, the van der Waals and surface tension forces are typically much larger than the electrostatic force.

The particle adhesive forces can be measured using methods such as centrifugal separation, air sweeping, surface vibration, and AFM. The magnitude of the adhesive forces depends on the particle type, the material of the adhering surface, the surface roughness, and probably the biological features of bioaerosols, with adhesive forces typically on the order of nano-Newtons (nN). For bioaerosols like *Aspergillus niger* spores, the detachment of older spores from the growing colonies generally requires less force than younger spores.

The force required for particle resuspension is typically about 1% of the force required for detachment of the particles from the stationary status. Resuspension is mainly subjected to air sweeping, mechanical disturbances, and the electrostatic and physical/chemical properties of both the particles and the surfaces. Commonly, the particle resuspension rate caused by jet impingement is higher, while far fewer particles are resuspended as a result of human walking or vacuum cleaning.

When solid particles impact a surface, they may rebound if the rebound energy exceeds the adhesive energy between the particles and the surface or if the impact velocity is higher than the critical velocity for bounce. The rebound speed may not always increase with the particle approach speed. Particle bounce is mainly affected by particle size and the roughness and coating conditions of the collision surface.

Problems

4.1 The charge of a particle with a diameter of 10 μm is 10 e ($e = 1.6 \times 10^{-19}$ C), and the distance between the particle and the conductive surface is 10 nm. Calculate the electrostatic force and van der Waals force acting on the particle, respectively. The value of the Hamaker constant is 8×10^{-19} J.
[Answer: Electrostatic force: 0.23 nN; van der Waals force: 6.7 nN]

4.2 In a ventilation duct, a spherical particle (density $\rho_p = 1{,}000\ \text{kg/m}^3$, diameter $d_p = 5\ \mu\text{m}$) is attached to the duct wall. The adhesive force between the particle and the wall is 10 nN. Calculate the minimum air velocity that can detach the particle from the duct wall.
[Answer: 11.7 m/s]

4.3 A particle with a diameter of 2.5 μm and a density of 800 kg/m^3 is attached to a stainless steel surface. Assuming the sweeping air velocity around the particle is 2 m/s and the relative humidity is 40%, will the particle be detached from the surface?
[Answer: No]

4.4 A particle is placed on a vibrating surface. When the peak acceleration of the surface vibration is 1,000 times the gravitational acceleration, what is the size of the particle that can be just detached from the surface? Assume the density of the particle is 1,000 kg/m^3 and the air is dry.
[Answer: 110 μm]

4.5 A certain type of impactor uses a special coating that has $\beta = 3 \times 10^{-6}\ m^2/s$. To ensure that the solid particles with a diameter of $10\,\mu m$ do not rebound during sampling, please calculate the allowed maximum jet velocity of the impactor.

[Answer: 0.3 m/s]

4.6 Please explain why smaller particles are more difficult to detach from the contacting surface.

4.7 Please imagine variations of the drag force on particles and the particle resuspension rate with the increase of the turbulence intensity of the sweeping airflow. And also tell the reasons.

References

Aylor, D.E. (1975). Force required to detach conidia of *Helminthosporium maydis*. *Plant Physiology* 55: 99–101.

Boskovic, L., Agranovski, I.E., and Braddock, R.D. (2007). Filtration of nanosized particles with different shape on oil coated fibres. *Journal of Aerosol Science* 38: 1220–1229.

Braaten, D.A., Paw, U.K.T., and Shaw, R.H. (1990). Particle resuspension in a turbulent boundary layer – observed and modeled. *Journal of Aerosol Science* 21: 613–628.

Brauer, M., Hirtle, R., Lang, B., and Ott, W. (2000). Assessment of indoor fine aerosol contributions from environmental tobacco smoke and cooking with a portable nephelometer. *Journal of Exposure Analysis and Environmental Epidemiology* 10: 136–144.

Cheng, Y.-S. and Yeh, H.-C. (1979). Particle bounce in cascade impactors. *Environmental Science and Technology* 13: 1392–1396.

Corn, M. (1961). The adhesion of solid particles to solid surfaces II. *Journal of the Air Pollution Control Association* 11: 566–584.

Corn, M. and Stein, F. (1965). Re-entrainment of particles from a plane surface. *American Industrial Hygiene Association Journal* 26: 325–336.

Dahneke, B. (1971). The capture of aerosol particles by surfaces. *Journal of Colloid and Interface Science* 37: 342–353.

Feng, J. and Hays, D.A. (2003). Relative importance of electrostatic forces on powder particles. *Powder Technology* 136: 65–75.

Ferro, A.R., Kopperud, R.J., and Hildemann, L.M. (2004). Source strengths for indoor human activities that resuspend particulate matter. *Environmental Science and Technology* 38: 1759–1764.

Geagea, L., Huber, L., and Sache, I. (1997). Removal of urediniospores of brown (*Puccinia recondita f.sp. tritici*) and yellow (*P. striiformis*) rusts of wheat from infected leaves submitted to a mechanical stress. *European Journal of Plant Pathology* 103: 785–793.

Guo, D., Xie, G., and Luo, J. (2014). Mechanical properties of nanoparticles: basics and applications. *Journal of Physics D: Applied Physics* 47: 013001.

Hein, K., Hucke, T., Stintz, M., and Ripperger, S. (2002). Analysis of adhesion forces between particles and wall based on the vibration method. *Particle & Particle Systems Characterization* 19: 269–276.

Henry, C., Minier, J.-P., and Brambilla, S. (2023). Particle resuspension: challenges and perspectives for future models. *Physics Reports* 1007: 1–98.

Hinds, W.C. and Zhu, Y. (2022). *Aerosol Technology: Properties, Behavior, and Measurement of Airborne Particles*, 3e. Wiley.

Li, X., Zhang, T., and Wang, S. (2018). Measuring detachment of *Aspergillus niger* spores from colonies with an atomic force microscope. *Indoor Air* 28: 1–10.

Mei, X., Zhang, T., and Wang, S. (2016). Experimental investigation of jet-induced resuspension of indoor deposited particles. *Aerosol Science and Technology* 50: 230–241.

Tsai, C.-J., Pui, D.Y.H., and Liu, B.Y.H. (1991). Elastic flattening and particle adhesion. *Aerosol Science and Technology* 15: 239–255.

Ranade, M.B. (1987). Adhesion and removal of fine particles on surfaces. *Aerosol Science and Technology* 7: 161–176.

Wall, S., John, W., Wang, H.-C., and Goren, S.L. (1990). Measurements of kinetic energy loss for particles impacting surfaces. *Aerosol Science and Technology* 12: 926–946.

Wang, J., Niu, L., Zhang, T., and Wang, S. (2019). Detachment of adhered particles from a cloth surface subjected to a rod strike. *Aerosol Science and Technology* 53: 1–14.

Zhang, T., Li, N., Xu, X. et al. (2024a). Measurement and prediction of the detachment of *Aspergillus niger* spores in turbulent flows. *Journal of Hazardous Materials* 475: 134910.

Zhang, T., Xu, X., and Lv, M. (2024b). Measurement and prediction of the *Aspergillus niger* spore detachment from a vesicle unit subjected to air-blowing. *Science of the Total Environment* 912: 168714.

Ziskind, G. (2006). Particle resuspension from surfaces: revisited and re-evaluated. *Reviews in Chemical Engineering* 22 (1–2): 1–123.

5

Brownian Motion and Diffusion and Thermophoresis

5.1 Introduction

Brownian motion refers to the irregular wiggling movement of aerosol particles due to continuous, random bombardment of molecules of the carrier gases. This random motion results in wiggling in the particles' trajectories, as shown in Figure 5.1. Diffusion is the main mechanism for the transport of particles with diameters less than 0.1 μm. For instance, when the transport distance of ultrafine particles is short, as in filters and human lungs, the collection of these particles mainly relies on diffusion. In addition, diffusion often occurs in combination with other mechanisms such as agglomeration and sedimentation. This chapter addresses Brownian motion and the diffusion of particles, including the diffusion coefficient, particle mean thermal velocity, particle mean free path, Brownian displacement, and thermophoresis.

5.2 Diffusion Coefficient

In cases in which the migration of particles is mainly driven by Brownian motion, the particles move from regions of higher concentration to regions of lower concentration, a process known as diffusion. Diffusion can be characterized by the diffusion coefficient, D. A larger diffusion coefficient indicates more intense Brownian motion and a faster rate of mass transfer within a concentration gradient. In fact, the diffusion coefficient is a proportional constant that relates the flux J of aerosol particles (defined as the net number of particles passing through a unit area per unit time) to the concentration gradient $\mathrm{d}n/\mathrm{d}x$. The relationship between the two parameters is described by Fick's first law of diffusion. In the absence of external forces, this law is expressed as:

$$J = -D\frac{\mathrm{d}n}{\mathrm{d}x} \tag{5.1}$$

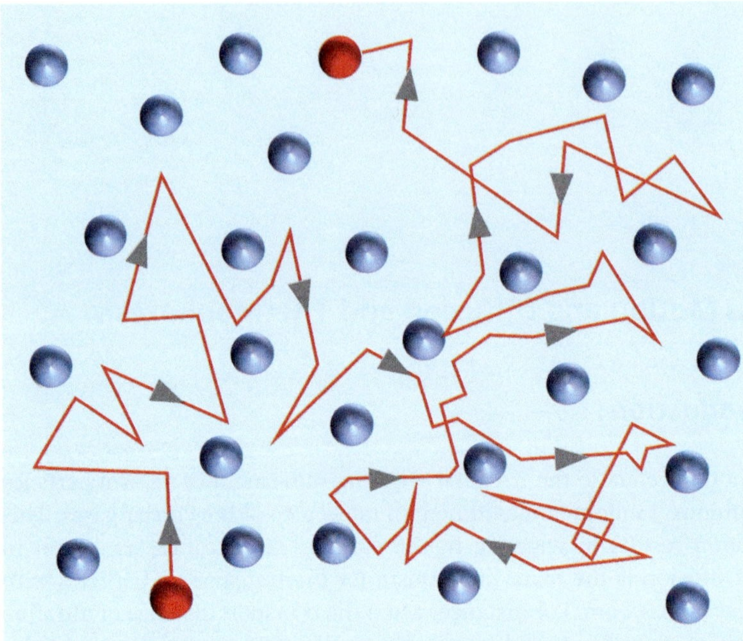

Figure 5.1 Schematic diagram of Brownian motion pathlines of a particle in tracking (red sphere).

where the negative sign indicates that diffusion occurs in the direction of decreasing concentration.

To derive the diffusion coefficient of particles, the concept of diffusion force is used. The diffusion force is the force acting on particles that drives them to move in the direction of decreasing concentration. The diffusion force is equal to the drag force exerted by the gases. According to Stokes's law, as described in Chapter 2, the diffusion force can be expressed as:

$$\text{Diffusion force} = F_{\text{Diff}} = \frac{3\pi\eta V d_p}{C_c} \tag{5.2}$$

Furthermore, the diffusion force acting on a particle is equal to the net osmotic pressure force acting on that particle (Hinds and Zhu 2022). A way to understand the osmotic pressure is by considering the molecules dissolved in a liquid. As shown in Figure 5.2, a membrane allows liquid molecules to pass through but prevents the passage of dissolved molecules. This membrane in the figure can move freely to the left or to the right. If the membrane is moved from the initial position, as shown in Figure 5.2a, to the position shown in Figure 5.2b, a force

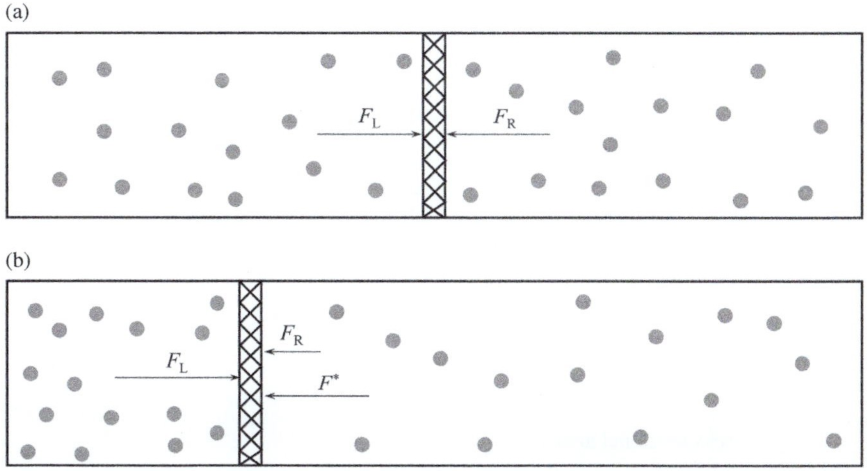

Figure 5.2 Schematic diagram of the generation of osmotic pressure force: (a) force F_L = force F_R initially; (b) an additional force F^* must be imposed to counterbalance the osmotic pressure force in order to keep the membrane in position. *Source:* Hinds and Zhu (2022) / with permission of John Wiley & Sons.

must be applied toward the left to keep the membrane in place. This force is equal to the net osmotic pressure acting toward the right. In fact, osmotic pressure can be regarded as the pressure generated by the higher concentration of dissolved molecules on the left side of the membrane. This is simply due to the higher chance of molecular collision with the membrane in the left-hand region, where the concentration of dissolved molecules is higher. The osmotic pressure p_0 for n suspended particles per unit volume can be expressed using van't Hoff's law as:

$$p_0 = \mathrm{k}Tn \tag{5.3}$$

where k is Boltzmann constant and T is the absolute temperature.

In a scenario where aerosol particles are diffusing from a region of higher concentration to a region of lower concentration, as shown in Figure 5.3, the net osmotic pressure generates a diffusion force F_{Diff}, which pushes the particles from surface E to surface E'. This force can be expressed as:

$$F_{\text{Diff}} = \mathrm{k}Tn_{\mathrm{E}} - \mathrm{k}Tn_{\mathrm{E'}} = -\mathrm{k}T\mathrm{d}n \tag{5.4}$$

When a unit cross-sectional area from surface E to surface E' is considered, as presented in Figure 5.3, the volume of gas between these two surfaces is dx. The number of particles within this volume is given by ndx. Consequently, we can

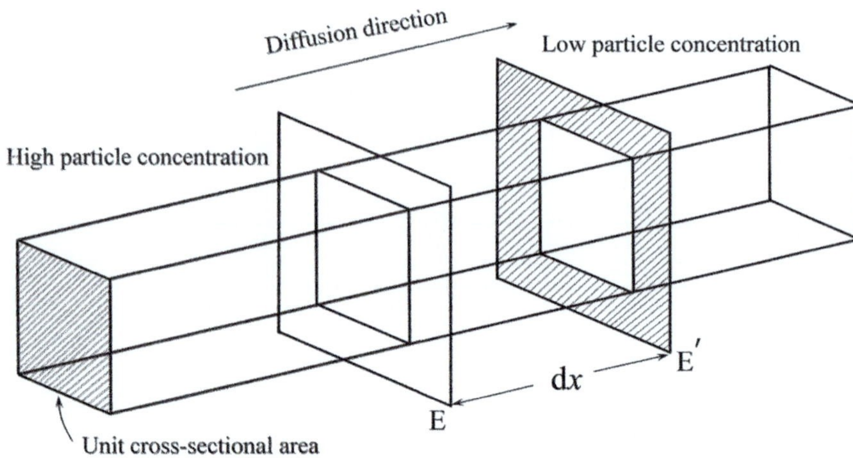

Figure 5.3 Schematic diagram of particle diffusion between two surfaces. *Source:* Hinds and Zhu (2022) / with permission of John Wiley & Sons.

derive the diffusion force acting on each particle within the volume $\mathrm{d}x$:

$$F_{\mathrm{Diff}} = -\frac{\mathrm{k}T}{n}\frac{\mathrm{d}n}{\mathrm{d}x} \tag{5.5}$$

This diffusion force results in the net motion of particles. Substituting Eq. (5.5) into Eq. (5.2) yields:

$$nV = -\frac{\mathrm{k}TC_{\mathrm{c}}}{3\pi\eta d_{\mathrm{p}}}\frac{\mathrm{d}n}{\mathrm{d}x} \tag{5.6}$$

where nV represents the flux of particles per unit area per second, which is equal to J in Eq. (5.1). Thus, the diffusion coefficient D can be expressed as:

$$D = \frac{\mathrm{k}TC_{\mathrm{c}}}{3\pi\eta d_{\mathrm{p}}} \tag{5.7}$$

Eq. (5.7) is also known as the Stokes–Einstein equation, and it can be further formulated in terms of the particle mobility B as:

$$D = \mathrm{k}TB \tag{5.8}$$

The unit of the diffusion coefficient D is m^2/s. The diffusion coefficient of particles increases with temperature, as is the case for gases. Additionally, the diffusion coefficient is subject to the particle size. For coarse particles, whose slip correction can be neglected, D is inversely proportional to the particle size. In contrast, for small particles, whose slip effect is significant, D is nearly inversely proportional to the square of the particle size.

5.3 Particle Mean Thermal Velocity

Aerosol particles continuously exchange energy with the carrier gas molecules, resulting in a shared average kinetic energy given by $KE = \frac{3}{2}\mathrm{k}T$. This relationship allows us to derive the root mean square (rms) velocity of a spherical particle with mass m and diameter d_p:

$$c_{\mathrm{rms}} = \left(\frac{3\mathrm{k}T}{m}\right)^{1/2} = \left(\frac{18\mathrm{k}T}{\pi \rho_\mathrm{p} d_\mathrm{p}^3}\right)^{1/2} \tag{5.9}$$

Similarly, the mean thermal velocity is defined as:

$$\bar{c} = \left(\frac{8\mathrm{k}T}{\pi m}\right)^{1/2} = \left(\frac{48\mathrm{k}T}{\pi^2 \rho_\mathrm{p} d_\mathrm{p}^3}\right)^{1/2} \tag{5.10}$$

The diffusion coefficient is independent of particle density, but the thermal velocity is not. This distinction arises because the forces acting on the particles, such as osmotic pressure and Stokes drag forces, are not reliant on particle density, whereas kinetic energy is directly related to the particle mass. The average net motion of a particle during diffusion results from a random combination of wiggling movements, while the mean thermal velocity represents the average net forward velocity in each step.

Table 5.1 presents the diffusion coefficients of particles of varying sizes. Notably, a particle with a diameter of 0.01 μm has a diffusion coefficient nearly three orders of magnitude smaller than that of an air molecule. Meanwhile, a 0.01 μm particle diffuses over 20,000 times faster than a 10 μm particle.

Table 5.1 Diffusion-related properties of standard-density spherical particles at 293.15 K (20 °C).

Particle diameter (μm)	Mobility (m/N·s)	Diffusion coefficient ($\mathrm{m^2/s}$)	Mean thermal velocity (m/s)
0.00037 (air molecule)	N/A	2.0×10^{-5}	460
0.01	1.3×10^{13}	5.4×10^{-8}	4.4
0.1	1.7×10^{11}	6.9×10^{-10}	0.14
1.0	6.8×10^{9}	2.7×10^{-11}	0.0044
10	6.0×10^{8}	2.4×10^{-12}	0.00014

Source: Hinds and Zhu (2022) / with permission of John Wiley & Sons.

Example 5.1
Calculate the diffusion coefficient of a particle with a diameter of 0.5 μm under standard conditions. From Chapter 2, the slip correction factor for a 0.5 μm particle is 1.337.

Solution

$$D = \frac{kTC_c}{3\pi\eta d_p} = \frac{1.38 \times 10^{-23} \times 293.15 \times 1.337}{3\pi \times 1.81 \times 10^{-5} \times 0.5 \times 10^{-6}} = 6.34 \times 10^{-11}\ \text{m}^2/\text{s}$$

5.4 Particle Mean Free Path

To further quantify the diffusion behavior of a particle, the concept of particle mean free path, denoted as λ_p, is introduced. This term, originally developed to describe the motion of gas molecules, can be adapted to characterize the Brownian motion of aerosol particles.

The mean free path of an aerosol particle is defined as the average straight-line distance that the particle's center travels before it completely changes direction, as illustrated in Figure 5.4. This concept is analogous to the mean free path of gas molecules, which describes the average distance a gas molecule travels between collisions. The motion of aerosol particles, whose sizes are significantly larger than gas molecules, is affected by billions of collisions with molecules of the carrier gases. While each collision may be imperceptible, the cumulative effect is that the particle proceeds in a meandering trajectory.

The mean free path λ_p can be expressed as the product of the particle's mean thermal velocity in Eq. (5.10) and the average time required for particles to change direction:

$$\lambda_p = \bar{c}\tau \tag{5.11}$$

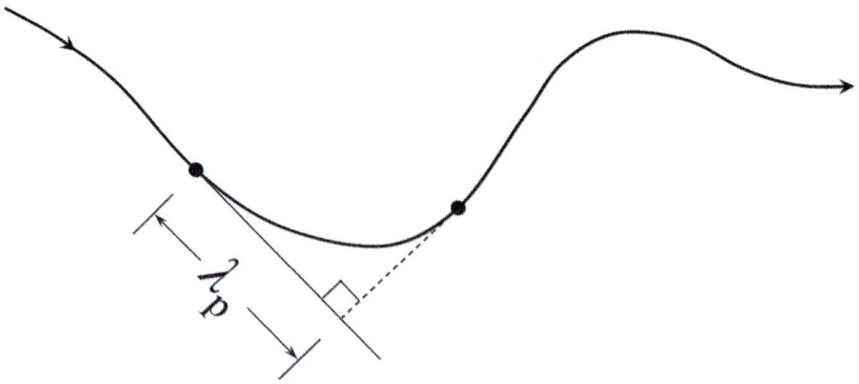

Figure 5.4 Schematic diagram of the trajectory and mean free path of a particle.

Table 5.2 Particle mean free path of standard-density spherical particles.

Particle diameter (μm)	Diffusion coefficient (m^2/s)	τ (s)	$\bar{c}$ (m/s)	λ_p (μm)
0.00037 (air molecule)	2.0×10^{-5}	N/A	460	0.066
0.01	5.4×10^{-8}	6.8×10^{-9}	4.4	0.030
0.1	6.9×10^{-10}	8.8×10^{-8}	0.14	0.012
1.0	2.7×10^{-11}	3.6×10^{-6}	0.0044	0.016
10	2.4×10^{-12}	3.1×10^{-4}	0.00014	0.044

Source: Hinds and Zhu (2022) / with permission of John Wiley & Sons.

Table 5.2 summarizes the values of λ_p, $\bar{c}$, and τ for standard-density spherical particles at 293.15 K. It is interesting to observe that while both $\bar{c}$ and τ vary significantly with the particle size, their product, λ_p, increases by less than fourfold, even when the particle size increases 1,000 times. This indicates that the mean free path of particles is not highly dependent on the particle size. Although larger particles have greater inertia than smaller ones, their lower average thermal velocity compensates for the increased inertial effect. Consequently, the Brownian motion scale is nearly the same for aerosol particles of various sizes, approximately 1/5 to 2/3 of the mean free path of gas molecules.

As is the case with gas molecules, the diffusion coefficient of a particle can also be expressed in terms of the mean thermal velocity $\bar{c}$ and mean free path λ_p:

$$D = \frac{\pi}{8}\lambda_p\bar{c} \tag{5.12}$$

This equation indicates that the diffusion coefficient is directly proportional to both the mean free path and the mean thermal velocity of the particle. Given that λ_p remains relatively constant across a variety of particle sizes, the diffusion coefficient is mainly dependent on the mean thermal velocity $\bar{c}$.

5.5 Brownian Displacement

The mean free path of a particle describes the micro-scale motion of the individual aerosol particle due to Brownian motion. In this section, the macro-scale motion of particles over a longer time interval, i.e., Brownian displacement, is analyzed. Due to the random nature of particle motion, a particle may be displaced in one direction at a given moment, while a collision from a different angle shortly afterward pushes it in the opposite direction. The net displacement of a particle over a given period is attributed to numerous small random movements in various directions. Through statistical analysis of these random motions, the net displacement over a longer time scale can be predicted.

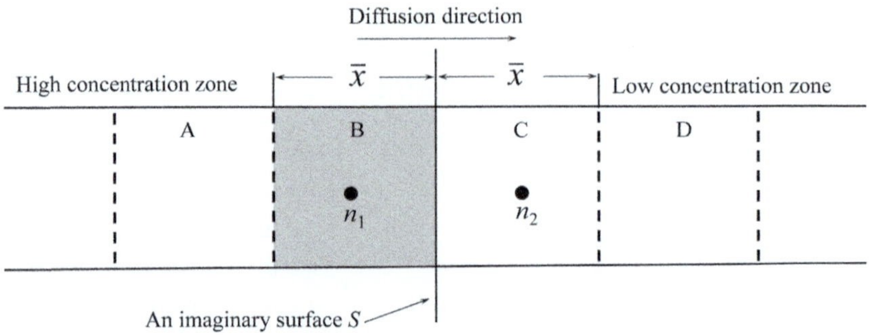

Figure 5.5 Schematic diagram of Brownian motion of particles from the high concentration zone to the lower concentration zone, for the derivation of Brownian displacement. *Source:* Hinds and Zhu (2022) / with permission of John Wiley & Sons.

Figure 5.5 displays the side view of a horizontal cylinder with a unit cross-sectional area. Particles diffuse from regions of high concentration to regions of low concentration. The net flux J of particles through an imaginary plane S from left to right is calculated. In a time interval t, larger than τ, each particle moves a net distance x_i, assuming that particles move only in the x-direction. This displacement is generally to the left for some particles and to the right for others. For the sake of simplicity, the individual displacements are replaced by their root mean square average $\bar{x}$.

Let us define two regions, B and C, as shown in Figure 5.5, characterized by particle number concentrations n_1 and n_2, respectively, at their midpoints. The concentration in region C in relation to the concentration in region B can be expressed from Taylor's expansion as:

$$n_2 = n_1 + \bar{x}\frac{dn}{dx} \tag{5.13}$$

where $\frac{dn}{dx}$ is negative because the concentration decreases along the diffusion direction, as shown in Figure 5.5.

The net transfer rate J of particles through surface S of unit area from left to right can be expressed as:

$$J = \frac{\bar{x}}{2t}n_1 - \frac{\bar{x}}{2t}n_2 = \frac{\bar{x}}{2t}(n_1 - n_2) \tag{5.14}$$

Substituting n_2 in Eq. (5.13) into Eq. (5.14) yields:

$$J = \frac{\bar{x}}{2t}\left(-\bar{x}\frac{dn}{dx}\right) \tag{5.15}$$

Rearranging gives:

$$J = -\frac{\bar{x}^2}{2t}\frac{dn}{dx} \tag{5.16}$$

Table 5.3 Net displacement in 1 second due to Brownian motion and gravitational settling for standard-density particles under standard conditions.

Particle diameter (μm)	Brownian displacement (x_{BM}) (m)	Gravitational settling (x_{Grav}) (m)	x_{BM}/x_{Grav}
0.01	3.3×10^{-4}	6.9×10^{-8}	4,800
0.1	3.7×10^{-5}	8.8×10^{-7}	42
1.0	7.4×10^{-6}	3.5×10^{-6}	0.21
10	2.2×10^{-6}	3.1×10^{-3}	7.1×10^{-4}

Source: Hinds and Zhu (2022) / with permission of John Wiley & Sons.

Eq. (5.16) presents another formulation of Fick's first law. When we compare Eqs. (5.16) and (5.1), we can express the diffusion coefficient D as:

$$D = \frac{\bar{x}^2}{2t} \tag{5.17}$$

Consequently, the root mean square average $\bar{x}$ along any axis after a time t can be cast as:

$$\bar{x} = \sqrt{2Dt} \tag{5.18}$$

The preceding derivation reveals the relationship between Brownian displacement and the particle diffusion coefficient and time. Particles undergoing Brownian motion move randomly in all directions. However, when a concentration gradient exists, particles tend to move from regions of higher concentration to regions of lower concentration. In the absence of a concentration gradient, Brownian motion still exists. However, the diffusion flux remains balanced in both directions, resulting in zero net mass transfer.

Table 5.3 presents the relative significance of Brownian motion and gravitational settling for particles of different sizes. Notably, the ratio of Brownian displacement to gravitational settling (x_{BM}/x_{Grav}) indicates that for extremely small particles (0.01 μm), Brownian displacement is approximately 4,800 times the gravitational settling. As the particle size increases, the gravitational settling gradually becomes dominant. In summary, both diffusion and Brownian motion are vital for the movement of aerosol particles with diameters less than 0.1 μm. Gravitational settling becomes increasingly significant for larger particles, particularly those exceeding 1 μm.

5.6 Diffusion Deposition

Deposition by diffusion refers to the process through which aerosol particles settle onto surfaces due to the random Brownian motion of particles. Unlike gas molecules, which tend to rebound from surfaces and generate a pressure difference, ultrafine aerosol particles often adhere to surfaces upon collision. Consequently, the concentration of aerosol particles near the surface decreases, creating a concentration gradient in the surrounding air. This gradient drives the continuous diffusion of particles from regions of higher concentration in the air toward the surface, leading to a gradual reduction in aerosol concentration.

Let us consider the simplest scenario, namely, a vertical plane surface immersed in an infinitely large volume of aerosols with a uniform initial particle concentration, denoted as n_0. Initially, there is no air movement near the surface, and our objective is to calculate the rate at which particles are removed from the air through Brownian deposition. Let x represent the distance from the surface, and let $n(x,t)$ denote the particle concentration at distance x from the surface at time t. The evolution of this concentration is governed by Fick's second law of diffusion:

$$\frac{\partial n}{\partial t} = D\frac{\partial^2 n}{\partial x^2} \text{ and } \begin{cases} n(x,0) = n_0, & \text{for } x > 0 \\ n(0,t) = 0, & \text{for } t > 0 \end{cases} \tag{5.19}$$

The solution to Eq. (5.19) gives the particle concentration at any distance x from the surface at time t:

$$n(x,t) = \frac{n_0}{(\pi D t)^{1/2}} \int_0^x \exp\left(\frac{-p^2}{4Dt}\right) \mathrm{d}p \tag{5.20}$$

where p is a dummy variable of integration that ranges from 0 to x. Although the derivation of Eq. (5.20) assumes an infinite aerosol volume, it remains valid for finite volumes as long as the concentration gradient from the other side walls does not interfere.

Figure 5.6 illustrates the concentration profiles of aerosol particles with a diameter of 0.05 μm near the surface over various time intervals. As time passes, the concentration gradient becomes less steep while simultaneously extending further away from the surface. Notably, even after 1,000 seconds, the aerosol concentration remains largely unaffected up to 6 mm from the wall. Consequently, particle deposition by Brownian diffusion does not result in a significant decrease in aerosol concentration unless the geometric scale is less than 1 cm, or the particle size is much less than 0.05 μm.

To obtain the rate of particle deposition on a wall surface, let us calculate the concentration gradient at the surface ($x = 0$) using Fick's first law. From Eq. (5.20), the concentration gradient at the surface is:

$$\left.\frac{\mathrm{d}n}{\mathrm{d}x}\right|_{x=0} = \frac{n_0}{\sqrt{\pi D t}} \tag{5.21}$$

Substituting Eq. (5.21) into Eq. (5.1), we obtain the rate at which particles deposit onto a unit area of surface at any given time t:

$$J|_{x=0} = -D\left.\frac{\mathrm{d}n}{\mathrm{d}x}\right|_{x=0} = -n_0\sqrt{\frac{D}{\pi t}} \tag{5.22}$$

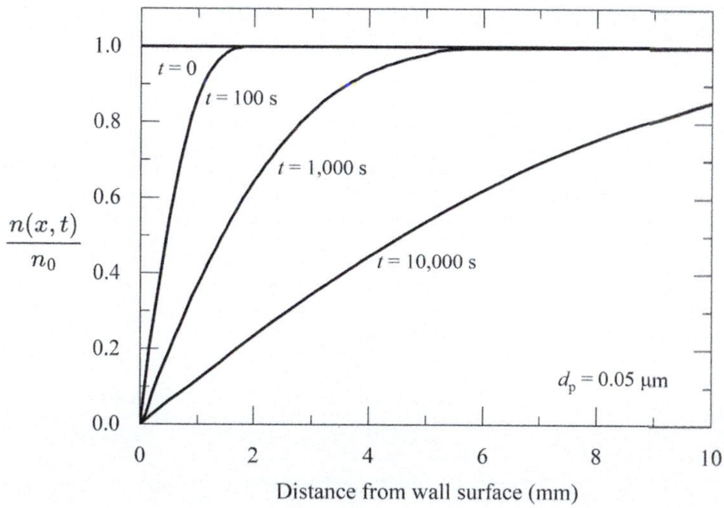

Figure 5.6 Concentration profiles over different time intervals for 0.05 μm particles in close proximity to a wall surface. *Source:* Hinds and Zhu (2022) / with permission of John Wiley & Sons.

To determine the total number of particles deposited on the surface during a given period, the deposition rate over time can be integrated. The cumulative number of particles deposited per unit area during time t is:

$$N(t) = \int J|_{x=0}\, dt = -\int_0^t n_0 \sqrt{\frac{D}{\pi t}} dt = -2n_0 \sqrt{\frac{Dt}{\pi}} \tag{5.23}$$

The preceding derivation assumes the concentration of stationary aerosols in an infinite volume remains at n_0 outside the gradient zone. However, Eq. (5.23) can still be used to predict the upper limit of the deposition loss of particles onto the walls of a finite-sized container (Hinds and Zhu 2022).

Table 5.4 compares particle deposition on a horizontal surface due to Brownian diffusion and gravitational settling. It can be seen that deposition by Brownian diffusion dominates when the particle size is less than 0.1 μm. However, as the particle size increases, gravitational settling gradually becomes the dominant mechanism.

The deposition rate can also be expressed by the deposition velocity, denoted as V_{Dep}, which is defined as the deposition flux divided by the uniform initial particle concentration n_0:

$$V_{Dep} = \frac{J}{n_0} = \sqrt{\frac{D}{\pi t}} \tag{5.24}$$

Table 5.4 Cumulative deposition of standard-density particles within 100 seconds by Brownian diffusion and gravitational settling (assuming that the concentration outside the gradient region is 1 particle/cm^3).

	Cumulative deposition		
Particle diameter (μm)	**Diffusion deposition (particles/m^2)**	**Gravitational deposition (particles/m^2)**	**Ratio of diffusion to gravity**
0.001	2.6×10^4	0.68	3.8×10^4
0.01	2.6×10^3	6.9	380
0.1	3.0×10^2	88	3.4
1.0	59	3,500	1.7×10^{-2}
10	17	3.1×10^5	5.5×10^{-5}
100	5.5	2.5×10^7	2.2×10^{-7}

Source: Hinds and Zhu (2022) / with permission of John Wiley & Sons.

The deposition velocity represents the effective speed at which particles migrate to the surface due to diffusion, similar to the way in which settling velocity indicates the speed of particle settling due to gravity.

The diffusion deposition described earlier is relatively simplified. The diffusion deposition of aerosol particles in ventilation ducts or particle sampling tubes, are more complex. Particle loss due to diffusion deposition in a cylindrical tube is addressed in Section 6.4.

5.7 Thermophoresis

Thermophoresis is the motion of aerosol particles due to random bombardment by molecules of the carrier gases with a temperature gradient. When there is a temperature gradient within the carrier gases, the particles are subjected to a thermal force that drives them to the low-temperature region and away from the high-temperature region.

The initial studies of thermophoresis originated from observations of heated objects, such as metal rods, immersing in smoke (Hinds and Zhu 2022). Researchers observed the formation of a dust-free layer around these heated objects, as illustrated in Figure 5.7. The smoke particles appeared to be repelled by the heated surface, creating a distinct, particle-free zone with a boundary layer often less than 1 mm thick. These early studies suggested that the thermal forces associated with the temperature gradient near the heated surface were responsible for displacing the particles. Furthermore, the studies established that the thickness of this particle-free layer is proportional to the square root of the temperature difference between the object and the surrounding gases, irrespective of the particle material.

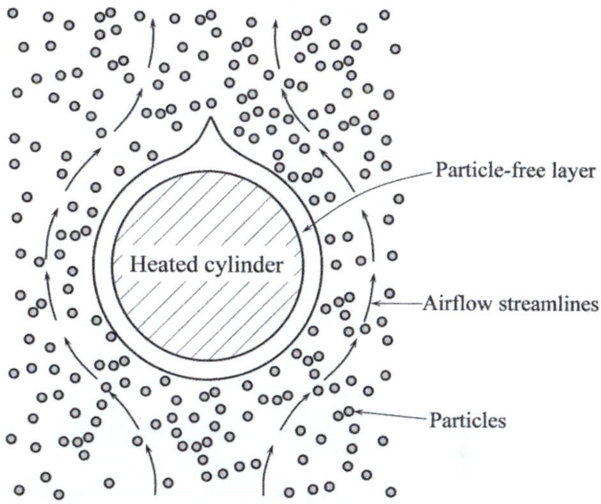

Figure 5.7 Schematic diagram of the particle-free layer around a heated horizontal cylinder due to a temperature gradient. *Source:* Hinds and Zhu (2022) / with permission of John Wiley & Sons.

For particles with a diameter d_p less than the gas mean free path λ, the thermophoretic force arises from an imbalance in momentum transfer. Specifically, gas molecules on the hot side of particles move faster and transfer more momentum to particles than do the molecules on the cold side. This differential momentum transfer generates a net force that drives particles toward the cooler region. The force can be described by the following equation (Hinds and Zhu 2022):

$$F_{\mathrm{Th}} = -\frac{p_g \lambda d_p^2 \nabla T}{T}, \quad \text{for } d_p < \lambda \tag{5.25}$$

where p_g is the gas pressure, λ is the mean free path of the carrier gases, ∇T is the temperature gradient, and T is the absolute temperature of the particle. The negative sign indicates that the direction of the force is toward the cooler region. The velocity with which the particle moves due to thermophoresis can be expressed as (Waldmann and Schmitt 1966):

$$V_{\mathrm{Th}} = -\frac{0.55\eta \nabla T}{\rho_g T}, \quad \text{for } d_p < \lambda \tag{5.26}$$

where η is the dynamic viscosity of the carrier gases and ρ_g is the gas density. Notably, for small particles ($d_p < \lambda$), the thermophoretic velocity is independent of the particle size and is directly proportional to the temperature gradient. This implies that a steeper temperature gradient results in faster particle movement toward the cooler region.

For air at standard pressure and a particle at 293.15 K (20 °C), the thermophoretic velocity for particles smaller than 0.1 μm is (Hinds and Zhu 2022):

$$V_{\mathrm{Th}} = -2.8 \times 10^{-8} \nabla T \tag{5.27}$$

For larger particles ($d_p > \lambda$), the thermophoretic mechanism becomes more complex. In these cases, a temperature gradient also exists within the particle itself, influencing the temperature distribution of the surrounding gases. However, similar to smaller particles, larger particles still experience a greater force from the hot side of the gas molecules than from the cold side, driving them toward the cooler region. The thermophoretic force for large particles can be described by the following equation (Hinds and Zhu 2022):

$$F_{\mathrm{Th}} = \frac{-9\pi d_{\mathrm{p}}\eta^2 H \nabla T}{2\rho_{\mathrm{g}} T}, \quad \text{for } d_{\mathrm{p}} > \lambda \tag{5.28}$$

where the coefficient H accounts for the temperature gradient inside the particle, while ∇T represents the overall temperature gradient in the gases in the absence of the particle. The value of H is subject to the ratio of the thermal conductivity of the surrounding gases k_{a} to that of the particle k_{p}. The expression for H is (Brock 1962):

$$H \cong \left(\frac{1}{1 + 6\lambda/d_{\mathrm{p}}}\right)\left(\frac{k_{\mathrm{a}}/k_{\mathrm{p}} + 4.4\lambda/d_{\mathrm{p}}}{1 + 2k_{\mathrm{a}}/k_{\mathrm{p}} + 8.8\lambda/d_{\mathrm{p}}}\right) \tag{5.29}$$

The thermophoresis velocity for larger particles can be obtained by equating the thermophoretic force to the Stokes drag force, which yields the following expression:

$$V_{\mathrm{Th}} = \frac{-3\eta C_{\mathrm{c}} H \nabla T}{2\rho_{\mathrm{g}} T}, \quad \text{for } d_{\mathrm{p}} > \lambda \tag{5.30}$$

Table 5.5 compares terminal settling velocities due to gravity with thermophoretic velocities in a temperature gradient of 100 K/m. For particles smaller than 0.1 μm, the thermophoretic velocity exceeds the terminal settling velocity, indicating the dominance of thermophoresis for small particles under moderate temperature gradients. Table 5.6 compares the particle deposition via diffusion and thermophoresis. In a temperature gradient of 100 K/m over 100 seconds, the cumulative deposition under thermophoresis is greater than that by diffusion for particles larger than 0.2 μm (Hinds and Zhu 2022).

Example 5.2

Calculate the thermophoretic velocity of a sodium chloride particle between vertical plates at 30 °C and 10 °C. The particle size is 5 μm, and the distance between the two plates is 2 mm. The thermal conductivity of air at 20 °C is 0.026 W/m·K, while that of the particle is 6.6 W/m·K. The mean free path of air at 20 °C is 0.066 μm.

Solution

$$k_{\mathrm{a}}/k_{\mathrm{p}} = 0.026/6.6 = 0.0039$$

$$\lambda/d_{\mathrm{p}} = 0.066/5 = 0.0132$$

Table 5.5 Terminal settling and thermophoretic velocities with a temperature gradient of 100 K/m at 293.15 K.

Particle diameter (μm)	Terminal settling velocity due to gravity (m/s)	Thermophoretic velocity[a)] (m/s)
0.01	6.7×10^{-8}	2.8×10^{-6}
0.1	8.6×10^{-7}	2.0×10^{-6}
1.0	3.5×10^{-5}	1.3×10^{-6}
10	3.1×10^{-3}	7.8×10^{-7}

a) For the calculation of H in Eq. (5.29), k_p=10 k_a.
Source: Hinds and Zhu (2022) / with permission of John Wiley & Sons.

Table 5.6 Cumulative deposition within 100 seconds by Brownian diffusion and thermophoresis from a unit aerosol concentration (1 particle/cm^3) in a temperature gradient of 100 K/m at 293.15 K.

Particle diameter (μm)	Cumulative deposition	
	Diffusion (particles/m^2)	Thermophoresis[a)] (particles/m^2)
0.01	2,600	280
0.1	290	200
1.0	59	130
10	17	78

a) For the calculation of H in Eq. (5.29), $k_p = 10\ k_a$.
Source: Hinds and Zhu (2022) / with permission of John Wiley & Sons.

From Eq. (5.29), we can obtain the value of H as:

$$H = \left(\frac{1}{1+6\lambda/d_p}\right)\left(\frac{k_a/k_p + 4.4\lambda/d_p}{1+2k_a/k_p+8.8\lambda/d_p}\right)$$

$$= \left[\frac{1}{1+6(0.0132)}\right]\left[\frac{0.0039+4.4(0.0132)}{1+2(0.0039)+8.8(0.0132)}\right] = 0.051$$

Then the thermophoretic velocity can be solved from Eq. (5.30) as:

$$V_{Th} = \frac{-3\eta C_c H \nabla T}{2\rho_g T}$$

$$= \frac{-3\times1.81\times10^{-5}\times1.034\times0.051\times((10-30)/0.002)}{2\times1.205\times293.15}$$

$$= 4.05\times10^{-5}\ \text{m/s}$$

5.8 Summary

This chapter addresses the Brownian motion and diffusion of aerosol particles. The main points are as follows:

The relationship between Brownian diffusive flux and the particle concentration gradient can be described using Fick's first law. The diffusive flux of particles is proportional to both the concentration gradient and the diffusion coefficient. The particle diffusion coefficient is subject to the air temperature and particle diameter. As the particle diameter increases, the diffusion coefficient of the particles decreases sharply.

The micro-scale migration of individual aerosol particles due to Brownian motion can be characterized by the particle mean free path, which represents the average straight-line distance that the particle's center travels before it completely changes direction. The mean free path is very close for particles of different diameters and is approximately 1/5 to 2/3 of the mean free path of gas molecules.

The macro-scale motion of particles over a long time interval in a diffusion process can be described by Brownian displacement, which is positively correlated with the particle diffusion coefficient and the time scale. For particles smaller than 0.1 μm in diameter, the Brownian displacement is much greater than the displacement caused by gravitational settling. In contrast, for particles larger than 1 μm, gravitational settling dominates.

The particle deposition due to Brownian motion can be significant for particles smaller than 0.1 μm. For particles larger than 1 μm, the deposition due to diffusion is negligible, as compared with the gravitational settling. In the presence of a temperature gradient in the carrier gases, the particles are subject to the thermophoretic force in the direction toward the low-temperature side. Under a temperature gradient of 100 K/m, the deposition by Brownian diffusion is much greater than that by thermophoresis for particles smaller than 0.1 μm. In contrast, for particles larger than 0.2 μm, thermophoretic deposition becomes dominant over Brownian diffusion.

Problems

5.1 A spherical particle with a diameter of 0.2 μm is suspended in air at 293.15 K. The slip correction factor C_c for this particle is 1.15. Please calculate the diffusion coefficient of the particle.
[Answer: 1.4×10^{-10} m^2/s]

5.2 A 0.05 μm particle has a diffusion coefficient of 5.4×10^{-8} m^2/s. Please calculate its root mean square displacement after 10 minutes.
[Answer: 8.0×10^{-3} m]

5.3 A 0.1 μm particle has a mean thermal velocity of 0.14 m/s and a direction-change time of 8.8×10^{-8} seconds. Please calculate its mean free path.
[Answer: 1.2×10^{-8} m]

5.4 A 0.01 μm particle is subjected to a temperature gradient of $\nabla T = 200$ K/m. Please calculate the thermophoretic velocity.
[Answer: -5.6×10^{-6} m/s]

5.5 An aerosol with a fixed initial concentration of 1×10^{12} particles/m^3. Assume the diffusion coefficient of particles is 6.9×10^{-10} m^2/s. Please calculate the deposition flux J onto a wall surface after 1 hour.
[Answer: -2.5×10^5 particles/(m$^2 \cdot$ s)]

5.6 Explain why the diffusion coefficient decreases sharply with increasing particle size. Additionally, discuss how the slip effect of particles influences the diffusion coefficient.

5.7 Provide an example on how thermophoresis is applied in engineering practice, and describe the particle size range that is most effectively affected by thermophoresis.

5.8 The derivation of Eq. (5.23) is based on the assumption of an infinite aerosol volume. What are the practical limitations when applying this equation to finite volumes or containers?

References

Brock, J.R. (1962). On the theory of thermal forces acting on aerosol particles. *Journal of Colloid Science* 17: 768–780.

Hinds, W.C. and Zhu, Y. (2022). *Aerosol Technology: Properties, Behavior, and Measurement of Airborne Particles*, 3e. Wiley.

Waldmann, L. and Schmitt, K.H. (1966). *Thermophoresis and Diffusiophoresis of Aerosol*. Academic Press.

6

Sampling and Measurement of Particle Concentrations

6.1 Introduction

Aerosol samples are extracted from the air for various purposes: observing the particle morphology, measuring geometric sizes, analyzing chemical components, quantifying mass or number concentrations, and so on. Sampling aerosols involves the extraction of aerosol-laden air from the ambient environment into a tube and eventually the transmission of the air to measuring devices. During the extraction process, there may be variations in the air motion that is created from the ambient environment to the sampling head and also inside the sampling tube. Due to the inertia of aerosol particles, the concentration of aerosols extracted into the sampling system may be different from that in the ambient environment. Some of the aerosols may deposit inside the sampling tube and thus are lost during transmission. The biases or uncertainties during particle extraction and transmission should be adequately evaluated. This chapter addresses the principle of particle sampling efficiency, the impact of sampling-head shape on the sampled particle concentration, the particle transport loss inside the sampling tube, and the basic principle for measuring the particle mass and number concentrations.

6.2 Particle Sampling

An aerosol sampling system collects aerosols from the ambient environment and carries them to a testing instrument via a sampling tube, as illustrated by the schematic diagram in Figure 6.1. The proportion of the ambient aerosols that enter the sampling tube through the sampling head is called the *inlet efficiency*. Small particles can closely trace the airflow and make their way into the sampling head. Meanwhile, large particles are not sensitive to changes in airflow direction, due to inertia; thus, some large particles may over- or under-enter the sampling head, which results in measurement bias.

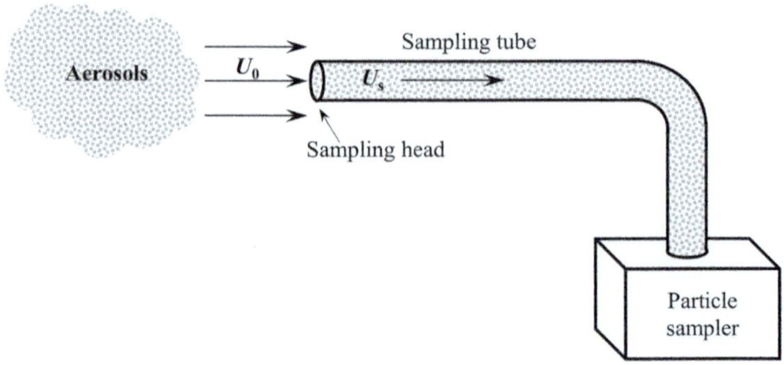

Figure 6.1 Schematic diagram of an aerosol sampling system.

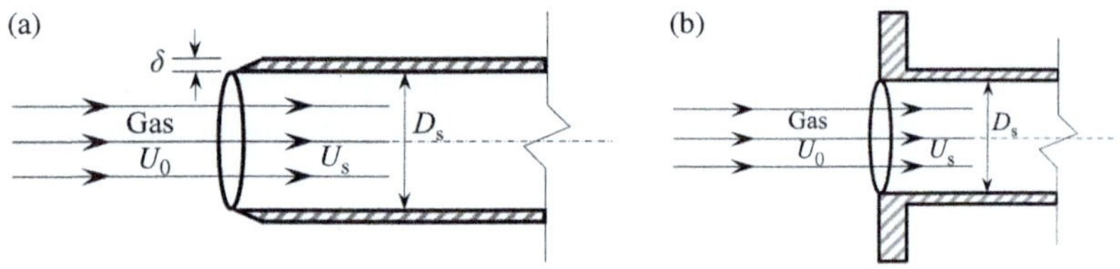

Figure 6.2 Two typical airborne particle sampling heads: (a) thin-wall sampling head; (b) blunt-wall sampling head.

Figure 6.2 shows two typical particle sampling heads: a thin-wall sampling head and a blunt-wall sampling head. Thin-wall sampling heads are commonly used in connection with particle measuring instruments to measure particle concentration. The wall thickness of the sampling head, δ, is much smaller than the sampling tube diameter, D_s, typically at a ratio of $D_s/\delta > 10$ (Zhang 2004). To minimize disturbance to the airflow, many thin-wall sampling heads are designed with sharp edges. However, to simulate the human mouth or nose, a blunt-wall sampling head is occasionally used, as shown in Figure 6.2b. As compared with the surrounding airflow, the airflow around a blunt-wall sampling head is significantly disrupted, which may result in a difference in particle concentration between the free stream and the sampled gas stream.

6.2.1 Isokinetic Sampling

As shown in Figure 6.3, when the axis of the sampling head (or sampling probe) is parallel to the gas streamlines, and the velocity of the gas entering the probe is equal to the free-stream velocity near the inlet, the sampling process is *isokinetic*. During isokinetic sampling, the direction of airflow at the interface between the sampling probe and the free stream remains unchanged, thereby preventing the

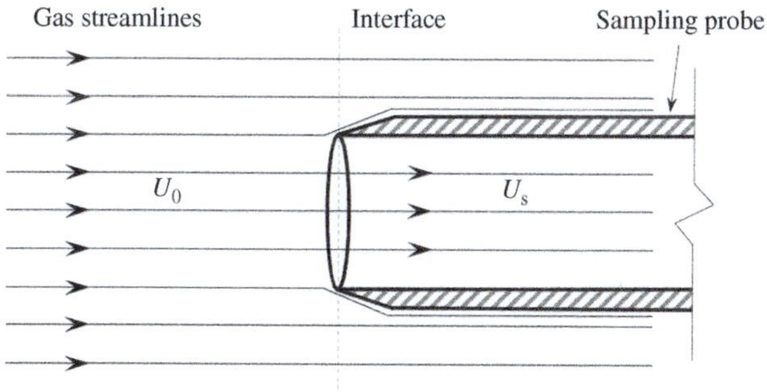

Figure 6.3 Schematic diagram of isokinetic sampling.

loss of particles from the sampled air. Thin-wall sampling probes cause minimal disruption to the airflow, making them more suitable than other types of samplers for achieving isokinetic sampling. Isokinetic sampling ensures that the collected particles are sufficiently representative in terms of particle concentration and size distribution.

To ensure isokinetic sampling in a free stream with a velocity of U_0 and a gas velocity of U_s in the probe, these velocities must be identical in magnitude and direction:

$$U_s = U_0 \tag{6.1}$$

When conducting isokinetic sampling in a duct, the flow rates in the duct and sampling probe are proportional to the areas of their respective cross-sections, as follows:

$$\frac{Q_s}{Q_0} = \left(\frac{D_s}{D_0}\right)^2 \tag{6.2}$$

where Q_s is the sampling flow rate and D_s is the diameter of the sampling probe, which is inside a circular duct of diameter D_0 with a flow rate Q_0.

Example 6.1

The airflow rate in a duct with a diameter of 0.04 m is 2 l/s. The diameter of the sampling probe is 0.01 m. What flow rate should be used to achieve isokinetic sampling?

Solution

From Eq. (6.2),

$$Q_s = Q_0 \times \left(\frac{D_s}{D_0}\right)^2 = 2\,\text{l/s} \times \left(\frac{0.01}{0.04}\right)^2 = 0.125\,\text{l/s}$$

6.2.2 Anisokinetic Sampling

Anisokinetic sampling refers to the failure to achieve isokinetic sampling. When sampling is anisokinetic, it results in a shifted particle size distribution and bias in the particle concentration. The inertia of large particles prevents their motion from closely tracing the air streams near the sampling head, and the anisokinetic sampling may lead to either an overrepresentation or underrepresentation of large particles. There are three general forms of anisokinetic sampling: misalignment sampling, superisokinetic sampling, and subisokinetic sampling.

6.2.2.1 Misalignment Sampling

Figure 6.4 shows the angle of misalignment θ between the axis of the probe and the free gas streamlines when the sampling probe extracts air at the isokinetic flow rate. In this situation, the particle concentrations are underestimated because particles with high inertia in the air are unable to turn quickly enough to enter the sampling probe. The maximum error in the sampled particle concentration can be estimated as:

$$\xi = \frac{C_s}{C_0} = \cos\theta, \text{ for Stk} > 6, 0^\circ < \theta < 90^\circ, \text{ and } U_0 = U_s \tag{6.3}$$

where C_s is the concentration inside the sampling probe, C_0 is the concentration in the free stream, and Stk is the Stokes number at the sampling head (probe inlet), which is defined as:

$$\text{Stk} = \frac{\tau U_0}{D_s} \tag{6.4}$$

When Stk < 0.01, the particle inertia is negligible and the concentration ratio $C_s/C_0 = 1$. Ultrafine particles can be regarded as closely following the gas streams into the probe inlet.

An empirical equation (Durham and Lundgren 1980) for evaluating the particle sampling efficiency as a function of the Stokes number at the sampling probe inlet

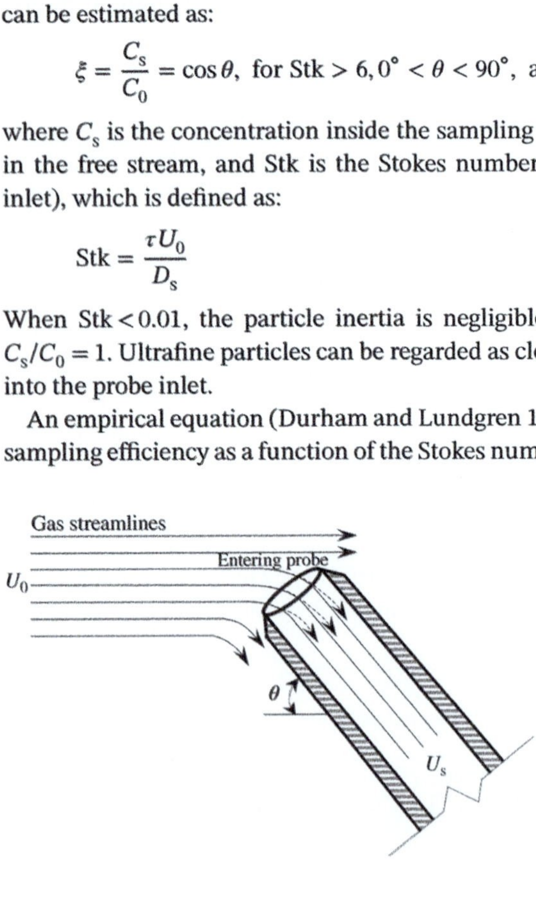

Figure 6.4 Schematic diagram of misalignment sampling, where $\theta \neq 0$.

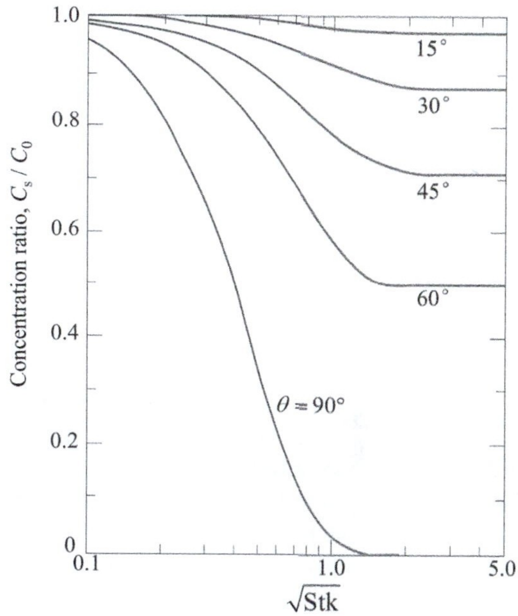

Figure 6.5 Effect of probe misalignment on the particle concentration ratio for $U_s = U_0$. *Source:* Hinds and Zhu (2022) / with permission of John Wiley & Sons.

with a misalignment angle ($0° \leq \theta \leq 90°$) is:

$$\xi = \frac{C_s}{C_0} = 1 + (\cos\theta - 1)\left(1 - \frac{1}{1 + 0.55(\text{Stk}') \cdot \exp(0.25\text{Stk}')}\right),$$
$$\text{for } 0.01 < \text{Stk} < 6 \text{ and } U_s = U_0 \tag{6.5}$$

where

$$\text{Stk}' = \text{Stk} \cdot \exp(0.022\theta) \tag{6.6}$$

and θ is expressed in degrees. Eq. (6.5) only applies when $U_s = U_0$. If Stk $\to 0$, $\xi \to 1$, which indicates that it is regardless of the misalignment angle. Figure 6.5 illustrates the effects of the Stokes number and misalignment angle on the concentration ratio C_s/C_0 when $U_s = U_0$.

When sampling conditions are such that $U_s \neq U_0$ and there is also a misalignment angle θ of the sampling probe, the particle concentration ratio in the extreme situation is:

$$\xi = \frac{C_s}{C_0} = \frac{U_0}{U_s}\cos\theta, \text{for Stk} > 6 \text{ and } 0° \leq \theta \leq 90° \tag{6.7}$$

Example 6.2

A thin-wall probe with a diameter of 11 mm is misaligned by 30° in a duct with an air speed of 7 m/s. What is the sampled concentration ratio for 5 μm particles of standard density? Assume that the sampling probe extracts air at the isokinetic flow rate.

Solution

For 5 μm particles, the slip correction factor $C_c \approx 1$.

$$\text{Stk} = \frac{\tau U_0}{D_s} = \frac{\rho_p d_a^2 C_c U_0}{18\eta D_s} = \frac{1000 \times (5 \times 10^{-6})^2 \times 1 \times 7}{18 \times 1.81 \times 10^{-5} \times 0.011} = 0.049$$

Then $0.01 < \text{Stk} < 6$, and $\theta = 30°$.

$$\text{Stk}' = \text{Stk} \cdot \exp(0.022\theta) = 0.049 \times \exp(0.022 \times 30) = 0.095$$

Then,

$$\xi = \frac{C_s}{C_0} = 1 + (\cos\theta - 1)\left[1 - \frac{1}{1 + 0.55\text{Stk}' \cdot \exp(0.25\text{Stk}')}\right]$$

$$= 1 + (\cos 30° - 1)\left[1 - \frac{1}{1 + 0.55 \times 0.095 \times \exp(0.25 \times 0.095)}\right] = 0.993$$

6.2.2.2 Superisokinetic Sampling

Superisokinetic sampling refers to a condition in which the sampling velocity U_s is higher than the free-stream velocity U_0. Under superisokinetic sampling conditions, the larger the particle inertia, the greater the loss. As shown in Figure 6.6, the gas streamlines entering the sampling tube converge when $U_s > U_0$. However, due to inertia, large particles cannot trace the gas streamlines well and are thus excluded from entering the sampling probe. The sampling efficiency for superisokinetic sampling is less than 1 for large particles with significant inertia.

The sampling error arises for large, heavy particles that continue along a straight line during the superisokinetic sampling, as shown in Figure 6.6 (and also applicable to Figure 6.7). The concentration ratio for these particles can be estimated as:

$$\xi = \frac{C_s}{C_0} = \frac{U_0}{U_s}, \text{ for Stk} > 6 \tag{6.8}$$

A generalized concentration ratio (Belyaev and Levin 1974) is proposed for a properly aligned probe, as follows:

$$\xi = \frac{C_s}{C_0} = 1 + \left(\frac{U_0}{U_s} - 1\right)\left(1 - \frac{1}{1 + \text{Stk}[2 + 0.62(U_s/U_0)]}\right) \tag{6.9}$$

In cases in which $\text{Stk} = 0$ or $U_0/U_s = 1$, $C_s/C_0 = 1$.

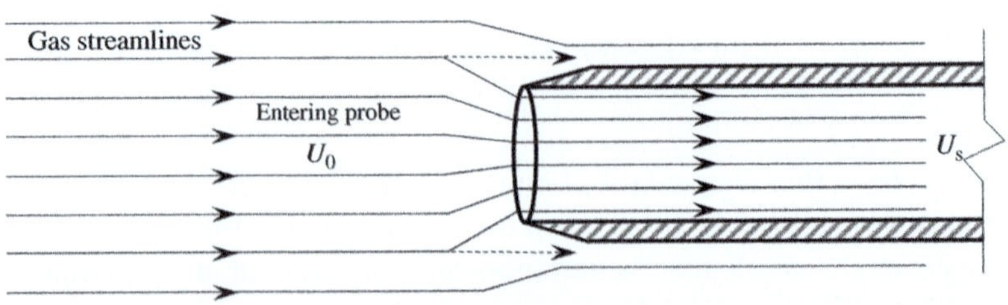

Figure 6.6 Schematic diagram of superisokinetic sampling in which $U_s > U_0$.

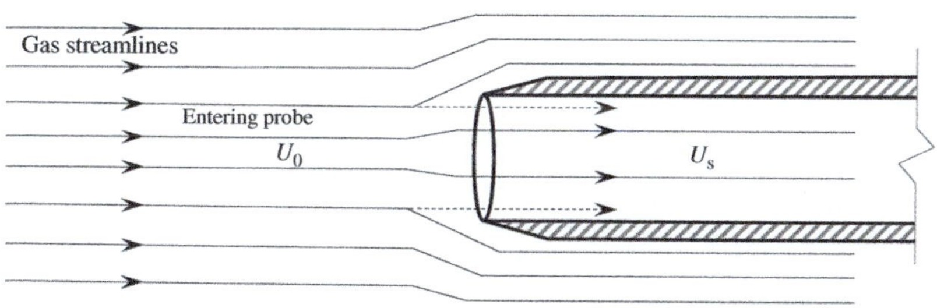

Figure 6.7 Schematic diagram of subisokinetic sampling in which $U_s < U_0$.

6.2.2.3 Subisokinetic Sampling

Subisokinetic sampling refers to a condition in which the sampling velocity U_s is lower than the free-stream velocity U_0. In this case, as shown in Figure 6.7, the free streamlines diverge, causing some particles with significant inertia that were not originally in the sampled air to be impacted into the sampler. Large particles are especially susceptible to such impaction. As a result, the sampling efficiency for subisokinetic sampling is greater than 1 for large particles. Eqs. (6.8) and (6.9) can also be employed to estimate the effect of subisokinetic sampling on the concentration ratio C_s/C_0.

In Figure 6.8, the impact of the velocity mismatch on the particle concentration ratio is depicted as a function of the square root of the Stokes number. When $\text{Stk} < 0.01$ and $0.2 < U_0/U_s < 5$, the anisokinetic loss is negligible and $C_s/C_0 \approx 1$.

Example 6.3

The sampling rate of a thin-wall probe with a diameter of 37 mm is 28.3 l/min. The probe is placed inside a duct in which the air velocity is 7 m/s. What is the concentration ratio for the sampled 6 μm particles of standard density? Assume that the probe is correctly aligned.

Solution

For 6 μm particles, the slip correction factor $C_c \approx 1$.

$$\text{Stk} = \frac{\tau U_0}{D_s} = \frac{\rho_p d_a^2 C_c U_0}{18\eta D_s} = \frac{1000 \times (6 \times 10^{-6})^2 \times 1 \times 7}{18 \times 1.81 \times 10^{-5} \times 0.037} = 0.0209$$

and

$$U_s = \frac{4Q_s}{\pi D_s^2} = \frac{4 \times 0.0283}{60 \times \pi \times (0.037)^2} = 0.439 \text{ m/s}$$

Then $0.01 < \text{Stk} < 6$, and $\theta = 0°$.

$$\xi = \frac{C_s}{C_0} = 1 + \left(\frac{U_0}{U_s} - 1\right)\left(1 - \frac{1}{1 + \text{Stk}(2 + 0.62(U_s/U_0))}\right)$$

$$= 1 + \left(\frac{7}{0.439} - 1\right)\left(1 - \frac{1}{1 + 0.0209(2 + 0.62(0.439/7))}\right) = 1.611$$

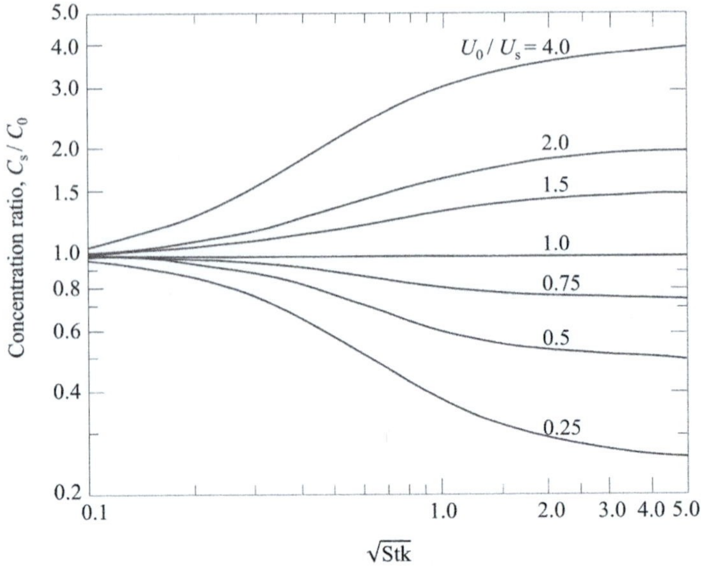

Figure 6.8 Particle concentration ratio versus the square root of Stokes number for various velocity ratios and $\theta = 0°$. *Source:* Hinds and Zhu (2022) / with permission of John Wiley & Sons.

6.3 Sampling from Still Air

There are two main sources of bias when particles are sampled from still air: one source is the settling velocity of the particles, and the other is related to particle inertia. The particle concentration is overestimated in situations in which the inlet velocity is low and the inlet tube is directed upward because additional particles that were not initially in the sampled air volume can settle into the inlet. On the other hand, the concentration is likely to be underestimated when the sampling tube is directed downward. Meanwhile, when the tube is horizontal, there should be no bias from the particle settling effect. Davies (1968) proposed the following criterion to minimize the sampling bias arising from particle settling for probes in any orientation:

$$U_s \geq 25 V_{TS} \tag{6.10}$$

or

$$D_s \leq \frac{2}{5}\left(\frac{Q_s}{\pi \tau g}\right)^{1/2} \tag{6.11}$$

where U_s is the air velocity in the probe, V_{TS} is the particles' terminal settling velocity, Q_s is the sample flow rate, D_s is the sampling probe diameter, and τ

is the particle relaxation time. For standard conditions, when slip correction is neglected, Eq. (6.11) can be rewritten as (Hinds and Zhu 2022):

$$D_s \leq 680\left(\frac{Q_s^{1/2}}{d_a}\right) \tag{6.12}$$

where D_s is expressed in mm, Q_s in m^3/h, and d_a in μm.

Regarding the bias due to particle inertia, let us consider a particle with significant inertia approaching a sampling probe inlet along a curved streamline. The stopping distance of the particle may be greater than the diameter of the probe inlet. Particles with high inertia may not enter the probe inlet, which results in collection loss. This bias becomes more pronounced for larger particles and higher inlet velocities. The following criterion (Davies 1968) was proposed for determining the minimum diameter of the sampling head (probe inlet):

$$D_s \geq 10\left(\frac{Q_s\tau}{4\pi}\right)^{1/3} \tag{6.13}$$

For standard conditions, when slip correction is neglected, Eq. (6.13) can be rewritten as:

$$D_s \geq 4.05{Q_s}^{1/3}d_a^{2/3} \tag{6.14}$$

where D_s is expressed in mm, Q_s in m^3/h, and d_a in μm.

When Eq. (6.11) is combined with Eq. (6.13), the range of sampling head diameter for particle sampling in still air is:

$$10\left(\frac{Q_s\tau}{4\pi}\right)^{1/3} \leq D_s \leq \frac{2}{5}\left(\frac{Q_s}{\pi\tau g}\right)^{1/2} \tag{6.15}$$

For convenience, the maximum allowable ambient air velocity for still-air sampling can be circumscribed as:

$$U_0 \leq \frac{1}{5}\left(\frac{Q_s}{4\pi\tau^2}\right)^{1/3} \tag{6.16}$$

In Figure 6.9, the maximum allowable air velocity U_0 for still-air sampling is plotted as a function of the sampling rate and aerodynamic particle diameter d_a. If the air velocity exceeds the maximum value shown in the figure, an isokinetic sampling head should be employed.

For an upward-facing probe, there is an error margin of less than 10% for still-air sampling, by adopting the following formula (Agarwal and Liu 1980):

$$D_s \geq 20\tau^2 g \tag{6.17}$$

This criterion is less restrictive than Davies's criteria because the former is less sensitive to the inlet velocity and the sampling rate.

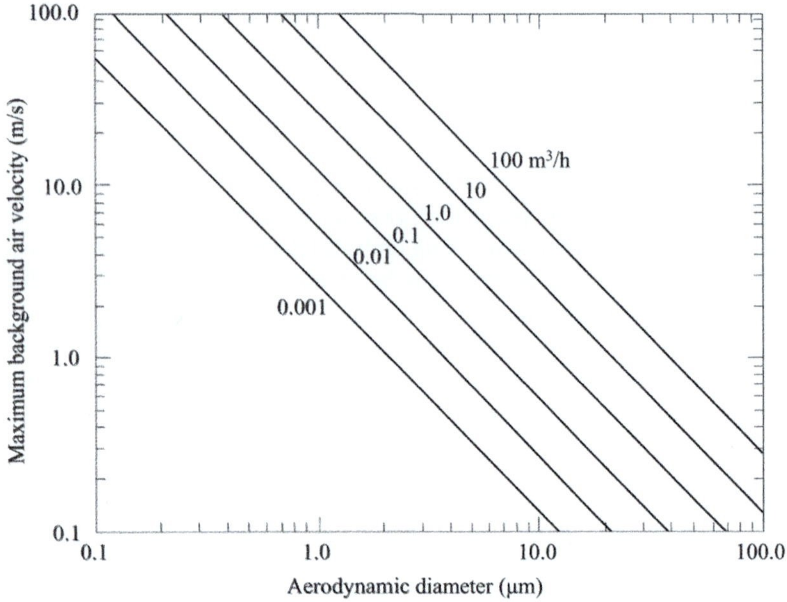

Figure 6.9 Criteria for still-air sampling with maximum ambient air velocity versus aerodynamic diameter under different sampling flow rates. *Source:* Hinds and Zhu (2022) / with permission of John Wiley & Sons.

Example 6.4

In ambient still air at 1.013×10^5 Pa and 293.15 K, particles with an aerodynamic diameter of 10 μm are sampled at 0.24 m³/h. What should the inlet diameter be in order to meet Davies's criteria?

Solution

From Eqs. (6.12) and (6.14),

$$D_{s,max} \leq 680\left(\frac{Q_S^{\frac{1}{2}}}{d_a}\right) = 680 \times \frac{(0.24)^{\frac{1}{2}}}{10} = 33\ \text{mm}$$

$$D_{s,min} \geq 4.05 Q_S^{\frac{1}{3}} d_a^{\frac{2}{3}} = 4.05 \times (0.24)^{\frac{1}{3}} \times (10)^{\frac{2}{3}} = 11.7\ \text{mm}$$

6.4 Particle Transport Losses

Particles extracted into a sampling tube may be lost in the tubing and fittings between the sampling probe inlet and the measuring device. Due to the converging or diverging streamlines at the entrance to the probe, particle losses can occur on the inside surfaces of the sampling probe near the inlet and along the sampling line as a result of the deposition mechanisms. These mechanisms include interception, inertial impaction, diffusion, gravitational settling, and electrostatic attraction. A shorter and straighter transport path line with the same probe diameter throughout may reduce the losses. Diluting the aerosol stream with clean air can help to reduce or prevent coagulation of particles and thus minimize particle loss.

6.4.1 Diffusion Deposition Loss

The loss of fine particles through diffusion deposition can be substantial, especially for ultrafine particles. The penetration rate, or the fraction of particles that exit the tube compared to the total entering, depends on a dimensionless parameter μ (Hinds and Zhu 2022):

$$\mu = \frac{4DL}{\pi D_s^2 \overline{U_s}} = \frac{DL}{Q_s} \tag{6.18}$$

where D is the diffusion coefficient of aerosol particles, L is the length of the sampling tube, D_s is the tube diameter, $\overline{U_s}$ is the average flow velocity inside the tube, and Q_s is the volumetric flow rate. Notably, for a given volumetric flow rate, the particle loss does not depend on the tube diameter. It is mainly because the increased diffusion distance in a wider tube is counterbalanced by the longer diffusion time due to the lower diffusion velocity, which is somehow shown in Eq. (5.24). Eq. (6.19) expresses the penetration rate P as a function of μ (Hinds and Zhu 2022), with an accuracy of 1% across all values of μ:

$$P = \begin{cases} \dfrac{n_{\text{out}}}{n_{\text{in}}} = 1 - 5.5\mu^{2/3} + 3.77\mu, & \text{for } \mu < 0.009 \\ \dfrac{n_{\text{out}}}{n_{\text{in}}} = 0.819\exp(-11.5\mu) + 0.0975\exp(-70.1\mu), & \text{for } \mu \geq 0.009 \end{cases} \tag{6.19}$$

The preceding equation addresses diffusion deposition in tubes of laminar flow. However, in some tubes with a large diameter and/or high velocity, the internal flow can be turbulent. Diffusion deposition in turbulent flow is more complex and cannot be easily represented by simple equations. It is generally assumed that, outside of a thin boundary layer adjacent to a surface, the aerosol concentration remains constant at n_0. Within this boundary layer, however, the flow transitions to the laminar type, and the aerosol concentration decreases linearly from n_0 to zero at the tube surface. Under these conditions, the deposition velocity can be expressed as:

$$V_{\text{Dep}} = \frac{D}{\delta} \tag{6.20}$$

where D is the diffusion coefficient of the aerosol particles, and δ is the thickness of the boundary layer. One of the challenges in applying this equation is the accurate determination of the value of δ. The thickness of the boundary layer depends on various factors, including the flow dynamics, the characteristics of the velocity boundary layer, and the size of the aerosol particles (Hinds and Zhu 2022).

Wells and Chamberlain (1967) proposed an equation to estimate the deposition velocity in turbulent flow within a tube:

$$V_{\text{Dep}} = \frac{0.04\overline{U_s}}{\text{Re}^{1/4}}\left(\frac{\rho_g D}{\eta}\right)^{2/3} \tag{6.21}$$

where $\overline{U_s}$ is the average velocity inside the tube, Re is the Reynolds number of the gas flow, ρ_g is the gas density, and η is the dynamic viscosity of the gas.

Example 6.5

Calculate the loss of 0.01 μm particles due to diffusion to the walls of a sampling tube. The sampling flow rate is 0.2 l/min (3.33×10^{-6} m³/s), and the length and diameter of the sampling tube are 1 m and 0.4 cm, respectively. The diffusion coefficient of particles with a diameter of 0.01 μm under standard conditions is 5.45×10^{-8} m²/s.

Solution

The air velocity inside the tube is:

$$\overline{U_s} = \frac{3.33 \times 10^{-6}}{\pi/4 \times (0.4 \times 10^{-2})^2} = 0.265\ \text{m/s}$$

The Reynolds number inside the tube is:

$$\text{Re} = \frac{1.205 \times 0.265 \times 0.004}{1.81 \times 10^{-5}} = 70.56$$

Thus, the flow inside the tube is laminar. From Eq. (6.18), we can calculate the dimensionless parameter μ:

$$\mu = \frac{DL}{Q_s} = \frac{5.45 \times 10^{-8} \times 1}{3.33 \times 10^{-6}} = 0.0164$$

Because $\mu > 0.009$, we can obtain the penetration rate P from Eq. (6.19):

$$P = 0.819\exp(-11.5\mu) + 0.0975\exp(-70.1\mu)$$
$$= 0.819\exp(-11.5 \times 0.0164) + 0.0975\exp(-70.1 \times 0.0164) = 71\%$$

Therefore, 29% of the particles are deposited on the wall.

6.4.2 Inertial Deposition Loss

For laminar flow in tubes with circular cross-sections, Fuchs et al. (1965) proposed a formula to estimate the particle settling loss as:

$$\text{Settling loss} = \frac{2}{\pi}\left[2k_1k_2 - k_1^{1/3}k_2 + \arcsin\left(k_1^{1/3}\right)\right] \tag{6.22}$$

where

$$k_1 = \left(\frac{3LV_{TS}}{4D_s U_s}\right) \cos\theta \text{ and } k_2 = \left(1 - {k_1}^{2/3}\right)^{1/2} \tag{6.23}$$

and L is the sampling tube length, θ is the inclination of the sampling tube from the horizontal, arcsin θ is expressed in radians, D_s is the sampling tube diameter, and V_{TS} is the terminal settling velocity.

For aerosol particles larger than 1 µm, the dominant deposition mechanism in turbulent flow is inertia. Inertial deposition occurs when a particle acquires sufficient lateral velocity toward the wall, enabling it to penetrate the laminar boundary layer and settle on the tube wall. The overall penetration of aerosol particles through the tube, accounting for losses due to deposition by diffusion or inertia, can be expressed as (Lee and Gieseke 1994):

$$P = \frac{n_{out}}{n_{in}} = \exp\left(\frac{-4V_{dep}L}{D_s\overline{U}_s}\right) \tag{6.24}$$

For the inertial deposition of particles in a tube bend with laminar flow, the following empirical equation was developed (Crane and Evans 1977):

$$\text{Bend loss} = (\text{Stk})\phi \tag{6.25}$$

where ϕ is the bend angle in radians, and $\text{Stk} = \tau U_s/D_s$.

Pui et al. (1987) proposed an empirical equation for turbulent flow in a tube bend as:

$$\text{Bend loss} = 1 - \exp[-2.88(\text{Stk})\phi] \tag{6.26}$$

6.4.3 Electrostatic and Thermophoretic Deposition Loss

Electrostatic and thermophoretic deposition inside the sampling tube can also lead to particle transport loss. Given the particle charges and the electric field strength, the electrostatic terminal settling velocity V_{TE} toward the wall can be calculated by means of the formulas presented in Section 11.3. One can also calculate the thermophoretic velocity according to the formulas in Section 5.7. The resulting particle loss by either electrostatic or thermophoretic deposition can then be roughly estimated by substituting the corresponding deposition velocity into Eq. (6.23) or (6.24).

6.5 Off-Line Measurement of Particles

There are two primary approaches for measuring aerosol concentration: off-line measurement and real-time measurement. Off-line measurement methods include filter sampling, inertial separation, gravity sedimentation, centrifugal separation, and thermal sedimentation. Filter sampling and the inertial separator are the most common off-line measurement methods, and they are addressed in this section. Meanwhile, real-time measurement is introduced in Section 6.6.

6.5.1 Filter Sampling

Measuring the increase in the mass of the filter is the most accurate and direct off-line method for determining aerosol mass concentration, if the volume of the carrier gases passing through the filter is known. High-efficiency filters can capture particles in virtually all size ranges because the filter's collection efficiency is nearly 100%. The mass concentration is calculated by dividing the collected particle mass by the sample volume. The particles collected on the filters can be used for gravimetric, chemical, biological, or radiological analysis. In addition, single particles collected on the filter can be used for various forms of microanalysis, spectral analysis, or morphological analysis.

Figure 6.10 displays a schematic of a filtration sampling system used in aerosol mass measurement. Airborne particles are drawn through a membrane filter on a filter holder from an optional sampling head. Meanwhile, the airflow rate is monitored by a flow measurement device. Measuring the mass concentration by gravimetric analysis requires accurate weighing of the filter before and after

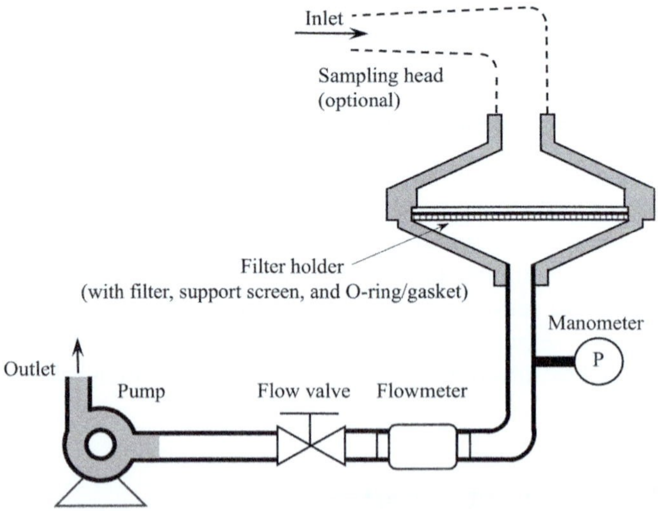

Figure 6.10 Schematic diagram of a filtration sampling system for aerosol mass concentration measurement. *Source:* Kulkarni et al. (2011) / with permission of John Wiley & Sons.

collection of particles and also accurate measurement of the sampling flow rate and sampling time.

The most frequently utilized filters for the collection of aerosol particles can be divided into three distinct categories: the fibrous filter, the membrane filter, and the capillary pore filter. The fibrous filter consists of a cluster of individual fibers, and particles are collected in various parts of the filter through interception, impaction, and diffusion. The membrane filter contains gels formed from colloidal solutions with a complex but relatively uniform microstructure on which particles are collected by adsorption. The capillary pore filter is a porous film composed of polycarbonate with a uniform pore size on the surface, and particles are collected on the pore walls by diffusion and around the pores by impaction and interception.

6.5.2 Inertial Classifier

Inertial separators sort particles of different aerodynamic sizes by changing the flow motion. Particles with greater inertia may deviate from the flow streamlines, but are eventually captured by walls, while particles with less inertia remain in the air stream. The relative velocity of the air (U_0) with respect to the impact body (impactor), the particle size (d_p), and the size of the impact body (d_b) determine the proportion of the particles that are collected by the impact body. The Stokes number is defined as the ratio of the particle's stopping distance to the geometry of the impact body, and it can be used to determine whether a particle is capable of hitting the impact body:

$$\mathrm{Stk} = \frac{\rho_p C_c d_p^2 U_0}{18\eta d_b} \tag{6.27}$$

where ρ_p is the density of the particles, C_c is the slip correction factor, U_0 is the relative velocity of the air with respect to the impact body, d_p is the diameter of the particles, η is the dynamic viscosity of air, and d_b is the size of the impact body. If the Stokes number is greater than 1, the particles are likely to hit the body.

Figure 6.11 presents four common inertial separators: the body impact sampler, the conventional impact sampler, the virtual impact sampler, and the cyclone. The body impact sampler utilizes inertia to separate airborne particles that are directed toward the impaction plate, as shown in Figure 6.11a. The impaction plate behaves as a collector, thereby facilitating the separation of large particles. The conventional impactor, as shown in Figure 6.11b, functions by directing the aerosol toward the impaction plate through the nozzle. It is difficult for large particles to trace the flow streamlines, and thus they collide with the impaction plate, whereas smaller particles proceed along the curved flow streamlines

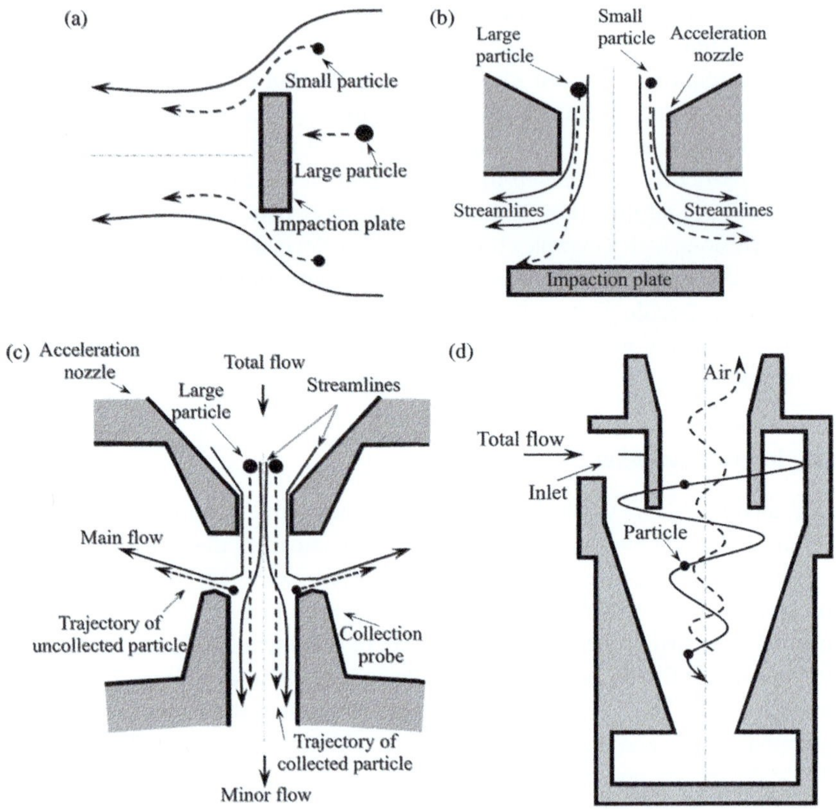

Figure 6.11 Sectional schematics of four typical inertial separation devices: (a) body impactor; (b) conventional impactor; (c) virtual impactor; (d) cyclone separator. *Source:* Kulkarni et al. (2011) / with permission of John Wiley & Sons.

without being collected. The virtual impactor uses a probe in lieu of an impaction plate, with a jet stream containing particles entering the collection probe through the nozzle and subsequently being separated, as shown in Figure 6.11c. The cyclone separator directs the free jet air to the inner surface of the cylinder, in the form of a spiral downward movement into the cone region of the cylinder, and the large particles drop down and hit the surfaces, as shown in Figure 6.11d. In the cone, the air and ultrafine particles undergo a change in direction, spiraling upward around the axis of the cyclone and then exiting.

Particles collected on the impact body/plate in certain size ranges can then be extracted for various measurements or analysis. Meanwhile, airborne particles residing in the air streams can be captured by filters for off-line measurement or can be sent directly for real-time measurement.

6.6 Real-Time Measurement of Particles

The off-line measurement of particles has a significant delay in the outputting of measurement results. A direct-reading instrument is capable of measuring particle concentration and particle size distribution in real time. However, real-time instruments, which make use of the association of particle concentrations with certain signals that can be easily measured, are generally not as precise as gravimetric analysis. The most common real-time measurement methods can be categorized in terms of their measuring principles as either semi-continuous mass measurement or optical measurement approaches.

6.6.1 Semi-Continuous Mass Measurement

There are several techniques for semi-continuous measurement of the aerosol mass concentration in near-real time. The quartz crystal microbalance (QCM) and the tapered element oscillating microbalance (TEOM) measure the particle mass concentration based on variations in the resonant frequency of a sensor surface on which particles are deposited. Another mature technique is the beta attenuation monitor (BAM).

The QCM employs a vibrating mass sensor to measure the mass concentration. Aerosol particles are collected by electrostatic precipitation or impaction on the surface of a piezoelectric quartz crystal connected to an electrode. The resonance frequency of the piezoelectric quartz crystal decreases with an increase in particle mass, and an electrical signal is produced in proportion to the mass of the particles. This method can measure particle mass directly and with high sensitivity. However, when a large number of particles adhere to the electrode surface, the measurement sensitivity may decline. Moreover, changes in the temperature and humidity of the measurement environment can impact the measurement results.

The TEOM is another type of vibrating mass sensor. Aerosols are drawn through a filter and along a hollow stalk at a rate of 0.5–5 l/min. The resonant frequency of the element decreases as the particle mass on the filter increases. This change is electronically sensed and then converted into a mass signal. A relationship has been established between the increase in mass, Δm, and the initial and final frequencies, f_i and f_f, respectively, as follows:

$$\Delta m = K_0 \left(\frac{1}{f_f^2} - \frac{1}{f_i^2} \right) \tag{6.28}$$

where K_0 is a calibration constant that is specific to each sensor. The use of a sufficiently small filter and holder is essential to the performance of the measurement.

The BAM is a simple instrument suitable for large-scale automatic monitoring. Since the beta ray intensity decreases nearly exponentially with the penetrating particle layer, the BAM quantifies particulate mass by measuring the change in this attenuation before and after the penetration, while compensating for the influence of temperature and pressure. The radioisotope source emits beta rays, and the ray intensity is measured by a specific electronic detector. To obtain accurate results, it is vital to choose an appropriate radioisotope source and detector. The half-life of the radioisotope source should be sufficiently long to avoid correction of the measurement time or frequent replacement of the source. The detector should be sensitive to the beta ray irradiation and capable of responding sufficiently fast.

6.6.2 Optical Measurement

Light scattering and extinction due to the presence of small particles are widely used to quantify the particle concentration and particle size distribution. A light beam incident on a particle gives rise to several phenomena, including the absorption of a fraction of the light as well as reflection, refraction, and diffraction of the light beam, as illustrated in Figure 6.12. This process leads to the dispersion of the incident light in multiple directions. Each particle becomes a source of scattered light, and the intensity of this light is contingent on the intensity of the incident beam, the wavelength of the light, the dimensions and morphology of the particles, the refraction index, the state and uniformity of the particle surfaces, and other related parameters.

6.6.2.1 Optical Particle Counter

An optical particle counter (OPC) can measure the particle concentration and particle size efficiently within a certain size range. The operation of the OPC is

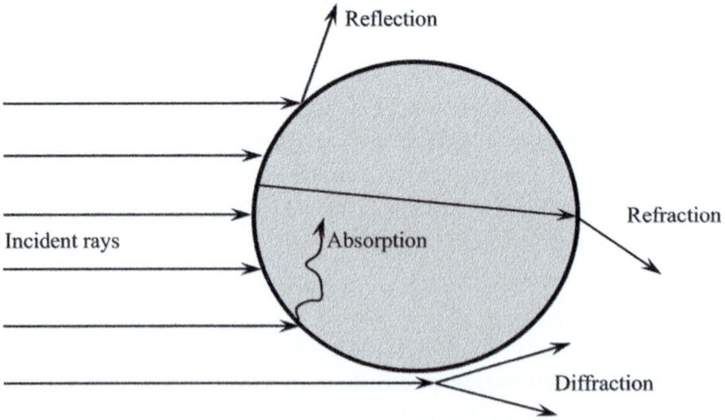

Figure 6.12 Schematics of a light beam incident on a single particle and the dispersion of the light. *Source:* Redmond et al. (2010) / with permission of Elsevier.

based on the interaction between an incident light beam and aerosol particles. It has the advantages of fast, continuous, and non-destructive detection of particles. However, the light source, the particle size and shape, the refractive index of the particles, etc., can affect the scattering of light, resulting in measurement uncertainty. Hence, the OPC can only provide an approximate analysis of the particle concentrations and the particle size distributions.

A schematic diagram of an OPC is displayed in Figure 6.13. A photodetector is used to detect particles passing through a sensing volume illuminated by the laser. A sheath flow exhausted by the pump keeps the particles focused across the laser light. As particles pass through the light sheath, they scatter light in the form of pulses that are simultaneously counted and sized. The measured concentration is sensitive to flow rate. The sheath flow remains entirely internal to the instrument, and the exhaust of the pump is connected to a HEPA filter. A portion of the HEPA-filtrated air is recirculated back to the chamber to maintain the sheath flow, while the rest of the filtrated air is exhausted. Two pressure transducers are

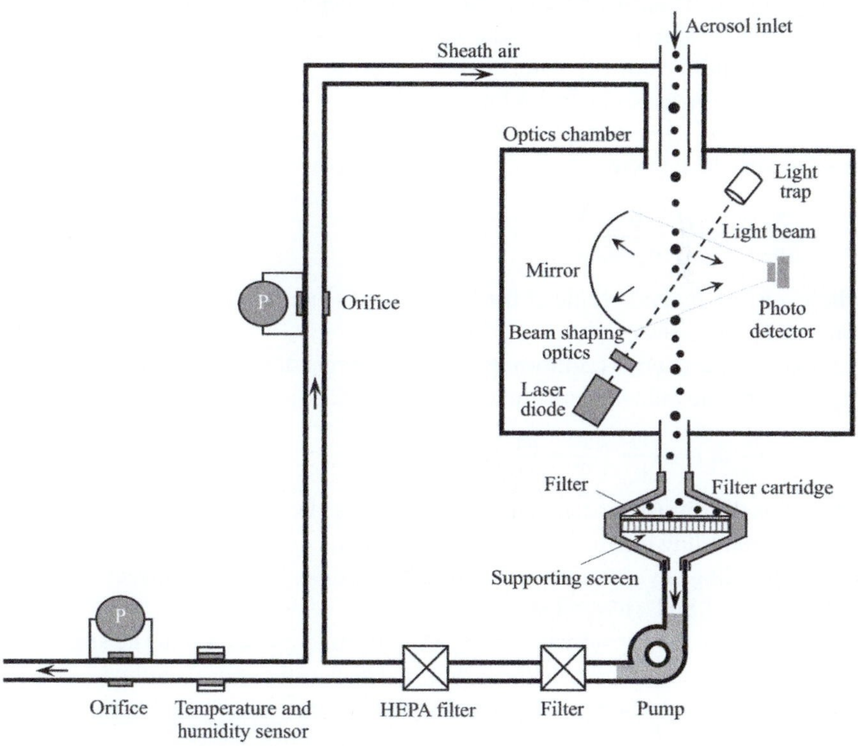

Figure 6.13 Schematic diagram of Optical Particle Sizer (model 3330). *Source:* Photo courtesy of TSI Incorporated, Shoreview, MN, USA.

employed to monitor the aerosol flow and the sheath flow, with the pump voltage adjusted as necessary to maintain the flows within their specified ranges. Particle pulses are counted individually and sorted into different channel bins according to pulse height.

6.6.2.2 Condensation Particle Counter

Conventional optical particle counters cannot sense particles smaller than 50 nm (Sang-Nourpour and Olfert 2019). The measurement of ultrafine particles may require the use of a condensation particle counter (CPC). The CPC is based on condensation technology, which involves the in-situ growth of particles to a specific size that can be measured using ordinary particle measurement techniques.

The particle measurements by CPC involve three distinct processes, namely, supersaturation of water or another working fluid, particle growth resulting from the vapor condensation, and particle detection and counting. The supersaturated environment fosters the condensation growth of particles using the ultrafine particles as the nuclei. This phenomenon can be attributed to the surface energy of the droplet, which is directly related to the degree of bending of the droplet surface. The vapor pressure at the curved droplet surface for maintaining a droplet in equilibrium is greater than the saturated pressure of the vapor near the flat liquid surface (p_s); this is known as the Kelvin effect. The saturated vapor pressure on the curved droplet surface (p_d) can be expressed as a function of the droplet particle diameter (d_p) (Bigelow and Trimble 1927):

$$\frac{p_d}{p_s} = \exp\left(\frac{4v\gamma}{R_u T d_p}\right) \tag{6.29}$$

where v is the molar volume of the liquid, γ is the surface tension, and R_u is the universal gas constant.

When a droplet grows uniformly due to condensation, the droplet size can be expressed by the following formula (Barrett and Clement 1988):

$$d_p \frac{d}{dt}(d_p) = 4\frac{vD}{R_u}\left(\frac{p}{T} - \frac{p_d}{T_d}\right) f(\mathrm{Kn}) \tag{6.30}$$

where D is the diffusion coefficient of the condensed vapor, p is the vapor pressure of the surrounding environment away from the particle, T is the temperature of the surrounding environment away from the particle, $f(\mathrm{Kn})$ is a correction factor, and Kn is the Knudsen number.

The CPC has the capacity to expand the ultrafine particles into the size range of 2–15 μm, which can be sensitively detected by certain optical methods, such as optical microscopy, photography, light extinction, and light scattering.

6.6.2.3 Fluorescence-based Bioaerosol Detection

Ultraviolet-induced fluorescence (UV-LIF) technology enables nearly real-time counting of certain bioaerosols, such as pollens, molds, and even some bacteria, in a manner similar to that of the OPC. Unlike the traditional culture method that counts only viable bioaerosols, the UV-LIF measurement can count both viable and non-viable bioaerosols. The measurement is based on the phenomenon by which the ultraviolet excitation of endogenous fluorophores in bioaerosols induces fluorescence emissions; hence, the bioaerosols can be distinguished from non-bioaerosols. Fluorophores include aromatic amino acids (such as phenylalanine, tyrosine, and tryptophan) and metabolic coenzymes (such as nicotinamide adenine dinucleotide [NADH], flavin adenine dinucleotide [FAD], and flavin mononucleotide [FMN]). Among these, the NADH/FAD/FMN serve as metabolic markers of biological activity and thus can indicate whether a microorganism is alive or dead. The fluorescence spectrum of a bioaerosol is the superposition of the fluorescence spectra of all present fluorophores (Hernandez et al. 2016).

A bioaerosol sensor used for fluorescence-based detection is depicted in the schematic diagram shown in Figure 6.14. The sampled aerosols are arranged in a single-file stream in the laminar flow regime passing through the optics chamber, which aligns the probable fibrous particles longitudinally along the stream. As

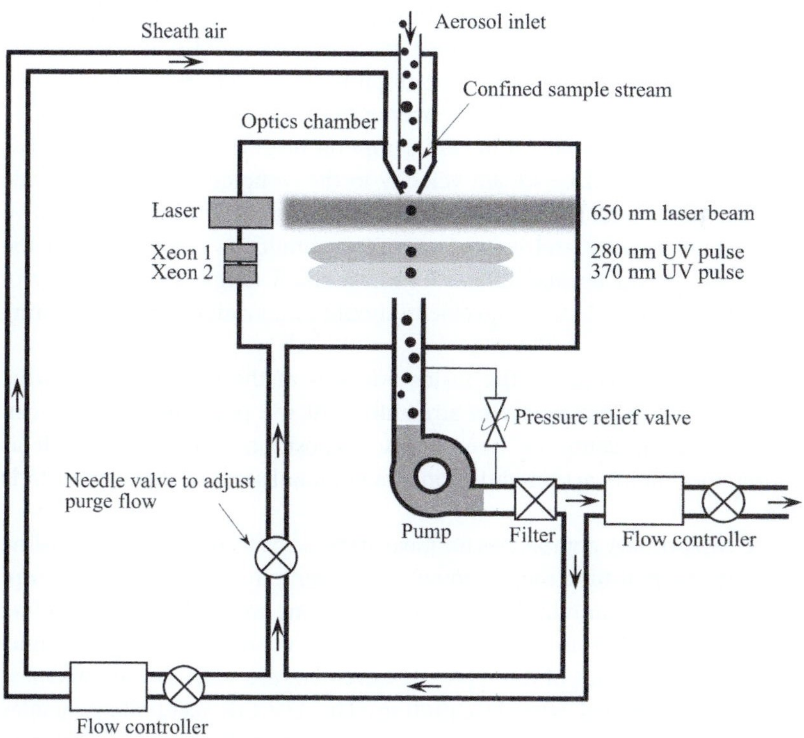

Figure 6.14 Schematic diagram of an ultraviolet-induced fluorescence bioaerosol sensor. *Source:* Adapted from Könemann et al. (2019).

a particle traverses the 635 nm laser beam, it scatters light in all directions. The scattered light is collected by two mirrors for detection by a photomultiplier tube (PMT). The PMT converts the scattered light into an electrical signal pulse, which triggers two xenon flashlamps for the generation of ultraviolet light centered at wavelengths of 280 and 370 nm, respectively. The bioaerosols' response to the 280 nm UV excitation is precisely timed for detection by both channels of the detector system, followed by detection of their response to the 370 nm UV excitation also by both channels. The first channel of the detector system senses the fluorescence emission between 310 and 400 nm, while the second channel not only senses the fluorescence emission between 420 and 650 nm but also acts as an OPC in response to the scattered light by the 635 nm laser beam. Consequently, both the total and fluorescence particle numbers and sizes are measured.

6.7 Summary

This chapter introduces the principles of particle sampling efficiency, the effects of sampling-head shape on the sampled particle concentrations, the particle transport losses that occur through the sampling tube, and basic principles for measuring particle mass and number concentrations.

The sampling efficiency of a thin-wall sampling head depends on the Stokes number, the ratio of the free-stream velocity to the sampled gas velocity inside the sampling tube, and the misalignment angle. To obtain a representative sample, constant velocity and coaxial sampling should be employed, and the Stokes number should be maintained within a small range. The influence of particle inertia and particle settling velocity should be considered when sampling in still air.

Particle losses can occur on the inside surfaces of the sampling probe due to the convergence or divergence of streamlines at the probe inlet. Losses can also occur along the sampling lines due to deposition mechanisms such as interception, inertial impaction, diffusion, gravitational settling, and electrostatic attraction.

There are two primary approaches for measuring aerosol concentration: off-line measurement and real-time measurement. Filter sampling and the inertial separator are the most common off-line measurement methods. Direct-reading instruments are capable of measuring mass or number concentration in real time. The two common methods of real-time measurement are semi-continuous mass measurement and optical particle counting. The UV-LIF technology enables nearly real-time counting of certain bioaerosols such as pollens, molds, and even some bacteria.

Problems

6.1 The air flow rate in a 30 cm duct is 1 m^3/s. A sampler probe with a diameter of 20 mm is placed inside the duct to sample particles. What sampling flow rate is required in the probe in order to attain isokinetic sampling?
[Answer: 0.0044 m^3/s]

6.2 A particle counter employs a sampling tube to draw air at a rate of 28.3 l/min from a duct. The air velocity in this duct is 10 m/s. What should the probe diameter be to sample isokinetically? Assume that the probe is in proper alignment.
[Answer: 7.7 mm]

6.3 A thin-wall probe with a diameter of 37 mm is misaligned by 20° in a duct with an air speed of 5 m/s. Assume the sampling probe extracts air at the isokinetic flow rate. What is the sampled concentration ratio for 10 μm particles of standard density?
[Answer: 0.998]

6.4 To ensure a representative sampling of the average particle concentration in a hallway under still-air conditions, a sampler with a diameter of 10 mm samples air at a rate of 5 l/min. Assuming that the sampler is aligned with the direction of air movement, what is the maximum walking speed that would be allowed?
[Answer: 3.82 km/h]

6.5 A 10 mm-diameter sampler is used to measure 5 μm particles of standard density in a duct with an air speed of 10 m/s. The sampling rate is 8 l/min. Assuming that the sampling probe is in alignment, what is the sampled particle concentration ratio?
[Answer: 1.68]

6.6 The sampling rate of a probe is 6 l/min. What range of probe diameters is required to satisfy Davies's criteria for still air? Assume that the particle size of concern is 5 μm.
[Answer: 8.4 mm $\leq D_s \leq$ 82 mm]

6.7 Determine the particle loss in a tube with a 60° bend. The tube has a diameter of 10 mm and is used to sample 8 μm particles of standard density in a duct with an air speed of 8 m/s.
[Answer: 37.7%]

6.8 Calculate the loss of 0.03 μm particles due to diffusion to the walls of a sampling tube under a sampling rate of 8×10^{-7} m^3/s. The diameter and length of the tube are 5 mm and 1.2 m, respectively. The diffusion coefficient of particles with a diameter of 0.03 μm under standard conditions is 6.39×10^{-9} m^2/s.
[Answer: 78%]

6.9 Please compare the filter sampling and inertial separation methods for off-line particle measurement. The measurement accuracy, applicability, and suitability for different particle size ranges should be analyzed.

6.10 Please evaluate the precision and constraints of real-time particle measurement techniques such as the quartz crystal microbalance (QCM), tapered element oscillating microbalance (TEOM), optical particle counter (OPC), and condensation particle counter (CPC) in the measurement of particle concentration and size distribution, particularly in relation to environmental conditions and particle properties.

References

Agarwal, J.K. and Liu, B.Y. (1980). A criterion for accurate aerosol sampling in calm air. *American Industrial Hygiene Association Journal* 41 (3): 191–197.

Barrett, J.C. and Clement, C.F. (1988). Growth rates for liquid drops. *Journal of Aerosol Science* 19 (2): 223–242.

Belyaev, S.P. and Levin, L.M. (1974). Techniques for collection of representative aerosol samples. *Journal of Aerosol Science* 5 (4): 325–338.

Bigelow, S.L. and Trimble, H.M. (1927). The relation of vapor p to particle size. *The Journal of Physical Chemistry* 31 (12): 1798–1816.

Crane, R.I. and Evans, R.L. (1977). Inertial deposition of particles in a bent pipe. *Journal of Aerosol Science* 8 (3): 161–170.

Davies, C.N. (1968). The entry of aerosols into sampling tubes and heads. *Journal of Physics D: Applied Physics* 1 (7): 921–932.

Durham, M.D. and Lundgren, D.A. (1980). Evaluation of aerosol aspiration efficiency as a function of stokes number, velocity ratio and nozzle angle. *Journal of Aerosol Science* 11 (2): 179–188.

Fuchs, N.A., Daisley, R.E., Fuchs, M. et al. (1965). The mechanics of aerosols. *Physics Today* 18 (4): 73.

Hernandez, M., Perring, A.E., McCabe, K. et al. (2016). Chamber catalogues of optical and fluorescent signatures distinguish bioaerosol classes. *Atmospheric Measurement Techniques* 9 (7): 3283–3292.

Hinds, W.C. and Zhu, Y. (2022). *Aerosol Technology: Properties, Behavior, and Measurement of Airborne Particles*, 3e. Hoboken, NJ: John Wiley & Sons.

Könemann, T., Savage, N., Klimach, T. et al. (2019). Spectral intensity bioaerosol sensor (SIBS): an instrument for spectrally resolved fluorescence detection of single particles in real time. *Atmospheric Measurement Techniques* 12 (2): 1337–1363.

Kulkarni, P.A., Baron, P., and Willeke, K. (ed.) (2011). *Aerosol Measurement: Principles, Techniques, and Applications*. Hoboken, NJ: John Wiley & Sons.

Lee, K.W. and Gieseke, J.A. (1994). Deposition of particles in turbulent pipe flows. *Journal of Aerosol Science* 25 (4): 699–709.

Pui, D.Y.H., Romay-Novas, F., and Liu, B.Y.H. (1987). Experimental study of particle deposition in bends of circular cross section. *Aerosol Science and Technology* 7 (3): 301–315.

Redmond, H.E., Dial, K.D., and Thompson, J.E. (2010). Light scattering and absorption by wind blown dust: theory, measurement, and recent data. *Aeolian Research* 2 (1): 5–26.

Sang-Nourpour, N. and Olfert, J.S. (2019). Calibration of optical particle counters with an aerodynamic aerosol classifier. *Journal of Aerosol Science* 138: 105452.

Wells, A.C. and Chamberlain, A.C. (1967). Transport of small particles to vertical surfaces. *British Journal of Applied Physics* 18 (12): 1793.

Zhang, Y. (2004). *Indoor Air Quality Engineering*. Boca Raton, FL: CRC Press.

7

Indoor Aerosol Sources

7.1 Introduction

Some of the aerosols found indoors have their origin in the outdoor environment. These aerosols are transported into indoor spaces by means of infiltration and/or ventilation. Aerosols are also generated indoors by activities such as cooking, burning of fuel, candles, or incense, smoking, ironing, and printing. When we examine aerosol sources, we are mainly concerned with the chemical substances of the aerosols, the aerosol release amount or rate, the aerosol size spectrum, etc. This chapter addresses some of the major particle emission sources in households, offices, and lavatories. Intrusion of outdoor aerosols into buildings, including the general outdoor aerosols and pollens and molds, are described as well.

7.2 Aerosol Sources in Households

7.2.1 Cooking Oil Fumes

In residential buildings, cooking is deemed to be one of the major sources of indoor aerosol emissions (Wallace and Ott 2011). It has been speculated that cooking generates 43.2% of indoor particulate matter and gives rise to 52.5% of indoor particulate exposure (Abdullahi et al. 2013). Available studies have identified serious adverse health outcomes from over-exposure to cooking aerosols. Epidemiological studies (Sun et al. 2007; Mu et al. 2013) have confirmed a significant association between cooking aerosols (cooking oil mist) and lung cancer in female non-smokers in Asia (for example, in China and Singapore). Exposure to cooking aerosols has also been associated with other cancers, such as cervical cancer. Cytotoxicity studies have shown that some chemicals in cooking aerosols may lead to intracellular peroxide production, DNA damage, cell proliferation, and release of pro-cytopathic factors associated with genotoxicity and tumor promotion (Wang et al. 2017). Combustion of fuel during cooking, heating, or burning of cooking oil and food, evaporation of moisture, and collision of pans and utensils are all sources of cooking aerosols.

7.2.1.1 Chemical Compounds in COFs

Cooking oil fumes (COFs) or particles are a primary source of volatile organic compounds (VOCs), particulate matter (PM), carbonyl compounds, polycyclic aromatic hydrocarbons (PAHs), and heavy metals in households (Huang et al. 2011b). Organic matter is the primary component of COFs. According to extant investigations, COFs contain more than 300 different organic compounds (Dai et al. 2018). In addition, black carbon (BC) is a constituent of particles emitted during cooking due to the incomplete combustion of carbon-based fuels. BC emissions interact with the emissions of the aforementioned organic compounds (Adhikari et al. 2020). More critically, BC can act as a carrier of toxic chemicals (Liu et al. 2022b). Fine particles that carry hazardous substances can penetrate the defense mechanisms of the human respiratory system and result in prolonged exposure.

7.2.1.2 Emission Dynamics of COFs

The Particle Team (PTEAM) study in the United States found that the particle concentrations inside homes during cooking were approximately 20 μg/m^3 higher than the concentrations present at other times (Wallace et al. 2006). These cooking-generated particles, primarily in the ultrafine (UFP, <100 nm) and fine size ($PM_{2.5}$) ranges, arise from oil droplets, combustion byproducts, and steam condensation. On a larger scale, it has been claimed that commercial cooking contributes significantly to PM emissions, with annual outputs of 79,300 tons of $PM_{2.5}$ and 85,500 tons of PM_{10}, whose amounts are comparable to half of highway traffic emissions in the United States (Roe et al. 2004).

COFs have mean diameters ranging from 22 to 69 nm. Approximately 80% of $PM_{2.5}$ and over 50% of particles in the size range of 16–674 nm have been identified as UFPs (Yeung and To 2008, Zhao et al. 2011, Zhao and Zhao 2018). During oil heating, the mode diameter of particles varies from 25 to 82 nm (Torkmahalleh et al. 2012). The overall particle size distribution is typically unimodal and lognormal, with median diameters ranging from 0.022 to 0.063 μm and 0.64 to 0.94 μm for particles sized from 0.015 to 0.685 μm and 0.5 to 20 μm, respectively (He et al. 2004), and varies with the ventilation conditions, stove types, and heating temperatures.

Additionally, cooking temperature significantly influences the particle diameter distribution. The submicron aerosol concentrations rise sharply as temperatures increase, particularly in the size of the accumulation mode (0.1–1.0 μm) (Yeung and To 2008). When oils are heated above their smoke point, particles in the 10–500 nm range are emitted, with the dominant size ranging from 25 to 230 nm.

The UFPs constitute 76–99% of total particles at temperatures between 131 and 197 °C (Tseng and Chen 2013). For PM_{10}, the particle sizes are mainly within the size range of 1–4 μm, exhibiting a mode diameter of approximately 2.8 μm (Gao et al. 2013b,c). These findings underscore the complex feature of emissions from cooking activities. Factors such as cooking method, food type, and temperature play a significant role in particle emissions. For example, grilling or deep-frying ingredients with high fat content usually produces more complex COFs. Through a systematic evaluation of these factors, some critical insights into the overall contribution of COFs to indoor air pollution can be obtained, with the aim of proposing effective strategies for reducing exposure. These factors are reviewed and summarized in Table 7.1.

In domestic kitchens, cooking methods and ventilation conditions are determining factors. PM emission rates are generally higher in Chinese kitchens than in Western kitchens. This is because Chinese cooking favors prolonged stir-frying. The emission rates of UFPs for Chinese and Western cooking are $4.7 \times 10^{11} - 148.3 \times 10^{11}$ particles/min (Wu et al. 2012) and $8 \times 10^{11} - 11 \times 10^{11}$ particles/min (Xiang et al. 2021), respectively.

Oil-based cooking techniques, such as stir-frying, pan-frying, deep-frying, and grilling, produce more PMs than water-based methods like steaming and boiling. Among the cooking methods, stir-frying also results in high UFP emissions, at a rate of 148.3×10^{11} particles/min, followed by pan-frying, boiling, and steaming at

Table 7.1 COF emissions in various cooking methods.

Cooking method	$PM_{2.5}$ emission rate (mg/s)	PM_{10} emission rate (mg/s)	UFP emission rate (10^{11} particles/min)	References
Western cooking	0.001–0.046	N/A	8–11	He et al. (2004) Xiang et al. (2021)
Chinese cooking	0.053–0.134	N/A	4.7–148.3	Liao et al. (2006), Wu et al. (2012)
Stir-frying	N/A	N/A	148.3	Wu et al. (2012)
Pan frying	N/A	N/A	104.4	
Boiling	N/A	N/A	5.3	
Steaming	N/A	N/A	4.7	
Frying vegetables	N/A	N/A	15–76	Buonanno et al. (2011)
Frying meats	N/A	N/A	18–96	
Oil heating	0.05–0.9	0.51–2.33	3.3–34.98	Torkmahalleh et al. (2012), Gao et al. (2013a)

104.4×10^{11}, 5.3×10^{11}, and 4.7×10^{11} particles/min, respectively (Wu et al. 2012). Thus, UFP emissions from stir-frying are about 28 times that from boiling and 31.6 times that from steaming.

Differences in the composition of oils also affect the temperature at which they decompose and release visible smoke fumes. Particles are emitted when oils release semi-volatile compounds that condense and form liquid aerosols. The usage of a particular oil also plays a key role in the production of COFs. The emission rates and concentrations of PM_{10} from the heating of oil have been found to range from 0.51 to 2.33 mg/s and 7.4 to 30 mg/m^3, respectively, depending on the type of oil (Gao et al. 2013a). Among the most commonly used edible oils, olive oil reportedly produces the highest emissions of both $PM_{2.5}$ and PM_{10}. Coconut oil yields the highest UFP concentration at 2.43×10^6 particles/cm^3, while peanut oil has the highest emission rate at 34.98×10^{11} particles/min. Meanwhile, soybean oil results in the lowest UFP concentration (0.58×10^6 particles/cm^3) and emission rate (3.3×10^{11} particles/min) (Torkmahalleh et al. 2012). Generally, oils with a higher smoke point produce lower particle concentrations within the measured range of 131–197 °C. Both wok preheating and oil heating significantly contribute to particle emissions; therefore, minimizing the duration and intensity of oil preheating is advisable for reducing exposure to COFs (Zhao and Zhao 2018).

7.2.2 Particle Emission from Heat Sources

7.2.2.1 Cooking Heat Sources

In residential environments, household heat sources are used for cooking and sometimes for heating the indoor space. Solid fuels, natural gas, and electric stoves, as shown in Figure 7.1, are three common heat sources for home cooking. In developed countries, typical heat sources for cooking are gas and electric stoves. Meanwhile, solid fuels are still used for cooking in many developing countries (Kang et al. 2009; Zhou et al. 2014; Hankey et al. 2015). Cooking with

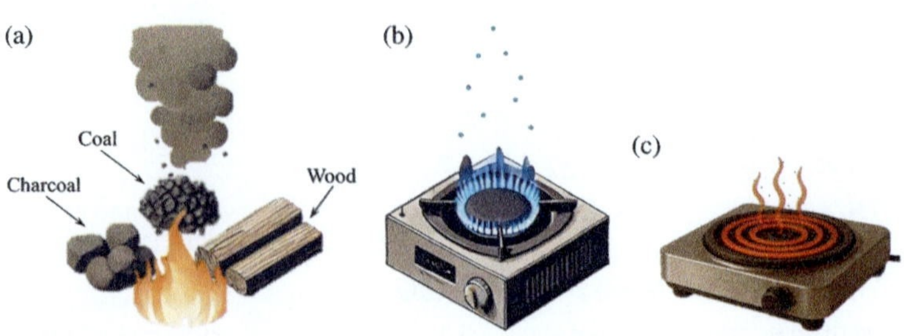

Figure 7.1 Three primary heat sources for household cooking: (a) solid fuel combustion; (b) natural gas stove; (c) electric stove. *Source:* Generated with AI using ChatGPT.

Table 7.2 Particle (UFP) emission rates from different cooking heat sources.

Cooking heat source	Peak particle diameter (nm)	Emission rate (10^{12} particles/ min)	Emission rate (10^2 μg/min)	References
Gas burner	5.2–6.4	10.6	N/A	Wallace et al. (2008)
Electric stove	6–8	5.4	N/A	
Gas grilling	10–20	1.5 ± 0.2	5.2 ± 0.2	Buonanno et al. (2009)
Electric grilling	10–20	1.2 ± 0.1	15 ± 2	

solid fuels can produce significantly higher PM concentrations than cooking with gas or electricity. Annually, there are an estimated four million deaths in developing countries as a result of indoor air pollution from solid fuel cooking (Ezzati et al. 2002).

Studies have found notably higher $PM_{2.5}$ concentrations in gas-fueled kitchens, averaging 48.8% (Huang et al. 2008) to 81% (Wong et al. 2011) more than in electric-cooking kitchens. Similarly, gas cooking increases PM_{10} levels by 30% compared to electric cooking in various frying methods (To and Yeung 2011). As shown in Table 7.2, the gas burner emits a higher particle number rate (10.6×10^{12} particles/min) than the electric stove (5.4×10^{12} particles/min), even in the absence of food, oil, or pans, likely due to the much smaller particle (5.2–6.4 nm) generation during combustion. Similarly, gas grilling produces a higher particle number emission rate (1.5 ($\pm$0.2)$\times 10^{12}$ particles/min) than electric grilling (1.2 ($\pm$0.1)$\times 10^{12}$ particles/min) (Buonanno et al. 2009), while the mass emission rate for electric grilling (15 ($\pm$2)$\times 10^2$ μg/min) is about three times that for gas grilling (5.2 ($\pm$0.2)$\times 10^2$ μg/min) (Wallace et al. 2008). These findings suggest that gas cooking tends to produce more numerous, smaller particles, whereas electric grilling results in higher mass emissions.

Solid fuels such as charcoal and coal remain the primary cooking energy source for approximately half of households worldwide (Gakidou et al. 2017). Particulate matter emissions from solid fuel combustion depend on fuel type, stove design, and combustion conditions. However, studies collectively indicate that the $PM_{2.5}$ emission ratios for these fuels generally fall within a range of 0.27–17 g/kg of fuel mass (Dang et al. 2019; Thompson et al. 2019; Liu et al. 2022a). A study conducted in Yunnan Province, China, surveyed emission rates from 37 residential stoves under uncontrolled real-world cooking conditions. These stoves primarily utilized coal, woody biomass, and agricultural waste as fuel sources. Under full combustion conditions, the observed $PM_{2.5}$ mass emission rate averaged 20 g/h (Thompson et al. 2019). However, detailed emission rate data specific to individual fuel types (e.g., coal versus biomass) remain lacking.

7.2.2.2 Building Heat Sources

In colder climates, small home heating units (boilers, fireplaces, etc.) are widely used. The energy sources for these heating devices are primarily coal, peat, or other biomass fuels (Molnár and Sallsten 2013). Although in-home combustion is harmful to human health, approximately 2.8 billion people worldwide still use solid fuels for space heating (Bonjour et al. 2013). Data on emission rates for some fuels are still unavailable. Table 7.3 shows the particle emission rates when various fuels are burned. The burning of solid fuels like wood and coal releases more particles than the burning of gases. The type of equipment used and the quality of the fuel affect the emission rate. For the fuels listed in the table, carbon monoxide and nitrogen oxides are also common byproducts of combustion, but they are not discussed here.

7.2.3 Cigarette Smoke Aerosols

Combustible cigarette smoke contains more than 7,000 different chemicals. Cigarette combustion produces potential toxicants in mainstream smoke (MS), side-stream smoke (SS), secondhand smoke (SHS), thirdhand smoke (THS), and discarded cigarette butts (CBs) (Rodgman and Perfetti 2008). These chemical compounds include argon, acetone, nitrogen oxides, and volatile sulfur compounds (Zhao et al. 2010a,b). The chemicals are bound to aerosol particles or are free in the gas phase. Chemical compounds in tobacco can evaporate into smoke

Table 7.3 Particle emission rates from different building heating sources.

Fuel type	Emission ratio	Particle size	Appliance	References
Wood logs	$9.5(\pm0.7)\times10^{14}$ particles/kg	PM_1 with a geometric mean diameter of 51 ± 2 nm	Heating stove	Tissari et al. (2008)
Coal	10–79 g/kg	$PM_{2.5}$	Heating stove	Butcher and Ellenbecker (1982)
Natural gas	0.6–3.2 mg/kg 0.9–3 mg/kg	$PM_{2.5}$	Furnace boiler	Traynor (1996)
Liquefied petroleum gas	4.5×10^{12} particles/kg	PM_1 with a peak diameter of around 20 nm	Open burner stove	Shen et al. (2017)
Charcoal	5.5×10^{15} particles/kg	PM_1 with a peak diameter of 30–35 nm	Natural draft stove	

or can react to form other chemicals. The particulate phase, which can be trapped in a filter for analysis, contains more than 3,500 individual chemical compounds, such as nicotine, PAHs, heavy metals, nitrosamines, carbonyls, phenols (Zhao et al. 2010a,b), BTEX (benzene, toluene, ethylbenzene, and xylene) (Dobaradaran et al. 2021), and other compounds.

Traditional combustible cigarettes predominantly release fine and ultrafine particles (UFPs), with a mass median aerodynamic diameter (MMAD) of approximately $0.5 \pm 0.01\,\mu m$ (Cui et al. 2023). The particles released from traditional cigarettes remain relatively stable in the particle phase. During the smoking process, the particle number concentration (PNC) from traditional cigarettes can reach 1.11×10^6 particles/cm^3 (Papaefstathiou et al. 2020), with a slow decay rate in the surrounding environment due to lower volatility.

Electronic cigarettes, often referred to as "e-cigarettes," are the most well-known type of electronic nicotine delivery device. These devices, typically shaped like traditional cigarettes, are designed to vaporize a blend of nicotine, propylene glycol, and various other chemicals (Fernández et al. 2015). Concern has been raised about the potential for passive exposure to the aerosols exhaled by e-cigarette users. Several studies have shown that the aerosols generated by e-cigarettes contain toxic compounds such as VOCs, aldehydes, nitrosamines, PAHs, glycols, and nicotine, although in lower amounts than found in conventional cigarettes (McAuley et al. 2012; Pellegrino et al. 2012; Schripp et al. 2013).

The aerosols generated by e-cigarettes are composed of volatile substances such as propylene glycol (PG) and vegetable glycerin (VG), resulting in a bimodal particle size distribution. The MMAD of particles from e-cigarettes ($1.06 \pm 0.14\,\mu m$) is generally larger than that from combustible cigarettes (Cui et al. 2023). However, the aerosols from e-cigarettes tend to evaporate or coagulate faster, which leads to a more rapid decrease in the PNC compared to traditional cigarettes. The PNC from e-cigarettes immediately after vaping can be around 1.18×10^6 particles/cm^3 (Papaefstathiou et al. 2020), but it decays quickly due to the volatile nature of the particles.

The emission rates of both types of cigarettes in offices and residential homes are compared in Table 7.4. The emission rate is a suitable assessment indicator if it can be assumed that, after being exhaled by users, the smoke or aerosols instantaneously mix with ambient air throughout the space. The emission rate can be expressed as (Li et al. 2020):

$$\mathrm{ER} = (C \times V \times \mathrm{ACH} \times t)/n \tag{7.1}$$

where ER (μg/puff) is the emission rate of chemicals or particles from the tested cigarettes or e-cigarettes, C is the time-averaged concentration of PM (μg/m^3) excluding the background particles, V is the volume of the test chamber (m^3),

Table 7.4 Typical particle concentrations during the smoking of cigarettes and the inferred particle emission rates.

Item	Location	Concentration (μg/m^3)	Emission rate (μg/puff)	References
E-cigarettes	Office	197	27.6	Schober et al. (2014)
	Home	1,330	274.9	Cheng et al. (2022)
Combustible cigarettes	Home	870	179.8	Cheng et al. (2022)

ACH is the air change per hour (h^{-1}), t is the emission duration (h), and n is the number of puffs during the smoking or vaping session.

7.2.4 Incense Stick and Candle Burning Aerosols

Incense burning releases fine particulate matter with a particle size smaller than 2.5 μm ($PM_{2.5}$). The PNC in the incense smoke can reach high levels, similar to those produced by other indoor combustion sources like cigarette smoking. The chemical composition of the particulate matter from incense burning includes both elemental carbon (EC) and organic carbon (OC), along with various metals such as aluminum (Al), iron (Fe), and toxic elements like arsenic (As) and lead (Pb) (See and Balasubramanian 2011). The PAHs can also be detected in incense smoke (Lin et al. 2001). The presence of these toxic metals, particularly in water-soluble forms, poses a significant health risk as they can more readily interact with biological tissues and contribute to respiratory and cardiovascular diseases.

Incense burning generates particles with a unimodal size distribution peaking at approximately 125 nm, exhibiting an average total emission rate of 3.1 $(\pm 0.3) \times 10^{12}$ particles/h (Manoukian et al. 2013). The emission rates of $PM_{2.5}$ from incense burning vary depending on the brand of incense used, with some brands emitting as much as 60.9 mg/h (See and Balasubramanian 2011). The smokeless incense emits the smallest amount of $PM_{2.5}$, comparable to the particle emission from gas cooking (See and Balasubramanian 2011).

Candles serve diverse purposes, including religious ceremonies, ambiance creation, and therapeutic applications. Numerous studies have measured their combustion emissions, identifying $C_{10}H_{16}$ (monoterpenes) as the dominant compound that may subsequently form secondary organic aerosols (SOAs) (Pagels et al. 2009). The compound emissions from candle combustion are lower than those from incense burning (Manoukian et al. 2013). However, candle combustion can significantly raise particulate concentrations, especially UFP concentrations. In the normal stage of candle burning, the emitted particles peak at 23 nm, and then the peak size is increased to 50 nm in the smoldering stage (Zai et al. 2006). The candle burning of 5 g for 1 hour may increase the particle concentrations from 0.6 ($\pm$0.5) $\times 10^4$ to 22 ($\pm$0.2) $\times 10^4$ particles/cm^3, yielding an average emission rate of 2.0 ($\pm$0.1) $\times 10^{13}$ particles/h, which is higher than the emission rate from the incense burning (Manoukian et al. 2013).

7.2.5 Ironing Aerosols

The particles emitted during ironing contain both organic and elemental carbon, as well as a range of inorganic elements like sodium, magnesium, aluminum, and chlorine. Notably, EC is only detected during ironing, likely from burned fabric fibers. Bromide concentrations are found to be 9–51 times higher than background levels, possibly due to the presence of brominated flame retardants in textiles (Vicente et al. 2021). PNC during ironing can reach high levels, especially under minimal ventilation. Steam ironing has been found to increase the PNC by factors ranging from 35 to 194 compared to the background levels. Peak PNC from the steam ironing alone and from the steam ironing with a boiler has been identified to 2.4×10^5 and 2.6×10^5 particles/cm^3, respectively. The Aitken mode (30–100 nm) of the size range accounts for 59–69% of the total particle count, indicating a high presence of UFPs during ironing (Vicente et al. 2021).

The majority of the particulate matter emitted during ironing consists of fine particles. $PM_{2.5}$ reportedly constitutes 85–94% of PM_{10}, while PM_1 contributes 85–93% of PM_{10}. The geometric mean diameter of the emitted particles ranges from 41.1 to 62.6 nm, suggesting that UFPs are a significant component of the emissions. A steam iron under minimal ventilation was found to produce emissions at a rate of 6.6 ± 1.4 μg/s, while ironing with an open door reduced the emission rate to 1.9 ± 1.6 μg/s. The highest emission rate (8.3 ± 3.1 μg/s) was measured for a steam iron with a boiler, highlighting the contribution of steam to particle emissions (Vicente et al. 2021).

7.2.6 Secondary Organic Aerosols (SOAs)

7.2.6.1 Formation of SOAs

While the aerosols emitted directly by indoor sources constitute the majority of particulate matter in typical indoor environments, the formation of SOAs by means of chemical reaction of VOCs with certain oxidants represents another pathway for indoor particle emissions (Pytel et al. 2022). In residential environments, SOA emission is due primarily to the ozone-initiated oxidation of various terpenes and terpenoid compounds. Their indoor concentrations are dependent much on the emissions of household cleaning products (HCPs), personal care products (PCPs), perfumes, and so on.

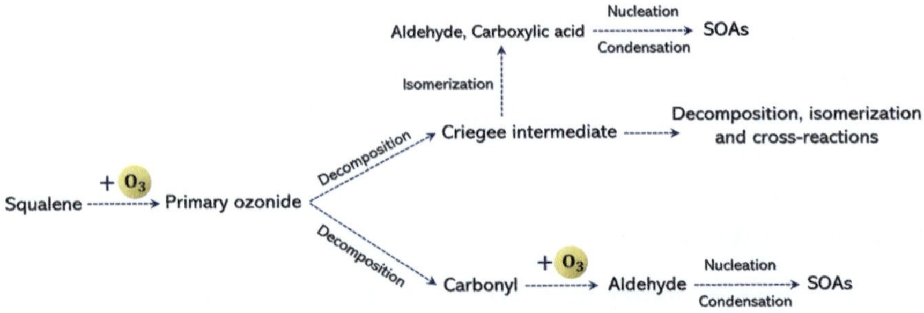

Figure 7.2 Mechanism of squalene ozonolysis (when exposed to ozone at a concentration of 100 ppb) leading to emission of SOAs (Fooshee et al. 2015). *Source:* Pytel et al. (2022) / with permission of Elsevier.

Figure 7.2 outlines the mechanism of squalene ozonolysis for the generation of SOAs. The gas-phase oxidation of reactive organic compounds by indoor oxidants, such as ozone, hydroxyl radicals, and nitrogen oxides, drives the formation of SOAs. This process generates organic radicals, followed by multiple oxidation steps or isomerization, ultimately producing low-volatility oxygenated compounds. These condensable species are then adsorbed or condensed on existing particles, contributing to the growth of the aerosols and also to the SOA mass (Ehn et al. 2014).

The emission rate of indoor SOAs is closely related to the indoor environmental conditions. First, relative humidity plays a significant role. An increase in the humidity level enhances the adsorption of polar compounds on surfaces, thus increasing the likelihood of surface reactions (Sarwar and Corsi 2007). Higher humidity also raises the probability of particle nucleation, thereby promoting SOA formation (Rösch et al. 2017). Temperature is another factor affecting the formation rate of SOAs. A temperature rise from 15 to 23 °C can increase the concentration of SOAs by approximately 58% (Huang et al. 2011a). However, a further temperature rise to 30 °C inhibits the formation of UFPs smaller than 40 nm (Huang et al. 2011a). This can be attributed to the volatilization of semi-volatile products at higher temperatures, which suppresses both nucleation and vapor phase condensation. Ventilation represents another critical factor. Higher ventilation rates dilute the content of VOCs in indoor environments, thus hindering the formation of UFPs in the presence of ozone (Huang et al. 2011a).

7.2.6.2 Spraying of Household Cleaning Products or Liquid Air Fresheners

The use of HCPs may lead to significant respiratory problems (Quirce and Barranco 2010), such as asthma (Walters et al. 2018; Clausen et al. 2020), skin irritation, and other undesirable health conditions, as a result of exposure to the HCP constituents through inhalation (Wolkoff et al. 1995). The terpenes and siloxanes contained in cleaning products can be oxidized into SOAs, thereby further increasing the concentration of potentially toxic UFPs in the air (Nazaroff and Weschler 2004). Additionally, chemical reactions among mixed cleaning agents (e.g., chemicals from different products) should also be considered, as these reactions may produce new, toxic chemical species (Siracusa et al. 2013).

The HCP users can be exposed to toxic components by means of several distinct mechanisms. Generally, these products are diluted with water (Shepherd et al. 1996), and then diffused or wiped on indoor surfaces. Some of the HCP ingredients remain on the surfaces, but the volatile constituents enter the ambient air and some of them are subsequently inhaled (Pommer 2003). Nonvolatile constituents, by means of primary aerosols or nuclei of SOAs, can also be inhaled during use, as the spraying process releases the chemicals in the form of gases and aerosols. Inhalation exposure may therefore be higher during and immediately after the spraying process as compared to other means of product application that do not involve spraying (Nazaroff and Weschler 2004).

Air fresheners can be in gaseous, liquid, or solid form. The largest amounts of aerosols may be released by liquid air fresheners due to the atomization of the liquid into aerosols when the products are sprayed and the formation of SOAs when the released VOCs are oxidized. Only the liquid air fresheners are discussed in this book. Table 7.5 summarizes three surveys of aerosol emissions during the spraying of liquid air fresheners. The measurement data indicate that the formed particles exhibit a size peak between 10 and 20 nm, with a release rate of up to 2.6×10^{12} particles/min and sizes smaller than 0.4 μm.

Another study measured the aerosols released during the spraying of nine different products, including air fresheners, deodorants, and cleaning agents (Kim et al. 2020). The spraying duration was 5 seconds, with both a manual-pump nozzle and a propellant-driven nozzle. The propellant-driven nozzle had a spraying rate of 1.68 ± 0.3 g/s during the 5-second spraying, while the manual pump had a rate of 2.32 ± 0.1 g/s. The propellant-driven spray exhibited a PNC peak in the 10–30 nm size range, with a concentration peak range of 3,000–8,000 particles/cm^3. In contrast, the manual-pump spraying resulted in concentrations lower than or equal to 100 particles/cm^3, with a mode particle size around 100 nm. These concentration differences can likely be attributed to the different spraying strengths,

Table 7.5 Aerosol emissions when a liquid air freshener is sprayed.

Item	Particle size (μm)	Emission rate (10^{11} particles/min)	References
Liquid air freshener spray	0.01–0.4	26	Wallace and Ott (2011)
	0.005–0.02	18	Géhin et al. (2008)
	<0.3	2.3	Afshari et al. (2005)

Source: Li et al. (2023) / with permission of Elsevier.

while the variations in particle size distribution primarily stem from the nozzle structure and the liquid product composition. Furthermore, due to enhanced coagulation under high initial nanoparticle concentrations (Riley et al. 2002; Hinds and Zhu 2022), the particle concentrations for the propellant-driven spraying decayed very slowly, with high particle concentrations persisting more than 90 minutes. In contrast, the nanoparticles generated by the manual-pump spraying were substantially diminished within approximately 30 minutes.

7.2.6.3 Personal Care Products Usage

PCPs contain various chemicals, which are mostly used as fragrances (Bickers et al. 2003), preservatives (Soni et al. 2001), and surfactants (Dodson et al. 2012). Generally, PCPs are directly smeared or sprayed onto the user's skin. Accompanied by the thermal plume rising from the human body, these chemicals and their reaction products can easily be lifted into the breathing zone. Most PCPs are aromatic formulations that do not cause immediate sensory irritation to users, and thus their health impacts are easily overlooked. However, certain fragrances in PCPs have been associated with indoor air quality concerns (Fung et al. 2014). Previous findings indicated the potential of monoterpenes to form secondary products like formaldehyde and the ultrafine SOAs. Limonene, a common ingredient in most PCPs, has been identified as the major VOC contributing to the aerosol generation in indoor environments after it is oxidized (Vartiainen et al. 2006).

Most studies on PCPs have focused on gaseous pollutants (Yeoman et al. 2021). The thermal decomposition of PCPs in a small reaction chamber (simulating human skin temperature) and the subsequent ozone injection into the chamber were found to facilitate VOC-to-SOA conversion (Vannucci and Nazaroff 2017). After a period of 20 minutes in this chamber test, tea tree oil fragrance and lavender body lotion exhibited the highest particulate emission rates and total emissions, as shown in Table 7.6. The tea tree oil and lavender released various terpenoid compounds (Weschler 2006; Hsu et al. 2012).

Table 7.6 Oxidation of personal care products (PCPs) by injected ozone in a test chamber to generate secondary organic aerosols (SOAs) in the ultrafine particle size range.

Product category	Fragrance	Total emission (particles)	Emission rate (particles/min)
Scalp/hair treatment	Tea tree oil	$5.5 (\pm 2.6) \times 10^9$	$6.2 (\pm 2.8) \times 10^7$
Body and face lotion	White lavender	$1.1 (\pm 1.2) \times 10^9$	$2.0 (\pm 1.4) \times 10^7$
Facial scrub/mask	Apricot	$1.5 (\pm 0.9) \times 10^7$	$5.7 (\pm 3.1) \times 10^5$
	Green tea scrub	$8.3 (\pm 3.5) \times 10^4$	$4.6 (\pm 1.9) \times 10^3$
Antiwrinkle cream	Grape seed	$1.4 (\pm 0.6) \times 10^7$	$5.0 (\pm 2.4) \times 10^5$

Source: Vannucci and Nazaroff (2017) / with permission of American Chemical Society.

7.3 Aerosol Sources in Offices

As shown in Figure 7.3, common sources of particulate matter in office spaces come from the operation of electronic devices like printers, photocopiers, and computers, which may also emit VOCs and ozone.

7.3.1 From Printers

Multifunction printers have been identified as a major source of particulate matter in smoke-free office environments. Notably, the emissions from laser printers can be comparable to those from indoor sources such as incense burning and smoking. Printer-emitted particles may contain trace amounts of heavy metals, including barium, mercury, and copper. Although the concentrations of these heavy metals are low, they still pose long-term exposure risks. Additionally, the presence of benzene and its derivatives in toner synthesis results in the release of various VOCs, such as toluene, xylene, and styrene, during the printing process.

Figure 7.4 summarizes data measured in chamber tests on the rates of particle emission from laser printing (He and Han 2021). The data demonstrate that the particle emission rates from laser printers vary widely across studies, and even data reported in the same study under the same settings can differ by more than two orders of magnitude (Salthammer et al. 2012; Sung et al. 2017). Laser printing with monochrome toner was found to result in emission rates ranging from 3.4×10^8 to 1.6×10^{12} particles/min per printer (Scungio et al. 2017). In contrast, under comparable experimental conditions, a range of $4.25\ (\pm 2.10) \times 10^9$–$3.3\ (\pm 0.64) \times 10^{12}$ particles/minute per printer was observed for 30 laser printers in another study (He et al. 2010). Addressing the ratio between the peak submicron PNC from the printer and the background concentration, another study (He et al. 2007) found that approximately 60% of the printers examined ($n = 62$) were

Figure 7.3 Three main types of equipment that emit aerosols in offices.

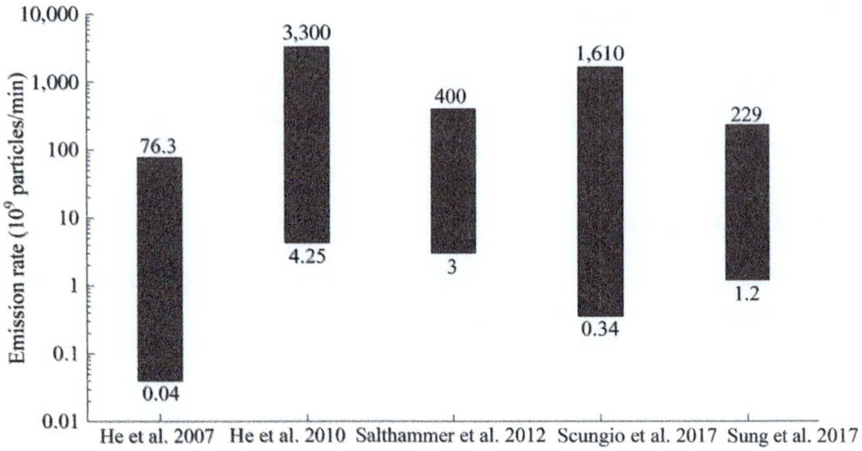

Figure 7.4 Ranges of particle emission rates from laser printing in different studies. *Source:* He and Han (2021) / with permission of Springer Nature.

non-emitters (ratios <1), 13% were low (ratios = 1–5) or median (ratios = 5–10) emitters, and 27% were high emitters (ratios >10).

The primary reason for particle emission from printers is the incomplete transfer of toner to the paper, with approximately 75% of the toner being effectively transferred, while the remaining particles are released into the surrounding air. The emitted mean mobility-equivalent particle size (50–244 nm) was found to be proportional to the printing speed (5–38 ppm). The printing speeds of 5, 24, and 38 pages per minute (ppm) reportedly correspond to particle mass emission rates of 0.007, 0.039, and 0.449 mg/h, respectively (Byeon and Kim 2012). The printer model, as well as the use of new versus old cartridges, can impact the particle size and release rate (Zou et al. 2022), as shown in Table 7.7. Specifically, the average sizes of emitted particles were found to range from 0.16 to 0.34 μm, with older toner cartridges producing larger particles and higher overall emissions. Notably, the emissions from laser printers were significantly higher than those from ink-jet printers (Lee et al. 2001).

7.3.2 From Photocopiers

During the operation of photocopiers and corona units, the discharge generates positive and negative ions that trigger chemical reactions with VOCs and indoor air, producing UFPs and ozone (Brown 1999). Early surveys indicated that the ozone emission rate ranges from 1.3 to 7.9 mg/h per device (Leovic et al. 1996).

According to a study of particulate emissions from 12 photocopy centers, the 24-hour averaged indoor $PM_{2.5}$ concentrations in two photocopy centers

Table 7.7 Particle sizes and emission rates of two types of printers with new or old cartridges.

Printer model	Toner cartridge	Average particle size (nm)	Emission rate (10^8 particles/min)
HP-1008	New	169	$PM_{0.1}$: 5.71 (1.57–21.71)
			$PM_{0.1-1}$: 7.98 (3.46–45.96)
	Old	264	$PM_{0.1}$: 7.69 (1.91–24.29)
			$PM_{0.1-1}$: 31.50 (17.57–69.30)
HP-104 W	New	176	$PM_{0.1}$: 12.80 (1.20–53.83)
			$PM_{0.1-1}$: 11.22 (6.20–24.47)
	Old	333	$PM_{0.1}$: 8.68 (3.60–15.96)
			$PM_{0.1-1}$: 71.74 (22.60–143.81)

Source: Adapted from Zou et al. (2022).

exceed 100 μg/m^3, which is the local guideline exposure limit (Lee and Hsu 2007). Figure 7.5 shows the PNC obtained in a particular photocopy center. The particulate matter emission rate from a printer can be inferred as below:

$$\frac{dC}{dt} = \frac{n \cdot \text{ER}}{V} - \frac{\text{ACH}}{60} \cdot C \tag{7.2}$$

where C is the concentration of particulate matter in the room (particles/cm^3) after deducting the background value, ER is the release rate of particulate matter (particles/min), V is the room volume (m^3), n is the number of copiers, and ACH is the air change per hour (h^{-1}).

Under steady-state conditions, the concentrations tend to be stable, so that the change rate $\frac{dC}{dt} = 0$. Therefore, the particle emission rate can be derived from Eq. (7.2) as:

$$\text{ER} = \frac{(C_r - C_{bg}) \cdot \text{ACH} \cdot V}{60 \cdot n} \tag{7.3}$$

where C_r is the steady-state particle concentration inside the room when the copiers are running, and C_{bg} is the background particle concentration in the room when the copiers are not running.

Thus, it is possible to obtain the release rate of a photocopier for a single particle size range at a given time. Within 40 minutes from the start of operation, the release rate of 20–50 nm particles was 1.61×10^7 particles/min per copier, while particles smaller than 20 nm were released at a rate of 1.38×10^7 particles/min per copier. Hence, the total emission rate of UFPs smaller than 50 nm amounted to 2.99×10^7 particles/min per copier.

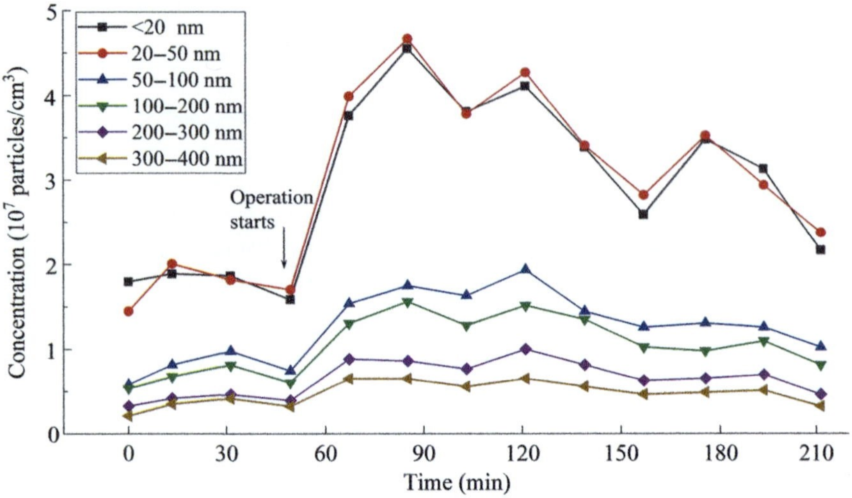

Figure 7.5 Measured temporal particle number concentrations in a photocopy center with three copiers, where the space volume is 30 m^3 and the ACH is 3 $hours^{-1}$. *Source:* He and Han (2021) / with permission of Springer Nature.

7.3.3 From Computers

Although particle emissions from office equipment (e.g., printers and photocopiers) are relatively well studied and supported by clear quantitative data, studies on particle emissions from computers are still scarce. Much of the existing literature focuses on the emission of VOCs and ozone, while particulate emissions have been mentioned in only a few studies (Nakagawa et al. 2003; Bako-Biro et al. 2004; Waye et al. 2013). Specific particle emission rates (e.g., μg/h) have not been systematically investigated, and further research is required in order to fill this gap.

7.4 Aerosol Sources in Lavatories

Lavatories, washrooms, restrooms, bathrooms, or water closets constitute a special type of indoor environment in both residential and office buildings. People typically need to use the lavatories every few hours. Even in settings where living or working space is not shared, people may need to access a shared lavatory. Consequently, the particle sources and hygiene status in lavatories are of particular concern to the general public.

Generally, the particle sources inside a lavatory are not very strong. However, even a low release of infectious bacteria, viruses, or fungi in a lavatory may lead to negative health outcomes. Therefore, significant care should be taken to maintain healthful, pleasant conditions within the lavatory. The particles inside the lavatory may originate from outdoors. People also release amounts of particles when they access the lavatory, through spitting, coughing, talking, and even breathing. The flushing of various toilets and urinals, cleaning of the lavatory, and use of hand dryers have all been reported to emit aerosols.

This section addresses the emission of liquid aerosols during the flushing of a sitting toilet, squat toilet, vacuum toilet, and male urinal, and also when hands are dried under a high-speed air-jet hand dryer. The released liquid aerosol numbers, their size spectrum, and the typical concentrations are summarized. However, due to the relative scarcity of pertinent research, this section primarily presents information about liquid aerosols overall, with limited mention of specific bioaerosols.

7.4.1 Aerosol Emission from Toilet Flushing

7.4.1.1 Surveyed Bioaerosol Emission after Flushing of a Toilet

Lavatories are sites of microbial growth, especially in public bathrooms. Table 7.8 summarizes a number of microbial aerosol surveys in real-world bathrooms. In hospital bathrooms, *S. epidermidis*, *C. difficile*, *Enterococci*, and human adenovirus were detected. In addition, Gram-negative bacteria such as *Klebsiella* and *Fusobacterium*, and the Norovirus were detected inside and outside toilets (Nazaroff 2016; Paddy et al. 2023).

Specifically, the measured concentration of *S. epidermidis* rose from 210 CFU/m^3 (prior to toilet flushing) to an average of 278 CFU/m^3 after flushing of the stool (Knowlton et al. 2018). The concentration of 240 CFU/m^3 was measured when the toilet was flushed without the presence of the stool matter (Knowlton et al. 2018). Similarly, the measured concentration of *C. difficile* rose from 1.8 to 5.8 CFU/m^3 after flushing of the toilet, and *enterococci* accounted for 54% of the sampled bacteria (Wilson et al. 2020). In a washroom at a nephrology clinic, the human adenovirus concentration was measured to be 64,444 PFU/m^3 and 293 CFU/m^3 of total bacterial genera after the toilet was flushed (Verani et al. 2014). In contrast, the concentrations of these two microorganisms in the washroom of an office building were much lower. The total bacteria count in the office building's washroom was half that in the clinic's washroom, while the human adenovirus concentration in the former washroom was at least two orders of magnitude lower (Verani et al. 2014).

Table 7.8 also presents the flushing-induced emission ratios of certain viruses and bacteria that were purposely released into a toilet in a laboratory. The emission ratio varies with the toilet type, the flushing water pressure, and the type of microorganism. The emission ratios of the viruses are higher than those of the bacteria. In general, the emission ratios are quite low.

Table 7.8 Surveyed bacterium and virus emission in washrooms.

Microorganism	Location	Toilet type	Bioaerosol concentration/emission ratio	References
S. epidermidis.	Hospital	Sitting toilet	210 CFU/m^3 (baseline), 278 CFU/m^3 (post flushing)	Knowlton et al. (2018)
C. difficile, Enterococci	Hospital	Sitting toilet	1.8 CFU/m^3 (baseline), 5.8 CFU/m^3 (post flushing)	Wilson et al. (2020)
Total bacterium count and human adenovirus	Hospital	Sitting toilet	293 CFU/m^3, 64,444 PFU/m^3	Verani et al. (2014)
	Office		147 CFU/m^3, 433 PFU/m^3	
Staphylococcus aureus	University lavatory (controlled)	Sitting toilet	855.15 ± 84.81 CFU/m^3 (poor ventilation), 466.43 ± 49.47 CFU/m^3 (natural and mechanical ventilation)	Ali et al. (2022)
P22	Laboratory	Squatting toilet	19.79 ± 10.81 PFU (with 10^9 PFU input)	Zhang et al. (2024)
MS2			8.72 ± 2.17 PFU (with 10^9 PFU input)	
E. coli			0.24 ± 0.10 CFU (with 10^9 CFU input)	
S. epidermidis	Laboratory	Sitting toilet with 400 kPa flushing pressure	266 ± 73 CFU (with 10^9 CFU input)	Lai et al. (2018)
E. coli			13 ± 4 CFU (with 10^9 CFU input)	

Although the available survey-based studies are limited, they demonstrate that flushing a toilet may induce emission of pathogenic bioaerosols. The following sections address the dynamics of toilet flushing in detail and the resulting liquid aerosol emission.

7.4.1.2 Flushing a Sitting Toilet

Sitting toilets usually employ a siphon for flushing. Figure 7.6a shows a classic sitting toilet. When the toilet is flushed, the flushing water flows out of water jet openings, forming jets that sweep the toilet's inner surface, and the angle of the water jet is commonly designed to make the water in the bowl form a vortex. The flushing water impacts the bowl and enters the "drain tunnel," which fills up and creates a siphon effect that empties the bowl of waste and water. At the end of the siphoning process, a certain amount of flushing water is stored in the bowl again.

Table 7.9 presents the emitted droplet/particle number and the percentage of the droplet nuclei in different size ranges. An increase in the flushing water pressure leads to an increase in the droplet/particle emission rate. The total emission rate under 400 kPa was measured at approximately 3.6, 21.0, and 19.8 times the rate under 200, 9.8, and 4.9 kPa, respectively (Lai et al. 2018). The flushing pressure

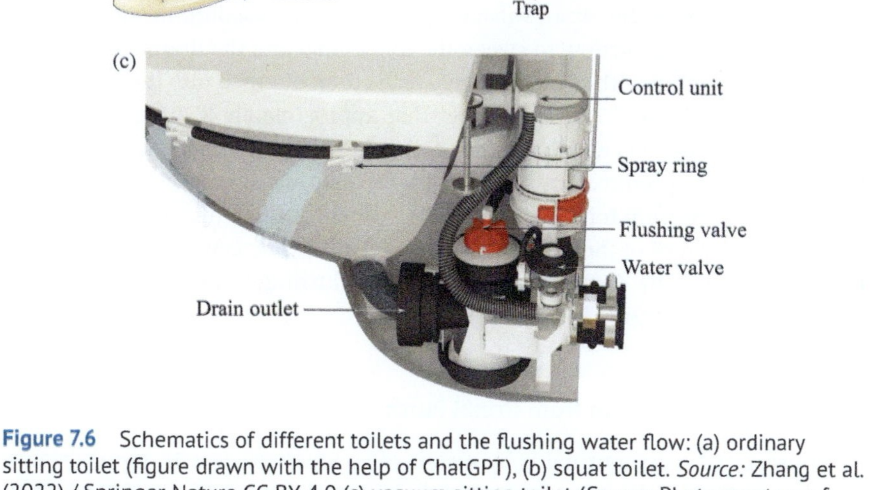

Figure 7.6 Schematics of different toilets and the flushing water flow: (a) ordinary sitting toilet (figure drawn with the help of ChatGPT), (b) squat toilet. *Source:* Zhang et al. (2022) / Springer Nature CC BY 4.0 (c) vacuum sitting toilet (*Source:* Photo courtesy of Evac company, Espoo, Finland).

of 9.8 kPa produced the fewest particles instead of 4.9 kPa. Most of the sampled particles were in the submicron range.

7.4.1.3 Flushing a Squatting Toilet

Squatting toilets also commonly utilize a siphon flushing mechanism. As shown in Figure 7.6b, the water jets in the squat pan create a flow to sweep the bottom pan and sides, ensuring that the waste and water flow are quickly removed.

Squatting toilets have no lid and may create a higher risk of bioaerosol exposure after flushing. A single flush was found to release 293,237 particles with diameters greater than 0.3 μm, more than 90% of which were in the submicron range (Zhang et al. 2022). In contrast, another study reported total droplet/particle emission of 5.72×10^5 particles/flush; 96% of the particles were smaller than 2 μm in diameter, and among the total particles, submicron particles comprised 68.2% (Zhang et al. 2024). The difference between the studies can likely be attributed to a significant difference in flushing water pressure. In the study by Zhang et al. (2022), the flushing water pressure was estimated to be 9.8 kPa because the water flow was driven by water stored in a tank 1.0 m high. The flushing water pressure in the latter study was approximately 275 kPa, which was much higher than in the former. Therefore, both the droplet/particle emission rate and the size range are highly subject to the flushing water pressure.

7.4.1.4 Flushing a Vacuum Toilet

Vacuum toilets are used on aircraft and trains to conserve flushing water. As shown in Figure 7.6c, during flushing, approximately 300 ml of water is released to rinse the toilet bowl. Next, the drain valve is opened, and the vacuum inside the waste tank creates a strong airflow to sweep the toilet bowl and the drainage duct. The fecal waste, together with the flushing water, is then dumped into the waste tank.

As shown in Table 7.7, it was found that approximately $8{,}498 \pm 1{,}918$ liquid aerosols are emitted in each flush (Li et al. 2022). The resulting aerosols are much fewer than those produced by the flushing of a sitting or squat toilet. One reason is that the vacuum toilet uses much less water for flushing. Submicron particles account for 69% of the aerosols produced, and this percentage is close to that found in the study by Zhang et al. (2024) for the flushing of a squat toilet with an extremely high pressure.

7.4.2 Aerosol Emission from Urinal Flushing

The flushing of a male urinal is by means of gravity flow or with the assistance of pressure. Once a urinal is flushed, stored water flows out of small holes/ports along the urinal's upper edge, as shown in Figure 7.7. The water flows across the

Table 7.9 Droplet/particle emission rate and size range per flush for different toilet types and flushing water pressures.

Toilet type /flushing pressure		Droplets/particles emitted per flush	Droplet/particle proportion in different size ranges				References
Sitting toilet with different flushing pressures			0.3–0.6 μm	0.6–10 μm	1.5–10 μm	2–10 μm	Lai et al. (2018)
	400 kPa	$2.87 \times 10^5 \pm 0.33 \times 10^5$	85.5%	14.5%	2.8%	0.7%	
	200 kPa	$8.02 \times 10^4 \pm 0.69 \times 10^4$	77.1%	22.8%	7.6%	1.6%	
	9.8 kPa	$1.37 \times 10^4 \pm 0.3 \times 10^4$	61.3%	38.7%	5.8%	2.1%	
	4.9 kPa	$1.45 \times 10^4 \pm 0.21 \times 10^4$	59.3%	40.7%	4.8%	0.5%	
Squat toilet		$2.93 \times 10^5 \pm 0.26 \times 10^5$	0.3–0.5 μm	0.5–1.0 μm	1.0–3.0 μm	3.0–5.0 μm	Zhang et al. (2022)
			73.8%	17.4%	7.5%	0.5%	
		$5.72 \times 10^5 \pm 1.93 \times 10^5$	0.3–0.6 μm	0.6–1.0 μm	1.0–2.0 μm	2.0–10.0 μm	Zhang et al. (2024)
			46.3%	21.9%	27.6%	4.2%	
Vacuum toilet		$8{,}498 \pm 1{,}918$	0.3–0.5 μm	0.5–1.0 μm	1.0–3.0 μm	3.0–5.0 μm	Li et al. (2022)
			46%	23%	23%	7.0%	
Urinal		$4.54 \times 10^5 \pm 0.51 \times 10^5$	0.3–0.5 μm	0.5–1.0 μm	1.0–3.0 μm	3.0–25.0 μm	
			69%	23.6%	7.4%	0	

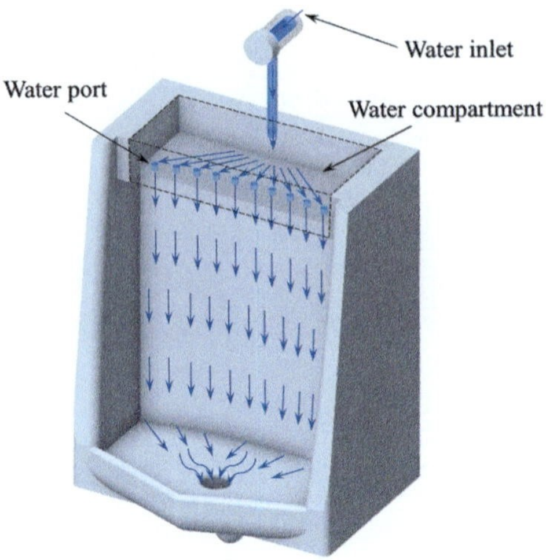

Figure 7.7 Schematic of the flushing water flow that sweeps a male urinal.

smooth wall surface, carrying waste (urine or saliva) into the drain. The flushing typically lasts for 5–6 seconds and consumes 2–3 l of water.

As shown in Figure 7.8a, at the beginning of the flushing cycle, marked by $t = 0$ seconds, the particle concentration was found to increase significantly from about 1.3 to 9.1 particles/ml in 9 seconds, reaching a peak concentration. At the end of the rinsing stage, the particulate emissions ceased, and the concentration began to gradually decrease, returning to its background level after about 4 minutes. Based on the measured temporal particle concentrations, it can be inferred that a single rinse releases approximately 453,536 particles in the 0.3–3 μm range, of which 312,418 are in the 0.3–0.5 μm range. Submicron particles account for 92.6% of the total particles, as shown in Figure 7.8b. No particles larger than 3 μm were detected.

7.4.3 Aerosol Emission from Handwashing

Handwashing results in the splashing of droplets, which in turn release aerosols. Once the faucet, sink, or storage water is contaminated, the emitted aerosols may contain contaminants. In addition, the microorganisms or chemical substances on the hands of individuals may be atomized into aerosols and subsequently transferred to the surrounding air, which may result in human exposure.

Water spray in the process of handwashing is mainly related to the faucet aerator and the water flow. Compared to the aerator, water flow has a relatively insignificant effect on aerosol production. Regardless of the type of aerator,

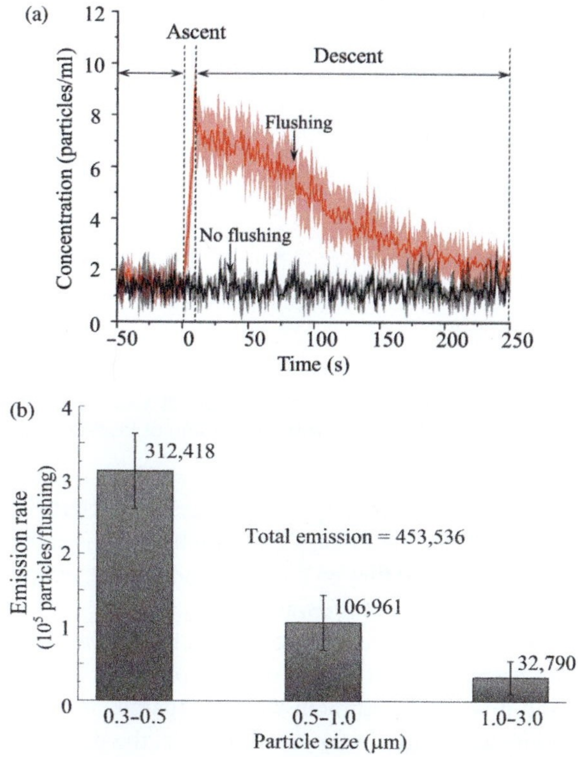

Figure 7.8 Measured particle concentration and inferred particle emission rate during the flushing of a male urinal: (a) temporal particle concentrations (0.3–3 μm) upon flushing; (b) particle emission (0.3–3.0 μm) rates.

the size distribution of aerosol particles is mainly below 1 μm (Benoit et al. 2021). Aerosol concentrations vary significantly across aerator types. Aerators with the aerated stream show the highest median concentration (59.9 μg/l), while the spray-stream in the laminar flow type corresponds to the lowest concentration (0.38 μg/l). When *S. epidermidis* was inoculated into a water tank for handwashing, the bacteria in the air were sampled in the size range of 2–3 μm (Hayward et al. 2025).

7.4.4 Aerosol Emission during Hand Drying under a Jet Dryer

High-speed air-jet dryers can dry hands quickly without the need for consumable materials. Water remaining on the hands after washing can be dried within a short time. When the fan inside the dryer is turned on, the surrounding air is drawn through the suction port located on the rear or underside of the device, as shown in Figure 7.9. The airflow is accelerated by the motor-driven fan and then discharged at a high speed through a series of small round apertures or a narrow-slit outlet.

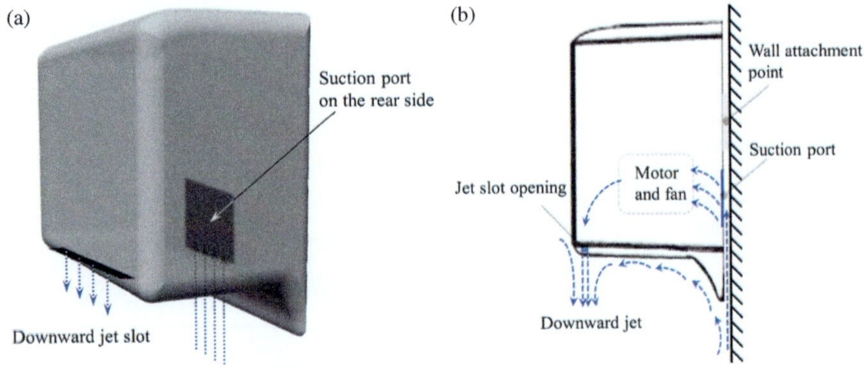

Figure 7.9 Schematics of a hand dryer: (a) appearance; (b) basic airflow pattern. *Source:* Gao et al. (2025) / with permission of Elsevier.

The speed of the discharged jets ranges from tens to hundreds of meters per second (Dyson United States 2024). The high-velocity jets impinge directly on the liquid droplets or liquid film on the palm and fingers, scraping water off the skin and also evaporating some of the water.

Two bacteria, *P. aeruginosa* and *S. aureus*, were measured by Taylor et al. (2000). The bacterial counts at the suction port and the discharge opening (outlet) of a hand dryer in a bathroom were compared, and a 30–75% reduction in bacterial counts was found in the discharge air; the discharge air had significantly fewer microorganisms than the air at the suction port (Taylor et al. 2000). When a warm air hand dryer is operated, the suction extracts air from the room and discharges it through the outlet to the user's hands. The air heating and filtration process within the hand dryer eliminates a portion of the microorganisms (Taylor et al. 2000). The aforementioned results were obtained in the context of good hand hygiene. In another experimental investigation, a subject's hands were immersed in a PR772 phage solution (6.5×10^8 gene copies/μl), and then the subject dried his/her hands under a jet air dryer while standing and wearing a mask. During the drying process, a viral titer of the user's mouthpiece reached 1.4×10^4 gene copies/μl (Moura et al. 2022). This finding indicates that pathogen transmission occurs during hand drying.

The temporal concentrations of droplet nuclei in the region near a hand dryer are shown in Figure 7.10a–c (Gao et al. 2025). The recorded concentrations are due only to the atomization of water remaining on the hands after washing. After the hand dryer is turned on, the aerosol concentration rises dramatically to 0.5 particles/ml and even higher. Most of the particles released are submicron particles. According to the particle number balance relationship, the liquid aerosol emission rate during hand drying is estimated to be 105,587 particles.

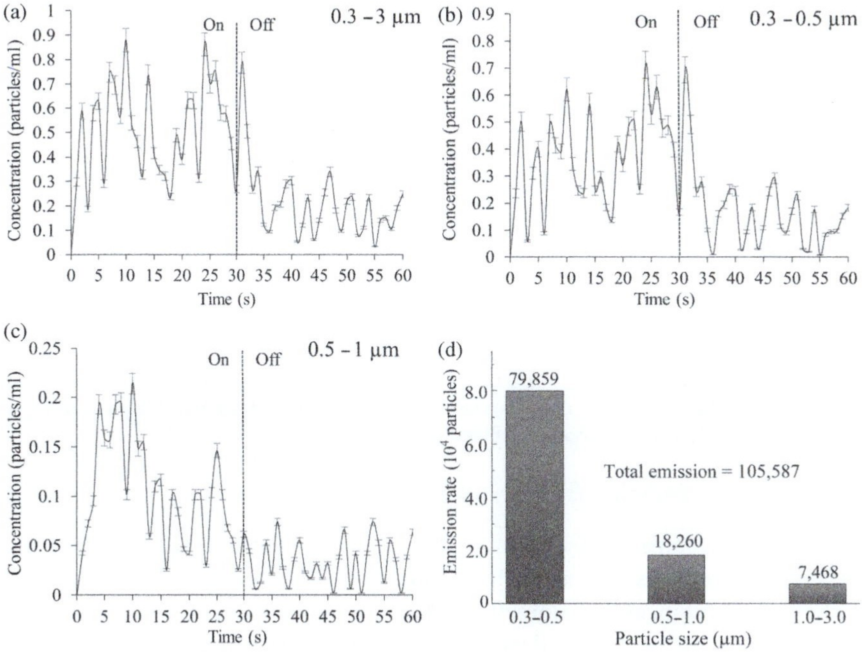

Figure 7.10 Measured liquid aerosol concentrations and emission rates for different particle diameter ranges in a test enclosure: (a) total aerosols in the range of 0.3–3 μm; (b) aerosols in the range of 0.3–0.5 μm; (c) aerosols in the range of 0.5–1 μm; (d) liquid particle numbers emitted from hand drying for 30 seconds. *Source:* Gao et al. (2025) / with permission of Elsevier.

The numbers of particles in the ranges of 0.3–0.5 and 0.5–1.0 μm are 79,859 and 18,260, respectively, as shown in Figure 7.10d, together constituting the largest share (92.93%) of the emitted aerosols.

In addition to bioaerosols and liquid aerosols, a hand dryer may release copper oxide (CuO) aerosols if the fan motor contains carbon brushes. The friction between the motor commutator, which is made of copper, and the carbon brushes produces fine and ultrafine CuO particles (Gao et al. 2024). As shown in Figure 7.11a, the PNC in an enclosure containing only a hand dryer increases immediately after the hand dryer is turned on, and decreases after the dryer is turned off. As shown in Figure 7.10b, the highest CuO particle emission rate is inferred to be 2.64×10^6 particles/s at the instant when the motor is started, at which point the friction between the commutator and brushes may be the most severe. Among the collected particles, approximately 99.7% of the emitted particulate matter is in the submicron size range, with particles smaller than 0.5 μm accounting for 96.5% of total emissions.

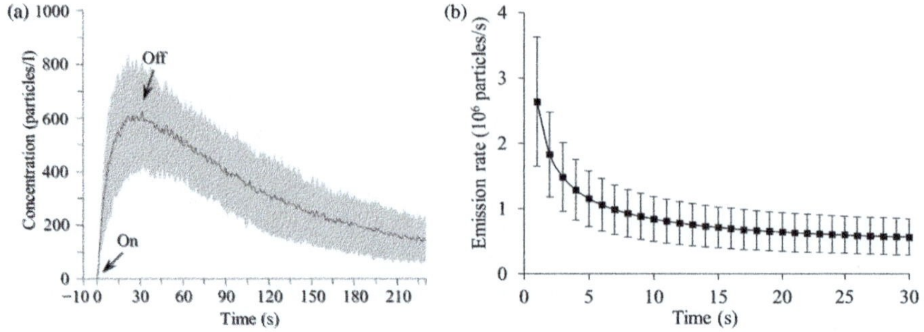

Figure 7.11 Measured CuO particle (0.3–3 μm) concentrations and inferred particle emission rates: (a) temporal mean particle concentration since the instant when the hand dryer is turned on; (b) particle emission rate during operation of the hand dryer. *Source:* Gao et al. (2024) / with permission of Elsevier.

7.4.5 Aerosol Emission from Showering

Showers are accompanied by substantial droplet splashing and the generation and dispersion of aerosols. The aerosolization process is likely to facilitate the spread of chemical substances (trihalomethane decontamination products, heavy metals, cyanobacterial toxins, etc.) (Falconer et al. 1994; Kerger et al. 2000) and microorganisms in water. Given that bathrooms are typically compact and enclosed spaces, understanding the mechanisms of aerosol formation during showering and quantifying the aerosol exposure pathways is critical.

Aerosol concentrations were observed to rise rapidly to their peak levels of 5–14 mg/m^3 within the first 1–2 minutes of showering (Zhou et al. 2007). The mechanisms of aerosolization during showers are primarily influenced by water temperature, water flow rate, and showerhead type (conventional head versus "rain" type). The showerhead type was found to have a relatively smaller effect on aerosolization (Tang et al. 2025). Meanwhile, an increase in water flow slightly increases the aerosol generation rate, and water temperature significantly impacts both the particle size and generation rate of the released aerosols. The aerosols released during showering and the main influencing factors are summarized in Table 7.10.

Showering with hot water produces aerosols at a much higher rate than cold-water showering. In a hot-water scenario, the production rate of peak-sized aerosols (5–6 μm) was found to be 7.7×10^4 μg/min, whereas in a cold-water scenario, the rate was only 19.62 μg/min for particles sized 1–2 μm (Tang et al. 2025), as shown in Table 7.10. In another hot-water scenario (43–44 °C), the count median diameter of generated aerosols was about 5 μm (Zhou et al. 2007). In a cold-water (24–25 °C) scenario, the average particle size is much smaller, around 1 μm. During hot showers, the higher water temperature leads to

Table 7.10 Aerosols emitted during showering.

Water temperature (°C)	Water flow rate (l/min)	Aerosol diameter (μm)	Emission or concentration rise rate	References
25.9	10	1–2	19.62 μg/min	Tang et al. (2025)
		2–3	25.88 μg/min	
41.0	10	5–6	7.7×10^4 μg/min	
		8–9	5.3×10^4 μg/min	
24–25	5.1	0.96	30.79 μg/m³/min	Zhou et al. (2007)
	9.0	0.97	87.15 μg/m³/min	
43–44	5.1	5.1	2,252 μg/m³/min	
	9.0	5.2	2,841 μg/m³/min	

increased humidity inside the shower stall, often approaching saturation. Such high humidity slows down the evaporation of droplets, allowing them to retain a larger size.

7.5 Intrusion of Outdoor Particles

7.5.1 Intrusion of General Outdoor Particles

Particles of outdoor origin constitute a significant contributor to indoor particulate pollution (Lee et al. 1997). There are three main pathways by which outdoor particles enter buildings, i.e., mechanical ventilation, natural ventilation, and infiltration through envelope leakage.

7.5.1.1 Mechanical Ventilation Pathway

Due to the limited particle-removal capability of mechanical ventilation systems, these systems can carry particles of outdoor origin into indoor spaces (Stephens and Siegel 2012). The airborne particles drawn into ventilation ducts may be intercepted, settle, and be partially removed by air filters if installed, but not completely. The interception and gravitational settling are only effective in removing coarse particles. The main determinant of the particle concentrations in the mechanical ventilation supply is the penetration ratio of the installed air filters. The removal efficiencies of 11 MERV-rated (minimum efficiency reporting values) filters for outdoor UFPs and $PM_{2.5}$ have been summarized by Azimi et al. (2014). Coarse filters (MERV <7) achieve geometric mean removal efficiencies of 5–12% for UFPs and only 1–8% for $PM_{2.5}$ (Azimi et al. 2014). In contrast, high-efficiency particulate air (HEPA) filters remove more than 99.97% of both types of aerosols. Particle filtration efficiency exhibits size-dependent mechanisms, the details of which are presented in Chapter 11.

7.5.1.2 Natural Ventilation Pathway

The majority of the outdoor-originating aerosols are carried into indoor spaces by natural ventilation through building openings such as open doors and windows. Most windows and some doors are equipped with screens. Conventional screens may only prevent insects from entering rooms and remove some dust in large sizes. Figure 7.12 presents common window screens with relatively large pores.

Screens with different pore sizes are used in different countries. The standard insect screens in Korea have a pore size of 75 μm, while nanofiber screens feature 0.5 μm pores. As shown in Table 7.11, particle-removal efficiency is dependent on the size of the particles and the screen type. The insect screens demonstrate a steady increase in the particle-removal efficiency with particle diameter,

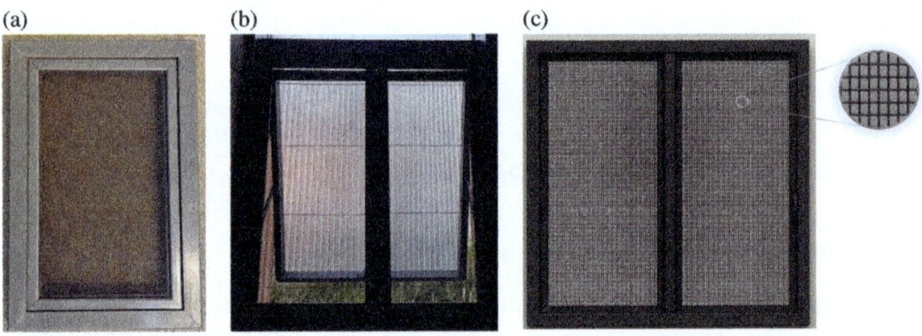

Figure 7.12 Window screens: (a) photograph of a single screen; (b) photograph of two screens installed on a window; (c) diagram of window screens and the enlarged mesh details, *Source:* Generated with AI using ChatGPT.

Table 7.11 Particle-removal efficiency of different types of window screens tested in Korea under a sweeping face velocity of 1 m/s.

	Particle-removal efficiency			
Particle size	**Insect netting with filtration**	**Non-woven fabric**	**Nanofiber**	**Electrostatic**
0.3–1.0 μm	15.7%	14.3%	43.0%	79.6%
1.0–3.0 μm	23.4%	54.2%	58.7%	81.9%
3.0–10.0 μm	30.3%	46.0%	57.5%	79.7%

Source: Adapted from Kang et al. (2020).

while the other three screen types remove the particles mainly in the size range of 1.0–3.0 μm, followed by particles in the 0.3–1.0 and 3.0–10.0 μm size ranges, respectively (Kang et al. 2020). It should be noted that the particle-removal efficiency decreases with an increase in wind speed.

7.5.1.3 Penetration Through Building Cracks

Building cracks, such as those around door and window frames (Figure 7.13), are found widely in building envelopes. These cracks form persistent entry routes for outdoor particles even when the doors and windows are closed. Buildings commonly have crack widths ranging from 0.5 to 7.5 mm (Hopkins and Hansford 1974), but normally less than 2.5 mm (Thomas and Dick 1953). The cracks around closed windows and doors can be classified in a straightforward manner according to shape: straight cracks (without a bend), L-shaped cracks (bend number = 1), or Z-shaped cracks (bend number = 2). The corresponding penetration routes or airflow directions are shown in the lower half of Figure 7.13, with the assumption that the direction of the leakage airflow is out from the screen or book page.

Both the shape of a crack and its geometric parameters, including crack length (L), width (W), and height (H), may affect the particle penetration ratio.

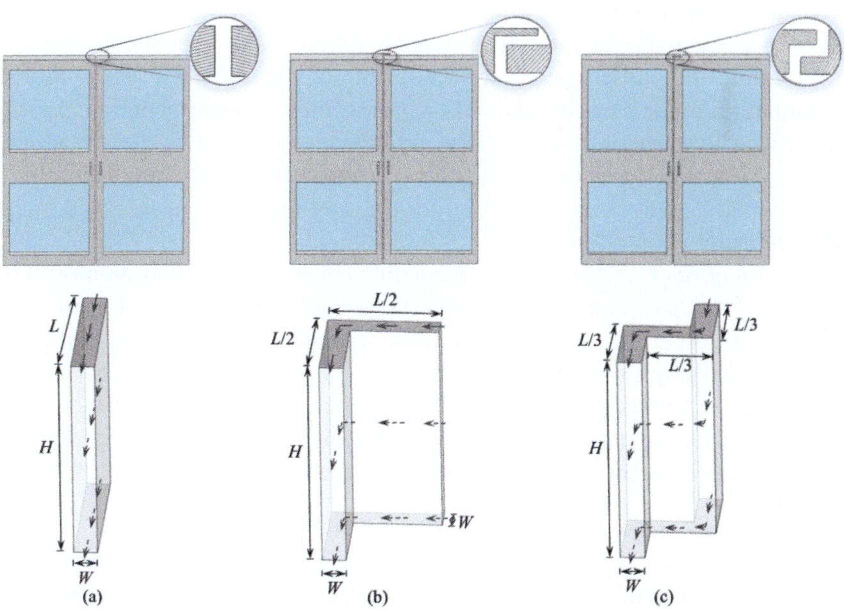

Figure 7.13 Typical cracks around window or door frames: (a) straight crack; (b) L-shaped crack (bend number = 1); (b) Z-shaped crack (bend number = 2).

The particle penetration ratio is defined as:

$$P = \frac{N_{\text{penetrated}}}{N_{\text{in}}} \tag{7.4}$$

where P is the particle penetration ratio, N_{in} represents the number of particles flowing into the crack, and $N_{\text{penetrated}}$ is the number of particles that penetrate through the crack. Due to gravitational deposition, Brownian diffusion, and inertial impaction, some of the airborne aerosols are deposited inside the crack. The overall particle penetration ratio can be estimated as the product of these three processes (Liu and Nazaroff 2001):

$$P = P_{\text{G}} \times P_{\text{D}} \times P_{\text{I}} \tag{7.5}$$

where P_{G} represents the ratio for penetration after gravitational settling, P_{D} stands for the penetration ratio associated with Brownian diffusion, and P_{I} is the penetration ratio after inertial impaction.

Figure 7.14 illustrates the particle penetration ratio through straight cracks under certain pressure differences. In accordance with common sense, both the ultrafine and coarse particles have relatively low penetration ratios. Longer cracks lead to lower penetration ratios, but not significantly. In contrast, crack width exerts a more pronounced effect, in that all the particles apparently have lower penetration ratios through 0.25 mm-wide cracks than through 1 mm-wide cracks, as shown in Figure 7.14a, b. Particles between 0.02 and 4 μm are almost able to penetrate ($P \approx 1$) 1 mm-wide cracks. Figure 7.14b also shows that for $W = 0.1$ mm, the particle penetration ratio through the crack is less than 7%. Accumulation-mode particles ($d_{\text{P}} = 0.1$–1 μm) exhibit a higher penetration ratio due to the reduced gravitational settling and Brownian diffusion in this size range. Notably, if H is much greater than W, then crack height (H) has virtually no effect on the particle penetration ratio. A larger pressure difference also results in an increased penetration ratio.

The number of crack bends has a negligible effect on the penetration ratio of submicron particles (<1 μm), whereas particles with diameters larger than 1 μm exhibit reduced penetration as the number of bends increases, as a result of a greater inertial impaction chance (Chen et al. 2012). Similar trends have been observed in automotive envelope penetration tests (Xu et al. 2010). Furthermore, the crack surface material and roughness exert a slight impact on the particle penetration through cracks under a small pressure difference (Liu and Nazaroff 2003).

7.5.1.4 Indoor/Outdoor Particle Concentration Ratio (I/O ratio)

The indoor/outdoor particle concentration ratio (I/O ratio) is widely used to indicate the contribution of outdoor-originating particles to indoor environments. The I/O ratio of particle concentrations is defined as:

$$\text{I/O ratio} = \frac{C_{\text{in}}}{C_{\text{out}}} \tag{7.6}$$

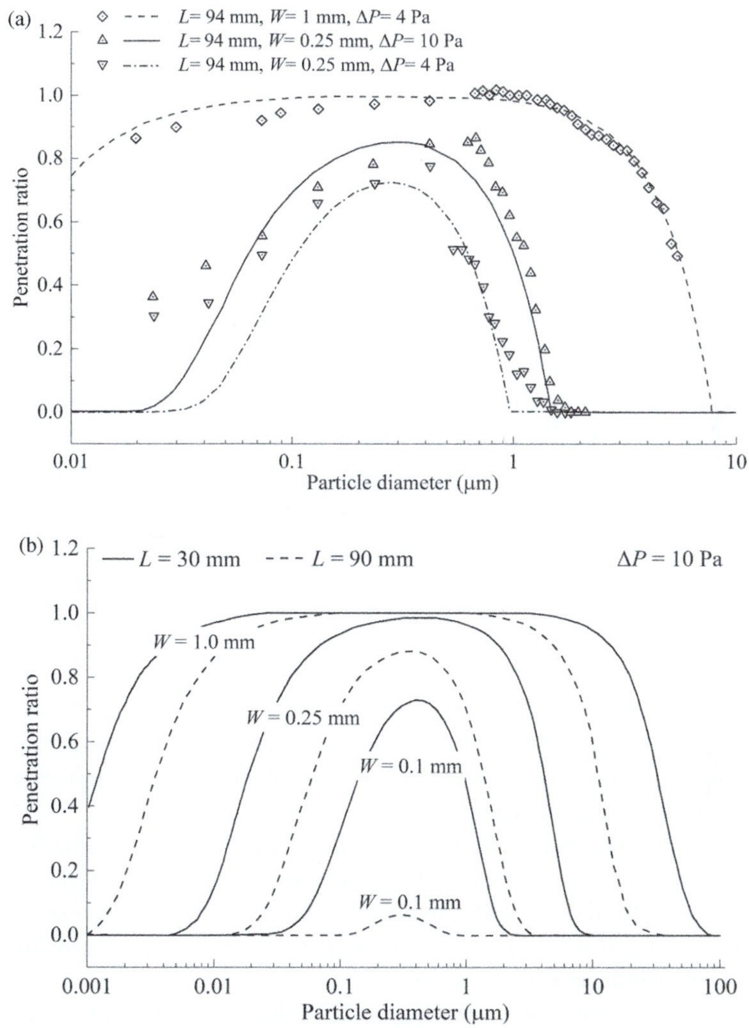

Figure 7.14 Particle penetration ratios through straight cracks for different crack dimensions and under different pressure differences: (a) comparison of the measurement data (symbols) with the model prediction (lines). *Source:* Adapted from Liu and Nazaroff (2003); (b) model prediction results. *Source:* Liu and Nazaroff (2003) / with permission of Elsevier.

where C_{in} and C_{out} are the indoor and outdoor particle concentrations, respectively.

Table 7.12 presents typical I/O ratios of particle concentrations in homes, a classroom, and an office. Ratios exceeding 1 signify that the indoor particles are predominantly from indoor PM sources. Considering the substantially higher

Table 7.12 Indoor/outdoor (I/O) particle concentration ratios in different studies.

Country/region	PM size	I/O ratio	Comment	References
USA	PM_{10}	1.15	With indoor emission sources	Abt et al. (2000)
	$PM_{2.5}$	1.19		
Hong Kong, China	PM_{10}	0.88	ACH > 3.5, with various human activities	Chao and Wong (2002)
	$PM_{2.5}$	0.92		
	PM_{10}	1.04	ACH < 3.5, with various human activities	
	$PM_{2.5}$	1.09		
China	$PM_{2.5}$	5.26	Smoking indoors	Zhou et al. (2016)
		0.6	No human activity	
China	$PM_{2.5}$	0.88	Natural ventilation	Wang et al. (2016)
		0.46	Mechanical ventilation	
Europe	$PM_{2.5}$	1.44	In the classroom during class	Kalimeri et al. (2019)
		0.62	In the office on weekdays	

emission rates of combustion-released UFPs, such as those from cooking and smoking, the $PM_{2.5}$ typically exhibits higher I/O ratios than do the PM_{10}. The coarse particles found indoors have typically been resuspended from surfaces due to disturbances caused by human activities.

For accumulation-mode particles (d_p = 0.1–1 μm), the I/O ratios approach 1 because indoor source emissions are minimal within this size range (Wallace 2006). Outdoor accumulation-mode particles readily infiltrate into indoor environments because of these particles' high penetration capability and low deposition velocities indoors (Long et al. 2001).

The room ventilation rate (or air exchange rate) presents greater complexity in its influence on the I/O ratios in scenarios without indoor sources. The I/O ratio generally increases with the ventilation rate (Kang et al. 2020). However, when high-emission indoor sources and low outdoor particle concentrations are present, the I/O ratio decreases with the ventilation rate, since a higher ventilation rate helps to dilute indoor particle concentrations (Chen and Zhao 2011). Under natural ventilation, the type of window screen also affects the I/O ratio. Measurements have revealed distinct I/O ratios when different window screen types are employed in Korea: the $PM_{2.5}$ I/O ratio is 0.24 when an electrostatic window screen is used, 0.51 for a nanofiber screen, 0.58 for a non-woven fabric screen, and 0.72 for a filtered insect netting (Kang et al. 2020).

7.5.2 Intrusion of Pollens and Molds

Pollen and mold may originate from both outdoor and indoor environments. The proteins and glycoproteins in pollen and mold act as allergens, exposure to which may induce allergic responses of the immune system in sensitive people (Traidl-Hoffmann et al. 2003). The indoor origin of pollens is in certain ornamental plants, while molds may grow on indoor surfaces in areas with high temperature and humidity. Commonly, outdoor environments contain higher concentrations of airborne pollens and molds, which make their way into the indoor spaces by means of ventilation or surface carriers.

7.5.2.1 Intrusion of Pollens

Pollen grain size is often denoted by the projected area diameter (Hirose and Osada 2016):

$$d_a = \sqrt{\frac{4A_p}{\pi}} \tag{7.7}$$

where d_a is the projected area diameter, while A_p is the projected surface area of the grain and can be measured by microscopy. Pollens have a size range of 5–100 μm, as shown in Table 1.8, and most pollens have a projected area diameter greater than 20 μm (Wang et al. 2022). For almost all anemophilous and entomophilous flowers, pollen diameters are larger than 20 μm (Lu et al. 2022). Table 7.13 lists the projected area diameters of some pollen grains and their aerodynamic diameters (Hirose and Osada 2016). The aerodynamic diameters of the pollen grains of Japanese cedar, Japanese cypress, and short ragweed are quite close to their corresponding projected area diameters, whereas the aerodynamic diameters of dry pine pollens are 30% smaller than the projected area diameters.

Since the aerodynamic diameters of most pollen grains are greater than 10 μm, these grains can only be carried by strong winds. However, pollen grains can be

Table 7.13 Projected area diameters and aerodynamic diameters of some typical plant pollens.

Plant type	Density (g/cm^3)	Projected area diameter (μm)	Aerodynamic diameter (μm)
Japanese cedar	1.06	28.4	27.2
Japanese cypress	0.82	27.6	27.2
Short ragweed	0.71	21.3	19.6
Japanese black pine	0.55	45.7	32.1
Japanese red pine	0.54	43.9	31.6

Source: Hirose and Osada (2016) / with permission of Springer Nature.

fractured into small fragments. Due to the smaller sizes of these pollen fragments, they can be more easily transported and suspended in the air. Notably, the pollen fragments can still contain allergens. Presently, the size spectrum of the pollen fragments remains unclear. Available studies have not identified the sampled pollens as being in the form of intact pollen grains, fragments, or a mixture, nor have the percentage ratios of these pollen forms been addressed.

Outdoor pollens can intrude into indoor spaces via openings in the building envelope, such as windows, doors, holes, cracks, and ventilating ducts. Because of the pollens' removal by air filters (if installed) in mechanical ventilation systems, and their interception and deposition inside ventilation ducts, the pollen concentration in the mechanical ventilation supply is typically low. The majority of the pollens found in rooms have been transported through building openings by natural ventilation. In addition, people's clothing, shoes, and pets can carry outdoor pollens into rooms (Wang et al. 2022).

Window screens may intercept the invading pollens, but not to a large extent. If doors and windows are closed, the pollens that infiltrate through cracks can be somewhat intercepted. Considering the relatively low indoor air speed, some of the pollens may be deposited on floors due to gravity. Consequently, the indoor pollen concentrations are commonly lower than the outdoor concentrations. Although the mechanisms of pollen invasion have not been thoroughly studied, the I/O ratios of pollen concentrations have been investigated for many years. The I/O ratio is affected by the pollen type, ventilation conditions, indoor plants, pollen monitoring locations, human and pet behavior, and the seasons (Wang et al. 2022).

Figure 7.15 presents the I/O ratios of pollen counts in different indoor environments. The average I/O ratios in most places are less than 0.5, which implies that outdoor environments contain higher concentrations of pollens. Moreover, in most spaces, the maximum I/O ratios are less than 1. In only three studies are the I/O ratios found to be greater than 1. The I/O ratios exceeding 1 may have been due to the presence of indoor ornamental plants that release pollens; outdoor pollens carried into the rooms by people or pets; or scouring by the precipitation that has settled the outdoor pollens. Since the pollens that enter via building openings are typically not mixed well with the indoor air, the I/O ratios in specific indoor positions decrease with the distance away from the building openings (Jantunen and Saarinen 2009). A room whose windows are larger or face the dominant wind direction corresponds to higher I/O ratios. Although pollen dynamics are strongly dependent on pollen size, available studies have not yet reported a clear correlation between I/O ratio and pollen component diameter.

7.5.2.2 Intrusion of Molds

Commonly, the outdoor air contains higher concentrations of mold aerosols than the indoor air, except in some very damp buildings. The mold aerosols include

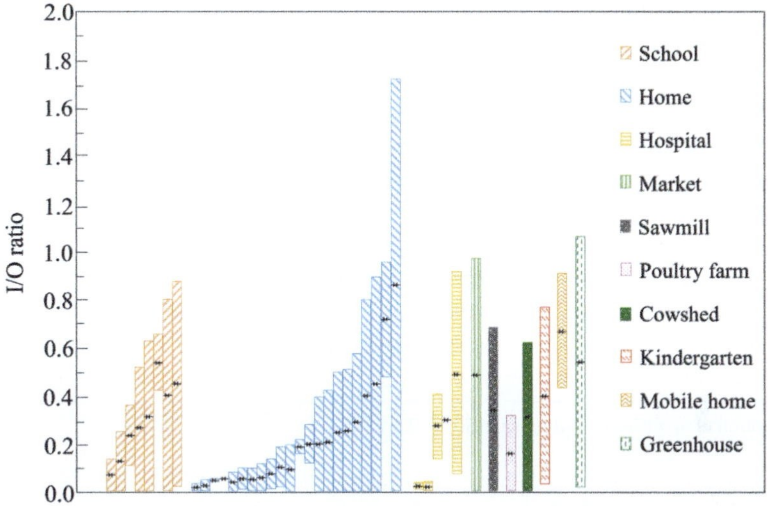

Figure 7.15 Range of I/O ratios of sampled pollen counts in different indoor environments, where the maximum, average, and minimum I/O ratios are displayed. *Source:* Adapted from Wang et al. (2022).

intact mold spores, ruptured spore pieces, and hyphal fragments. The aerodynamic diameters of most allergenic mold spores are between 2 and 6 μm (Reponen et al. 1996), whereas the ruptured spore pieces range from 0.3 to 1.3 μm, and the hyphal fragments may be in the submicron range. Consequently, the ruptured spore pieces and hyphal fragments may remain airborne much longer than the intact spores; however, the former retain similar allergenic potency to that of the intact spores (Górny et al. 2002). Table 7.14 presents the spore sizes and surface morphology of five common mold types.

The size spectrum of airborne mold spores in indoor spaces is strongly influenced by the sizes of outdoor molds. Field campaigns across diverse climates have reported a unimodal distribution of outdoor spore sizes peaking roughly in the range of 2–4 μm (Ye et al. 2021). Once the spores enter indoor spaces, the mode sizes of the spores become smaller. This shift is due to the more intense interception and settling effects on larger particles when the spores penetrate the building envelope. The indoor mold aerosols are often more abundant in the size range of 0.65–2.1 μm, whereas spores larger than 4.7 μm are quite few (Wu et al. 2020).

However, in moisture-damaged or mold-contaminated buildings, the spore size spectrum deviates from the aforementioned pattern. As shown in Figure 7.16, in mold-infested dwellings, airborne molds in the size range of 2.1–4.7 μm constitute 76% of the total mold aerosols, while the number ratio between 0.65 and 2.1 μm is 17%. In contrast, in normal dwellings without mold damage, the number ratio

Table 7.14 Common indoor mold spores.

Mold type	Spore size (μm)	Geometric shape	Surface morphology	References
Aspergillus fumigatus	2.5–3	Globose to subglobose	Rough-walled to echinulate	Samson et al. (1981)
Aspergillus versicolor	2–3.5	Globose	Echinulate	
Cladosporium cladosporioides	3–7 by 2–4	Ellipsoidal to lemon-shaped	Smooth	
Penicillium brevicompactum	3–4.5	Globose to subglobose	Smooth or slightly rough	
Penicillium melinii	5–6	Globose	Rough	Ramirez (1982)

Source: Reponen et al. (1996) / with permission of Elsevier.

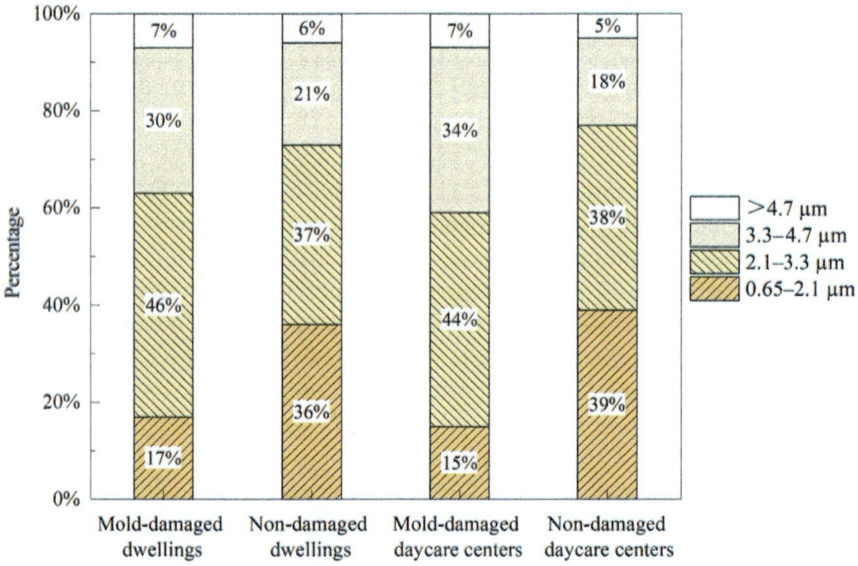

Figure 7.16 Comparison of the percentage ratios of mold aerosol counts in different size ranges between mold-damaged buildings and normally operated buildings. *Source:* Adapted from Reponen et al. (1994).

in the size range of 2.1–4.7 μm is only 58%, while the number ratio between 0.65 and 2.1 μm rises to 36%. Similar distributions are present in mold-damaged and non-damaged daycare centers, as shown in Figure 7.16. The mold-damaged daycare buildings also have slightly higher number ratios of airborne molds larger than 4.7 μm than do the non-damaged buildings. The shift in the mold aerosol

Table 7.15 Typical ranges of I/O ratios of fungal spore concentrations in various types of buildings.

Building type	Major mold genera	Human occupation[a]	I/O ratio	References
Typical urban residences	*Cladosporium*, *Penicillium*, *Aspergillus*	Yes	0.3–1	Shelton et al. (2002), Myers et al. (2021)
Hospital wards	*Aspergillus*, *Penicillium*, *Cladosporium*	No	0.36–0.68	Goudarzi et al. (2017)
		Yes	0.78–0.99	
Recreational facilities (bars and cafés)	*Cladosporium*, *Penicillium*, *Aspergillus*, *Alternaria*	Yes	0.5–1.8	Jo and Seo (2005)
Commercial office buildings	*Cladosporium*, *Penicillium*, *Aspergillus*	Yes	0.22	Tsai et al. (2007)
Flood-damaged residences	*Trichoderma*, *Penicillium*	Yes	1.3–13.5	Fabian et al. (2005)
Hospitals	*Penicillium*, *Aspergillus*, *Bjerkandera*, *Alternaria*, *Cladosporium*	Yes	<1	Sautour et al. (2009)
Suburban residences	*Aspergillus*, *Penicillium*, *Cladosporium*, non-sporing, *Alternaria*	Yes	0.9–1.2	Pei-Chih et al. (2000)
University libraries	*Penicillium*, *Cladosporium*, *Aspergillus*	Yes	<1	Hassan et al. (2021)
Retirement homes and school dormitories	*Penicillium*, *Cladosporium*, *Aspergillus*	Yes	0.57–1.77	Faridi et al. (2015)
Kindergartens	*Aspergillus*, *Cladosporium*, *Penicillium*	Yes	0.1–4.5	Chegini et al. (2020)

a) An occupation status of yes indicates the occupants were present during sampling.

size spectrum between the mold-damaged buildings and the normally operated buildings reflects either the removal of outdoor mold species when they invade buildings or a difference in growing status between the outdoor and indoor molds.

The indoor-to-outdoor (I/O) ratio of mold aerosol concentrations is widely used to diagnose potential indoor fungal contamination. An I/O ratio below 1 indicates that indoor mold aerosols originate mainly outdoors, while a ratio above 1 indicates the presence of indoor mold sources due to water intrusion, damp building materials, or inadequate ventilation (Chen and Zhao 2011). I/O ratios are strongly influenced by building function, ventilation strategy, and the degree of moisture damage.

Table 7.15 lists the I/O ratios of culturable airborne molds in 10 types of buildings. All these ratios were obtained from culture-based measurements, with impaction or passive settling of airborne molds on agar plates, followed by culture and quantification. In many of the buildings, the average I/O ratios remain below 1. Human occupation increases the I/O ratios. Commercial offices and university libraries have relatively lower I/O ratios as compared with other types of buildings. Recreational facilities have maximum I/O ratios as high as 1.8, which indicates the combined influence of outdoor infiltration and occupant activities. I/O ratios are higher in certain group-living settings; retirement homes and school dormitories show 0.57–1.77, and kindergartens span 0.1–4.5; however, the most pronounced amplification is linked to moisture damage, with flood-damaged residences soaring to 1.3–13.5. Table 7.15 also reveals that *Cladosporium* and *Penicillium* are dominant in buildings with a low I/O ratio, whereas *Trichoderma* (with *Penicillium* also common) is prominent in water-damaged buildings.

The I/O ratios of mold aerosols also vary noticeably with the seasons (Fan et al. 2021). The ratios are generally higher in summer and autumn than in winter and spring (Chen et al. 2024). This pattern corresponds to elevated ambient temperatures and relative humidity in warmer and humid seasons, which promote fungal growth, sporulation, and subsequent indoor accumulation, particularly in buildings with inadequate ventilation or moisture problems (Du et al. 2021). In contrast, colder and drier conditions in winter tend to suppress fungal activity and reproduction.

7.6 Summary

This chapter addresses the particle emission sources in households, offices, and lavatories, as well as the intrusion of outdoor aerosols into indoor spaces. The primary sources of indoor aerosol emissions include cooking, smoking, fuel and incense burning, printing, and operation of photocopiers. The main points are as follows:

7.6.1 In Households

Cooking may raise indoor PM concentrations by 20 μg/m^3, and the particle emissions vary with the cooking method and food type. The UFPs produced by stir-frying (148×10^{11} particles/min) are about 28 times that by boiling and 31.6 times that by steaming. The gas burners used for cooking (10.6×10^{12} particles/min, 5.2–6.4 nm) emit more particles than the electric stove, gas grilling, and electric grilling. For solid fuel stoves, the $PM_{2.5}$ emission rate reaches an average of 20 g/h. There are also particle emissions from the various heat sources used for house heating; notably, the burning of solid fuels, such as wood and coal, releases many more particles than the burning of gases.

The MMADs of the particles produced by combustible cigarettes and e-cigarettes are 0.5 ± 0.01 and 1.06 ± 0.14 μm, respectively. E-cigarettes (274.9 μg/puff) release more and larger particles than do combustible cigarettes (179.8 μg/puff).

The emission rate of incense burning varies with the brand of scent used, with some brands emitting $PM_{2.5}$ at a rate as high as 60.9 mg/h. Incense combustion produces unimodally distributed particles (peak: ~125 nm) at 3.1 ($\pm$0.3)$\times 10^{12}$ particles/h, while candle burning emits 2.0 ($\pm$0.1)$\times 10^{13}$ particles/h (peak: ~23 nm in the normal burning stage and 50 nm in the smoldering stage).

Most particles emitted during ironing are in the ultrafine size range, with the geometric mean diameter ranging from 41.1 to 62.6 nm. Specifically, the emission rate of a steam iron with a boiler amounts to 8.3 μg/s.

HCPs and PCPs release volatile constituents that form SOAs after the volatile constituents are oxidized. Spray of HCPs, particularly in the propellant-driven type, emits high concentrations ($\sim 2.6 \times 10^{12}$ particles/min) of fine particles (<0.4 μm). Terpene-based fragrances in PCPs (e.g., tea tree oil, lavender) can react with ozone, which rapidly produces SOAs (<100 nm, with the highest emission rate of 6.2 ($\pm$ 2.8)$\times 10^7$ particles/min).

7.6.2 In Offices

Laser printers emit fine particles with rates ranging from 3.4×10^8 to 3.3×10^{12} particles/min per printer, subject to the toner type, printing speed, and other conditions. Copiers also emit a significant amount of particles. UFPs less than 50 nm in diameter were found to dominate the PNC in a copy center, and the emission rate less than 50 nm was approximately 2.99×10^7 particles/min per copier.

7.6.3 In Lavatories

Approximately 90% of the aerosols produced by toilet flushing are in the submicron range, with the exception of squat toilet flushing, which is performed at an extremely high pressure, and vacuum toilet flushing. When a squatting toilet with an extremely high pressure or vacuum toilet is flushed, slightly less than 70% of the emitted particles are in the submicron sizes. Typical sitting toilets have an emission rate of 2.87×10^5 particles/flush, while the average emission rate for squatting toilets is generally higher, in the range of 2.93×10^5–5.72×10^5 particles/flush. Vacuum toilets have the lowest emission rate at 8,498 particles/flush. A single flushing of a male urinal releases approximately 4.5×10^5 particles. Toilet flushing also produces bioaerosols, and the rate of emission is related to the intensity of flushing and the type of microorganisms.

Handwashing produces aerosols that are typically less than 1 μm in diameter. The aerosol emission rate is influenced by the type of faucet aerator and water flow. The measured highest median concentration with certain aerators can reach 59.9 μg/l. Drying hands under a high-speed air-jet dryer may release 1.06×10^5 liquid aerosol particles, of which 92.93% are in the submicron size range. The motor of a hand dryer with a carbon brush also emits CuO particles (2.64×10^6 particles within 30 seconds, 99.7% of which are of submicron size).

Showering can raise the aerosol concentration to a peak of 5–14 mg/m^3 within 1–2 minutes. Hot showers (41 °C) release larger particles at a higher rate (5–6 μm, 7.7×10^4 μg/min), while cold showers release smaller particles at a lower rate (25.9 °C, 1–2 μm, 19.62 μg/min).

7.6.4 Intrusion of Outdoor Aerosols

The majority of the outdoor-originating aerosols are carried into indoor spaces by natural ventilation through building openings such as open doors and windows. The window screen used in some countries may remove some airborne particles in the size range of 0.3–10 μm. Both the shape of a crack and its geometric parameters, including crack length, width, and height, may affect the particle penetration ratio. Accumulation-mode particles (0.1–1 μm) have the highest penetration ratio through the cracks, while UFPs and coarse particles may be somewhat captured during the penetration process.

The indoor/outdoor particle concentration ratio (I/O ratio) is widely used to indicate the contribution of outdoor-originating particles to indoor environments. I/O ratios >1 signify dominant indoor sources, with UFPs exhibiting the highest ratios due to strong indoor combustion emissions (e.g., cooking). Coarse particles primarily originate from particle resuspension by human activities.

Commonly, outdoor environments contain higher concentrations of airborne pollens and molds, which make their way into the indoor spaces by means of ventilation or surface carriers. Most pollen grains (>10 μm in size) require strong winds for dispersal but may be fractured into allergen-containing pollen piece particles.

The pollen I/O ratio is affected by the pollen type, ventilation conditions, indoor plants, pollen monitoring locations, human and pet behavior, and the seasons.

The aerodynamic diameters of most allergenic mold spores are between 2 and 6 μm. Once the outdoor spores invade indoor spaces, the mode sizes of the spores become smaller due to the pronounced interception and settling effects of larger particles. Mold aerosol I/O ratios are strongly influenced by building function, ventilation condition, and the degree of moisture damage to buildings.

Problems

7.1 What are the primary chemical components found in cooking oil fumes (COFs)?

7.2 What is the mass median aerodynamic diameter (MMAD) of particles emitted by traditional combustible cigarettes versus e-cigarettes?

7.3 Why does incense burning pose a significant health risk, and what toxic elements are commonly detected in its particulate phase?

7.4 What is the primary mechanism by which photocopiers generate ultrafine particles during operation?

7.5 How does flushing water pressure influence aerosol emission rates from toilets?

7.6 Why do hot showers produce larger aerosol particles than cold showers?

7.7 Discuss the potential health implications of prolonged exposure to ultrafine particles (UFPs) emitted from cooking activities.

7.8 Compare the environmental and health impacts of burning solid fuels versus natural gas for household heating in developing countries.

7.9 Describe basic mechanisms for formation of indoor secondary organic aerosols (SOAs) and some strategies to reduce indoor SOA emissions.

7.10 Evaluate the effectiveness of ventilation systems in reducing aerosol concentrations generated by laser printers in office settings.

7.11 Propose strategies to mitigate bioaerosol exposure in public lavatories.

7.12 Describe how building cracks capture the invaded aerosols and the impacts to the particle size spectrum.

7.13 Describe indoor-to-outdoor (I/O) ratios of airborne pollen and mold concentrations in common indoor environments, analyze factors affecting the I/O ratios, and propose strategies to minimize indoor pollen and mold exposure.

References

Abdullahi, K.L., Delgado-Saborit, J.M., and Harrison, R.M. (2013). Emissions and indoor concentrations of particulate matter and its specific chemical components from cooking: a review. *Atmospheric Environment* 71: 260–294.

Abt, E., Suh, H.H., Allen, G., and Koutrakis, P. (2000). Characterization of indoor particle sources: a study conducted in the metropolitan Boston area. *Environmental Health Perspectives* 108: 35–44.

Adhikari, S., Mahapatra, P.S., Pokheral, C.P., and Puppala, S.P. (2020). Cookstove smoke impact on ambient air quality and probable consequences for human health in rural locations of southern Nepal. *International Journal of Environmental Research and Public Health* 17: 550.

Afshari, A., Matson, U., and Ekberg, L.E. (2005). Characterization of indoor sources of fine and ultrafine particles: a study conducted in a full-scale chamber. *Indoor Air* 15: 141–150.

Ali, W., Yang, Y.-F., Gong, L. et al. (2022). Emission characteristics and quantitative health risk assessment of bioaerosols in an indoor toilet after flushing under various ventilation scenarios. *Building and Environment* 207: 108463.

Azimi, P., Zhao, D., and Stephens, B. (2014). Estimates of HVAC filtration efficiency for fine and ultrafine particles of outdoor origin. *Atmospheric Environment* 98: 337–346.

Bako-Biro, Z., Wargocki, P., Weschler, C.J., and Fanger, P.O. (2004). Effects of pollution from personal computers on perceived air quality, SBS symptoms and productivity in offices. *Indoor Air* 14: 178–187.

Benoit, M.-È., Prévost, M., Succar, A. et al. (2021). Faucet aerator design influences aerosol size distribution and microbial contamination level. *Science of the Total Environment* 775: 145690.

Bickers, D.R., Calow, P., Greim, H.A. et al. (2003). The safety assessment of fragrance materials. *Regulatory Toxicology and Pharmacology* 37: 218–273.

Bonjour, S., Adair-Rohani, H., Wolf, J. et al. (2013). Solid fuel use for household cooking: country and regional estimates for 1980–2010. *Environmental Health Perspectives* 121: 784–790.

Brown, S. (1999). Assessment of pollutant emissions from dry-process photocopiers. *Indoor Air* 9: 259–267.

Buonanno, G., Morawska, L., and Stabile, L. (2009). Particle emission factors during cooking activities. *Atmospheric Environment* 43: 3235–3242.

Buonanno, G., Johnson, G., Morawska, L. et al. (2011). Volatility characterization of cooking-generated aerosol particles. *Aerosol Science and Technology* 45 (9): 1069–1077.

Butcher, S.S. and Ellenbecker, M.J. (1982). Particulate emission factors for small wood and coal stoves. *Journal of the Air Pollution Control Association* 32: 380–384.

Byeon, J.H. and Kim, J.-W. (2012). Particle emission from laser printers with different printing speeds. *Atmospheric Environment* 54: 272–276.

Chao, C.Y. and Wong, K.K. (2002). Residential indoor PM10 and PM2. 5 in Hong Kong and the elemental composition. *Atmospheric Environment* 36: 265–277.

Chegini, F.M., Baghani, A.N., Hassanvand, M.S. et al. (2020). Indoor and outdoor airborne bacterial and fungal air quality in kindergartens: seasonal distribution, genera, levels, and factors influencing their concentration. *Building and Environment* 175: 106690.

Chen, C. and Zhao, B. (2011). Review of relationship between indoor and outdoor particles: I/O ratio, infiltration factor and penetration factor. *Atmospheric Environment* 45: 275–288.

Chen, C., Zhao, B., Zhou, W. et al. (2012). A methodology for predicting particle penetration factor through cracks of windows and doors for actual engineering application. *Building and Environment* 47: 339–348.

Chen, Y., Liang, Z., Li, G., and An, T. (2024). Indoor/outdoor airborne microbiome characteristics in residential areas across four seasons and its indoor purification. *Environment International* 190: 108857.

Cheng, K.-C., Ott, W., Wallace, L. et al. (2022). PM2. 5 exposure close to marijuana smoking and vaping: a case study in residential indoor and outdoor settings. *Science of the Total Environment* 802: 149897.

Clausen, P.A., Frederiksen, M., Sejbæk, C.S. et al. (2020). Chemicals inhaled from spray cleaning and disinfection products and their respiratory effects. A comprehensive review. *International Journal of Hygiene and Environmental Health* 229: 113592.

Cui, T., Lu, R., Liu, Q. et al. (2023). PM1 exposure and spatial transmission of nicotine from the simulated second-hand vapor of pod-based electronic cigarettes. *Science of the Total Environment* 897: 165355.

Dai, W., Zhong, H., Li, L. et al. (2018). Characterization and health risk assessment of airborne pollutants in commercial restaurants in northwestern China: under a low ventilation condition in wintertime. *Science of the Total Environment* 633: 308–316.

Dang, J., Li, C., Li, J. et al. (2019). Emissions from solid fuel cook stoves in the Himalayan region. *Energies* 12: 1089.

Dobaradaran, S., Schmidt, T.C., Kaziur-Cegla, W., and Jochmann, M.A. (2021). BTEX compounds leachates from cigarette butts into water environment: a primary study. *Environmental Pollution* 269: 116185.

Dodson, R.E., Nishioka, M., Standley, L.J. et al. (2012). Endocrine disruptors and asthma-associated chemicals in consumer products. *Environmental Health Perspectives* 120: 935–943.

Du, C., Li, B., and Yu, W. (2021). Indoor mould exposure: characteristics, influences and corresponding associations with built environment – a review. *Journal of Building Engineering* 35: 101983.

Dyson United States (2024). Dyson airblade V (nickel) hand dryer. https://www.dyson.com/commercial/hand-dryers/airblade-v-nickel (accessed 1 September 2025).

Ehn, M., Thornton, J.A., Kleist, E. et al. (2014). A large source of low-volatility secondary organic aerosol. *Nature* 506: 476–479.

Ezzati, M., Lopez, A.D., Rodgers, A. et al. (2002). Selected major risk factors and global and regional burden of disease. *The Lancet* 360: 1347–1360.

Fabian, M., Miller, S., Reponen, T., and Hernandez, M.T. (2005). Ambient bioaerosol indices for indoor air quality assessments of flood reclamation. *Journal of Aerosol Science* 36: 763–783.

Falconer, I.R., Burch, M.D., Steffensen, D.A. et al. (1994). Toxicity of the blue-green alga (cyanobacterium) Microcystis aeruginosa in drinking water to growing pigs, as an animal model for human injury and risk assessment. *Environmental Toxicology and Water Quality* 9: 131–139.

Fan, G., Xie, J., Yoshino, H. et al. (2021). Investigation of fungal contamination in urban houses with children in six major Chinese cities: genus and concentration characteristics. *Building and Environment* 205: 108229.

Faridi, S., Hassanvand, M.S., Naddafi, K. et al. (2015). Indoor/outdoor relationships of bioaerosol concentrations in a retirement home and a school dormitory. *Environmental Science and Pollution Research* 22: 8190–8200.

Fernández, E., Ballbè, M., Sureda, X. et al. (2015). Particulate matter from electronic cigarettes and conventional cigarettes: a systematic review and observational study. *Current Environmental Health Reports* 2: 423–429.

Fooshee, D.R., Aiona, P.K., Laskin, A. et al. (2015). Atmospheric oxidation of squalene: molecular study using COBRA modeling and high-resolution mass spectrometry. *Environmental Science & Technology* 49: 13304–13313.

Fung, C.C., Shu, S., and Zhu, Y. (2014). Ultrafine particles generated from coloring with scented markers in the presence of ozone. *Indoor Air* 24: 503–510.

Gakidou, E., Afshin, A., Abajobir, A.A. et al. (2017). Global, regional, and national comparative risk assessment of 84 behavioural, environmental and occupational, and metabolic risks or clusters of risks, 1990–2016: a systematic analysis for the global burden of disease study 2016. *The Lancet* 390: 1345–1422.

Gao, J., Cao, C., Wang, L. et al. (2013a). Determination of size-dependent source emission rate of cooking-generated aerosol particles at the oil-heating stage in an experimental kitchen. *Aerosol and Air Quality Research* 13: 488–496.

Gao, J., Cao, C., Xiao, Q. et al. (2013b). Determination of dynamic intake fraction of cooking-generated particles in the kitchen. *Building and Environment* 65: 146–153.

Gao, J., Cao, C., Zhang, X., and Luo, Z. (2013c). Volume-based size distribution of accumulation and coarse particles (PM0. 1–10) from cooking fume during oil heating. *Building and Environment* 59: 575–580.

Gao, Z., Li, Y., Duanmu, L., and Zhang, T. (2024). Copper oxide particle emission and the spread in a public washroom from a high-speed jet air dryer. *Journal of Hazardous Materials* 462: 132808.

Gao, Z., Li, Y., Wang, F. et al. (2025). Emitted droplets and aerosols and their transmission when drying hands under an air-jet dryer. *Journal of Hazardous Materials* 482: 136508.

Géhin, E., Ramalho, O., and Kirchner, S. (2008). Size distribution and emission rate measurement of fine and ultrafine particle from indoor human activities. *Atmospheric Environment* 42: 8341–8352.

Górny, R.L., Reponen, T., Willeke, K. et al. (2002). Fungal fragments as indoor air biocontaminants. *Applied and Environmental Microbiology* 68: 3522–3531.

Goudarzi, G., Soleimani, Z., Sadeghinejad, B. et al. (2017). Visiting hours impact on indoor to outdoor ratio of fungi concentration at Golestan University Hospital in Ahvaz, Iran. *Environment and Pollution* 6: 62–68.

Hankey, S., Sullivan, K., Kinnick, A. et al. (2015). Using objective measures of stove use and indoor air quality to evaluate a cookstove intervention in rural Uganda. *Energy for Sustainable Development* 25: 67–74.

Hassan, A., Zeeshan, M., and Bhatti, M.F. (2021). Indoor and outdoor microbiological air quality in naturally and mechanically ventilated university libraries. *Atmospheric Pollution Research* 12: 101136.

Hayward, C., Ross, K.E., Brown, M.H., and Whiley, H. (2025). The impact of water flow rates on bioaerosol production from handwashing basins. *Building and Environment* 271: 112624.

He, S. and Han, J. (2021). Electrostatic fine particles emitted from laser printers as potential vectors for airborne transmission of COVID-19. *Environmental Chemistry Letters* 19: 17–24.

He, C., Morawska, L., Hitchins, J. et al. (2004). Contribution from indoor sources to particle number and mass concentrations in residential houses. *Atmospheric Environment* 38: 3405–3415.

He, C., Morawska, L., and Taplin, L. (2007). Particle emission characteristics of office printers. *Environmental Science & Technology* 41: 6039–6045.

He, C., Morawska, L., Wang, H. et al. (2010). Quantification of the relationship between fuser roller temperature and laser printer emissions. *Journal of Aerosol Science* 41: 523–530.

Hinds, W.C. and Zhu, Y. (2022). *Aerosol technology: properties, behavior, and measurement of airborne particles*. 3e. John Wiley & Sons.

Hirose, Y. and Osada, K. (2016). Terminal settling velocity and physical properties of pollen grains in still air. *Aerobiologia* 32: 385–394.

Hopkins, L. and Hansford, B. (1974). *Air Flow Through Cracks*. Elsevier.

Hsu, D.J., Huang, H.L., and Sheu, S.C. (2012). Characteristics of air pollutants and assessment of potential exposure in spa centers during aromatherapy. *Environmental Engineering Science* 29: 79–85.

Huang, Z.H., Li, S.C., and Yu, D.X. (2008). Investigation on environment of Chinese restaurants in Hong Kong. *Chinese Journal of Industrial Hygiene and Occupational Diseases* 26: 474–476.

Huang, Y., Ho, K.F., Ho, S.S.H. et al. (2011a). Physical parameters effect on ozone-initiated formation of indoor secondary organic aerosols with emissions from cleaning products. *Journal of Hazardous Materials* 192: 1787–1794.

Huang, Y., Ho, S.S.H., Ho, K.F. et al. (2011b). Characteristics and health impacts of VOCs and carbonyls associated with residential cooking activities in Hong Kong. *Journal of Hazardous Materials* 186: 344–351.

Jantunen, J. and Saarinen, K. (2009). Intrusion of airborne pollen through open windows and doors. *Aerobiologia* 25: 193–201.

Jo, W.K. and Seo, Y.J. (2005). Indoor and outdoor bioaerosol levels at recreation facilities, elementary schools, and homes. *Chemosphere* 61: 1570–1579.

K., W.L., Sheldon, L.S., Whitaker, D.A. et al. (1996). Measurement of indoor air emissions from dry-process photocopy machines. *Journal of the Air & Waste Management Association* 46: 821–829.

Kalimeri, K.K., Bartzis, J.G., Sakellaris, I.A., and de Oliveira Fernandes, E. (2019). Investigation of the PM2. 5, NO2 and O3 I/O ratios for office and school microenvironments. *Environmental Research* 179: 108791.

Kang, S., Li, C., Wang, F. et al. (2009). Total suspended particulate matter and toxic elements indoors during cooking with yak dung. *Atmospheric Environment* 43: 4243–4246.

Kang, K., Kim, T., Shin, C.W. et al. (2020). Filtration efficiency and ventilation performance of window screen filters. *Building and Environment* 178: 106878.

Kerger, B., Schmidt, C., and Paustenbach, D. (2000). Assessment of airborne exposure to trihalomethanes from tap water in residential showers and baths. *Risk Analysis* 20: 637–652.

Kim, T., Park, J., Seo, J. et al. (2020). Behavioral characteristics to airborne particles generated from commercial spray products. *Environment International* 140: 105747.

Knowlton, S.D., Boles, C.L., Perencevich, E.N. et al. (2018). Bioaerosol concentrations generated from toilet flushing in a hospital-based patient care setting. *Antimicrobial Resistance and Infection Control* 7: 1–8.

Lai, A., Tan, T., Li, W., and Ip, D.K.M. (2018). Emission strength of airborne pathogens during toilet flushing. *Indoor Air* 28: 73–79.

Lee, C.-W. and Hsu, D.-J. (2007). Measurements of fine and ultrafine particles formation in photocopy centers in Taiwan. *Atmospheric Environment* 41: 6598–6609.

Lee, H.S., Kang, B.W., Cheong, J.P., and Lee, S.K. (1997). Relationships between indoor and outdoor air quality during the summer season in Korea. *Atmospheric Environment* 31: 1689–1693.

Lee, S., Lam, S., and Fai, H.K. (2001). Characterization of VOCs, ozone, and PM10 emissions from office equipment in an environmental chamber. *Building and Environment* 36: 837–842.

Li, L., Lee, E.S., Nguyen, C., and Zhu, Y. (2020). Effects of propylene glycol, vegetable glycerin, and nicotine on emissions and dynamics of electronic cigarette aerosols. *Aerosol Science and Technology* 54: 1270–1281.

Li, P., Zhang, T., and Zhang, Y. (2022). Measuring the flushing-generated flow and aerosols in lavatory of commercial aircraft. *Building and Environment* 214: 108948.

Li, J., Fan, G., Ou, Y., Deng, Q. (2023). Characteristics and control strategies of indoor particles: An updated review. *Energy and Buildings* 294 (113232).

Liao, C.-M., Chen, S.-C., Chen, J.-W., and Liang, H.M. (2006). Contributions of Chinese-style cooking and incense burning to personal exposure and residential PM concentrations in Taiwan region. *Science of the Total Environment* 358: 72–84.

Lin, T.-C., Chang, F.-H., Hsieh, J.-H. et al. (2001). Environmental exposure to polycyclic aromatic hydrocarbons and total suspended particulates in a Taiwanese temple. *Bulletin of Environmental Contamination & Toxicology* 67: 332–338.

Liu, D.L. and Nazaroff, W.W. (2001). Modeling pollutant penetration across building envelopes. Atmospheric Environment 35: 4451–4462.

Liu, D.L. and Nazaroff, W.W. (2003). Particle penetration through building cracks. *Aerosol Science & Technology* 37: 565–573.

Liu, K., Wu, Q., Ren, Y., and Wang, S. (2022a). Air pollutant emissions from residential solid fuel combustion in the Pan-third pole region. *Environmental Science & Technology* 56: 15347–15355.

Liu, X., Xing, W., Xu, Z. et al. (2022b). Assessing impacts of additives on particulate matter and volatile organic compounds produced from the grilling of meat. *Food* 11: 833.

Long, C.M., Suh, H.H., Catalano, P.J., and Koutrakis, P. (2001). Using time-and size-resolved particulate data to quantify indoor penetration and deposition behavior. *Environmental Science & Technology* 35: 2089–2099.

Lu, X., Ye, X., and Liu, J. (2022). Morphological differences between anemophilous and entomophilous pollen. *Microscopy Research and Technique* 85: 1056–1064.

Manoukian, A., Quivet, E., Temime-Roussel, B. et al. (2013). Emission characteristics of air pollutants from incense and candle burning in indoor atmospheres. *Environmental Science and Pollution Research* 20: 4659–4670.

Mcauley, T.R., Hopke, P., Zhao, J. et al. (2012). Comparison of the effects of e-cigarette vapor and cigarette smoke on indoor air quality. *Inhalation Toxicology* 24: 850–857.

Molnár, P. and Sallsten, G. (2013). Contribution to PM 2.5 from domestic wood burning in a small community in Sweden. *Environmental Science: Processes & Impacts* 15: 833–838.

Moura, I.B., Bentley, K., and Wilcox, M.H. (2022). Assessment of potential for viral contamination of user and environment via aerosols generated during hand drying: a pilot study. *Frontiers in Public Health* 10: 1010802.

Mu, L., Liu, L., Niu, R. et al. (2013). Indoor air pollution and risk of lung cancer among Chinese female non-smokers. *Cancer Causes & Control* 24: 439–450.

Myers, N.T., Calderón, L., Pavilonis, B. et al. (2021). Presence and variability of culturable bioaerosols in three multi-family apartment buildings with different ventilation systems in the northeastern US. *Indoor Air* 31: 502–523.

Nakagawa, T., Wargocki, P., Tanabe, S., et al. (2003). Chemical emission rates from personal computers. *7th International Conference on Healthy Buildings 2003*, Singapore (7–11 December).

Nazaroff, W.W. (2016). Indoor bioaerosol dynamics. *Indoor Air* 26: 61–78.

Nazaroff, W.W. and Weschler, C.J. (2004). Cleaning products and air fresheners: exposure to primary and secondary air pollutants. *Atmospheric Environment* 38: 2841–2865.

Paddy, E.N., Afolabi, O.O., and Sohail, M. (2023). Toilet plume bioaerosols in health care and hospitality settings: a systematic review. *American Journal of Infection Control* 51: 324–333.

Pagels, J., Wierzbicka, A., Nilsson, E. et al. (2009). Chemical composition and mass emission factors of candle smoke particles. *Journal of Aerosol Science* 40: 193–208.

Papaefstathiou, E., Bezantakos, S., Stylianou, M. et al. (2020). Comparison of particle size distributions and volatile organic compounds exhaled by e-cigarette and cigarette users. *Journal of Aerosol Science* 141: 105487.

Pei-Chih, W., Huey-Jen, S., and Chia-Yin, L. (2000). Characteristics of indoor and outdoor airborne fungi at suburban and urban homes in two seasons. *Science of the Total Environment* 253: 111–118.

Pellegrino, R., Tinghino, B., Mangiaracina, G. et al. (2012). Electronic cigarettes: an evaluation of exposure to chemicals and fine particulate matter (PM). *Annali di Igiene* 24: 279–288.

Pommer, L. (2003). Oxidation of terpenes in indoor environments: a study of influencing factors. PhD thesis. Umeå University, Faculty of Science and Technology, Ecology and Environmental Science.

Pytel, K., Marcinkowska, R., Rutkowska, M., and Zabiegała, B. (2022). Recent advances on SOA formation in indoor air, fate and strategies for SOA characterization in indoor air-a review. *Science of the Total Environment* 843: 156948.

Quirce, S. and Barranco, P. (2010). Cleaning agents and asthma. *Journal of Investigational Allergology & Clinical Immunology* 20: 542–550.

Ramirez, C. (1982). *Manual and Atlas of the Penicillia*. Elsevier.

Reponen, T., Hyvärinen, A., Ruuskanen, J. et al. (1994). Comparison of concentrations and size distributions of fungal spores in buildings with and without mould problems. *Journal of Aerosol Science* 25: 1595–1603.

Reponen, T., Willeke, K., Ulevicius, V. et al. (1996). Effect of relative humidity on the aerodynamic diameter and respiratory deposition of fungal spores. *Atmospheric Environment* 30: 3967–3974.

Riley, W.J., Mckone, T.E., Lai, A.C. et al. (2002). Indoor particulate matter of outdoor origin: importance of size-dependent removal mechanisms. *Environmental Science & Technology* 36: 200–207.

Rodgman, A. and Perfetti, T.A. (2008). *The chemical components of tobacco and tobacco smoke*. CRC press.

Roe, S.M., Spivey, M.D., Lindquist, H.C., et al. (2004). National emissions inventory for commercial cooking. *13th International Emission Inventory Conference*, Clearwater, FL (26–29 September 2004).

Rösch, C., Wissenbach, D.K., Franck, U. et al. (2017). Degradation of indoor limonene by outdoor ozone: a cascade of secondary organic aerosols. *Environmental Pollution* 226: 463–472.

Salthammer, T., Schripp, T., Uhde, E., and Wensing, M. (2012). Aerosols generated by hardcopy devices and other electrical appliances. *Environmental Pollution* 169: 167–174.

Samson, R.A., Hoekstra, E.S., and Oorschot, C.V. (1981). *Introduction to Food-Borne fungi1981*. Centraalbureau voor Schimmelcultures.

Sarwar, G. and Corsi, R. (2007). The effects of ozone/limonene reactions on indoor secondary organic aerosols. *Atmospheric Environment* 41: 959–973.

Sautour, M., Sixt, N., Dalle, F. et al. (2009). Profiles and seasonal distribution of airborne fungi in indoor and outdoor environments at a French hospital. *Science of the Total Environment* 407: 3766–3771.

Schober, W., Szendrei, K., Matzen, W. et al. (2014). Use of electronic cigarettes (e-cigarettes) impairs indoor air quality and increases FeNO levels of e-cigarette consumers. *International Journal of Hygiene and Environmental Health* 217: 628–637.

Schripp, T., Markewitz, D., Uhde, E., and Salthammer, T. (2013). Does e-cigarette consumption cause passive vaping? *Indoor Air* 23: 25–31.

Scungio, M., Vitanza, T., Stabile, L. et al. (2017). Characterization of particle emission from laser printers. *Science of the Total Environment* 586: 623–630.

See, S.W. and Balasubramanian, R. (2011). Characterization of fine particle emissions from incense burning. *Building and Environment* 46: 1074–1080.

Shelton, B.G., Kirkland, K.H., Flanders, W.D., and Morris, G.K. (2002). Profiles of airborne fungi in buildings and outdoor environments in the United States. *Applied and Environmental Microbiology* 68: 1743–1753.

Shen, G., Gaddam, C.K., Ebersviller, S.M. et al. (2017). A laboratory comparison of emission factors, number size distributions, and morphology of ultrafine particles from 11 different household cookstove-fuel systems. *Environmental Science & Technology* 51: 6522–6532.

Shepherd, J.L., Corsi, R.L., and Kemp, J. (1996). Chloroform in indoor air and wastewater: the role of residential washing machines. *Journal of the Air & Waste Management Association* 46: 631–642.

Siracusa, A., De Blay, F., Folletti, I. et al. (2013). Asthma and exposure to cleaning products–a European academy of allergy and clinical immunology task force consensus statement. *Allergy* 68: 1532–1545.

Soni, M., Burdock, G., Taylor, S., and Greenberg, N.A. (2001). Safety assessment of propyl paraben: a review of the published literature. *Food and Chemical Toxicology* 39: 513–532.

Stephens, B. and Siegel, J.A. (2012). Comparison of test methods for determining the particle removal efficiency of filters in residential and light-commercial central HVAC systems. *Aerosol Science and Technology* 46: 504–513.

Sun, S., Schiller, J.H., and Gazdar, A.F. (2007). Lung cancer in never smokers – a different disease. *Nature Reviews Cancer* 7: 778–790.

Sung, G., Ha, S., Kwon, S., and Kim, T. (2017). Reduction of ultrafine particles emission from office laser printers. *Journal of Aerosol Science* 103: 15–23.

Tang, L., Eichelberg, A., Böni, F. et al. (2025). Modelling exposure to aerosols from showers: implications for microbial risk assessment. *Building and Environment* 275: 112825.

Taylor, J., Brown, K., Toivenen, J., and Holah, J.T. (2000). A microbiological evaluation of warm air hand driers with respect to hand hygiene and the washroom environment. *Journal of Applied Microbiology* 89: 910–919.

Thomas, D. and Dick, B. (1953). Air infiltration through gaps around windows. *Journal of the Institution of Heating and Ventilating Engineers* 21: 85–97.

Thompson, R.J., Li, J., Weyant, C.L. et al. (2019). Field emission measurements of solid fuel stoves in Yunnan, China demonstrate dominant causes of uncertainty in household emission inventories. *Environmental Science & Technology* 53: 3323–3330.

Tissari, J., Sippula, O., Kouki, J. et al. (2008). Fine particle and gas emissions from the combustion of agricultural fuels fired in a 20 kW burner. *Energy & Fuels* 22: 2033–2042.

To, W.M. and Yeung, L.L. (2011). Effect of fuels on cooking fume emissions. *Indoor and Built Environment* 20: 555–563.

Torkmahalleh, M.A., Goldasteh, I., Zhao, Y. et al. (2012). PM2. 5 and ultrafine particles emitted during heating of commercial cooking oils. *Indoor Air* 22: 483–491.

Traidl-Hoffmann, C., Kasche, A., Menzel, A. et al. (2003). Impact of pollen on human health: more than allergen carriers? *International Archives of Allergy and Immunology* 131: 1–13.

Traynor, G.W. (1996). *Pollutant Emission Factors from Residential Natural Gas Appliances: A Literature Review*. Lawrence Berkeley National Laboratory.

Tsai, F.C., Macher, J.M., and Hung, Y.Y. (2007). Biodiversity and concentrations of airborne fungi in large US office buildings from the BASE study. *Atmospheric Environment* 41: 5181–5191.

Tseng, L.-C. and Chen, C.-C. (2013). Effect of flow characteristics on ultrafine particle emissions from range hoods. *Annals of Occupational Hygiene* 57: 920–933.

Vannucci, M.P. and Nazaroff, W.W. (2017). Ultrafine particle production from the ozonolysis of personal care products. *Environmental Science & Technology* 51: 12737–12744.

Vartiainen, E., Kulmala, M., Ruuskanen, T.M. et al. (2006). Formation and growth of indoor air aerosol particles as a result of d-limonene oxidation. *Atmospheric Environment* 40: 7882–7892.

Verani, M., Bigazzi, R., and Carducci, A. (2014). Viral contamination of aerosol and surfaces through toilet use in health care and other settings. *American Journal of Infection Control* 42: 758–762.

Vicente, E.D., Evtyugina, M., Vicente, A.M. et al. (2021). Impact of ironing on indoor particle levels and composition. *Building and Environment* 192: 107636.

Wallace, L. (2006). Indoor sources of ultrafine and accumulation mode particles: size distributions, size-resolved concentrations, and source strengths. *Aerosol Science and Technology* 40: 348–360.

Wallace, L. and Ott, W. (2011). Personal exposure to ultrafine particles. *Journal of Exposure Science & Environmental Epidemiology* 21: 20–30.

Wallace, L., Williams, R., Rea, A., and Croghan, C. (2006). Continuous weeklong measurements of personal exposures and indoor concentrations of fine particles for 37 health-impaired North Carolina residents for up to four seasons. *Atmospheric Environment* 40: 399–414.

Wallace, L., Wang, F., Howard-Reed, C., and Persily, A. (2008). Contribution of gas and electric stoves to residential ultrafine particle concentrations between 2 and

64 nm: size distributions and emission and coagulation rates. *Environmental Science & Technology* 42: 8641–8647.

Walters, G., Burge, P., Moore, V., and Robertson, A.S. (2018). Cleaning agent occupational asthma in the west midlands, UK: 2000–16. *Occupational Medicine* 68: 530–536.

Wang, F., Meng, D., Li, X., and Tan, J. (2016). Indoor-outdoor relationships of PM2. 5 in four residential dwellings in winter in the Yangtze River Delta, China. *Environmental Pollution* 215: 280–289.

Wang, C., Liu, L., Liu, X. et al. (2017). Mechanisms of lung cancer caused by cooking fumes exposure: a minor review. *Chinese Medical Sciences Journal* 32: 193–197.

Wang, W., Ooka, R., Kikumoto, H. et al. (2022). Influence of various factors on indoor/outdoor pollen concentration ratio based on experimental research: a review. *Building and Environment* 219: 109154.

Waye, S.K., Anderson, A., Corsi, R.L., and Ezekoye, O.A. (2013). Thermal effects on polybrominated diphenyl ether mass transfer and emission from computer cases. *International Journal of Heat and Mass Transfer* 64: 343–351.

Weschler, C.J. (2006). Ozone's impact on public health: contributions from indoor exposures to ozone and products of ozone-initiated chemistry. *Environmental Health Perspectives* 114: 1489–1496.

Wilson, G.M., Jackson, V.B., Boyken, L.D. et al. (2020). Bioaerosols generated from toilet flushing in rooms of patients with Clostridioides difficile infection. *Infection Control & Hospital Epidemiology* 41: 517–521.

Wolkoff, P., Clausen, P.A., and Nielsen, P. (1995). Application of the field and laboratory emission cell "FLEC"-performance study, Intercomparison study, and case study of damaged linoleum in an office#. *Indoor Air* 5: 196–203.

Wong, T.W., Wong, A.H., Lee, F.S., and Qiu, H. (2011). Respiratory health and lung function in Chinese restaurant kitchen workers. *Occupational and Environmental Medicine* 68: 746–752.

Wu, C., Chao, C.Y., Sze-To, G. et al. (2012). Ultrafine particle emissions from cigarette smouldering, incense burning, vacuum cleaner motor operation and cooking. *Indoor and Built Environment* 21: 782–796.

Wu, D., Zhang, Y., Tian, Y. et al. (2020). On-site investigation of the concentration and size distribution characteristics of airborne fungi in a university library. *Environmental Pollution* 261: 114138.

Xiang, J., Hao, J., Austin, E. et al. (2021). Characterization of cooking-related ultrafine particles in a US residence and impacts of various intervention strategies. *Science of the Total Environment* 798: 149236.

Xu, B., Liu, S., and Zhu, Y. (2010). Ultrafine particle penetration through idealized vehicle cracks. *Journal of Aerosol Science* 41: 859–868.

Ye, J., Qian, H., Zhang, J. et al. (2021). Concentrations and size-resolved I/O ratios of household airborne bacteria and fungi in Nanjing, Southeast China. *Science of the Total Environment* 774: 145559.

Yeoman, A.M., Shaw, M., and Lewis, A.C. (2021). Estimating person-to-person variability in VOC emissions from personal care products used during showering. *Indoor Air* 31: 1281–1291.

Yeung, L.L. and To, W.M. (2008). Size distributions of the aerosols emitted from commercial cooking processes. *Indoor and Built Environment* 17: 220–229.

Zai, S., Zhen, H., and Jia-Song, W. (2006). Studies on the size distribution, number and mass emission factors of candle particles characterized by modes of burning. *Journal of Aerosol Science* 37: 1484–1496.

Zhang, T., Yao, L., Gao, Z., and Wang, F. (2022). Particle exposure risk to a lavatory user after flushing a squat toilet. *Scientific Reports* 12: 21088.

Zhang, H., Su, X., Xiao, Q., and Lai, A.C.K. (2024). Experimental study on droplet and bioaerosol emissions from flushing a squat toilet. *Building and Environment* 250: 111162.

Zhao, Y. and Zhao, B. (2018). Emissions of air pollutants from Chinese cooking: a literature review. *Building Simulation* 11: 977–995.

Zhao, P., Yu, K.P., and Lin, C.C. (2011). Risk assessment of inhalation exposure to polycyclic aromatic hydrocarbons in Taiwanese workers at night markets. *International Archives of Occupational and Environmental Health* 84: 231–237.

Zhao, J., Zhang, N., Qu, C. et al. (2010a). Cigarette butts and their application in corrosion inhibition for N80 steel at 90 C in a hydrochloric acid solution. *Industrial & Engineering Chemistry Research* 49: 3986–3991.

Zhao, J., Zhang, N., Qu, C. et al. (2010b). Comparison of the corrosion inhibitive effect of anaerobic and aerobic cigarette butts water extracts on N80 steel at 90 C in hydrochloric acid solution. *Industrial & Engineering Chemistry Research* 49: 12452–12460.

Zhou, Y., Benson, J.M., Irvin, C. et al. (2007). Particle size distribution and inhalation dose of shower water under selected operating conditions. *Inhalation Toxicology* 19: 333–342.

Zhou, Z., Dionisio, K.L., Verissimo, T.G. et al. (2014). Chemical characterization and source apportionment of household fine particulate matter in rural, peri-urban, and urban West Africa. *Environmental Science & Technology* 48: 1343–1351.

Zhou, Z., Liu, Y., Yuan, J. et al. (2016). Indoor PM2. 5 concentrations in residential buildings during a severely polluted winter: a case study in Tianjin, China. *Renewable and Sustainable Energy Reviews* 64: 372–381.

Zou, C., Jiang, M., Huang, H. et al. (2022). Size distribution, emission rate, and decay characteristics of particles emitted by printers. *Air Quality, Atmosphere & Health* 15: 1427–1438.

8

Indoor Aerosol Transport Modeling

8.1 Introduction

Both aerosols of outdoor origin and aerosols emitted indoors can be transported in indoor spaces. These aerosols are carried by air streams, spreading them inside rooms. Some of the aerosols may remain suspended in the air for a long time. Meanwhile, a portion of the aerosols may be deposited on surfaces, subsequently resuspended due to disturbances, exhausted out of the rooms by ventilation, or removed by air cleaners. When we model particle dynamics, the focus is primarily on the particle concentrations and their spatial distribution, the particle motion trajectories, and the final fates of the particles. This chapter addresses the modeling of particle release, particle transport within the air, particle deposition, and particle removal by various interception mechanisms or by filtration using air cleaners. The simplified multizone model and sophisticated CFD models are introduced. Table 8.1 presents an overview of common numerical models for solving indoor aerosol transport, the pros and cons of each model type, their suitable application areas, and some example software that adopts each model type.

8.2 Multizone Model

8.2.1 Basic Principles

The multizone model treats a room in a building as a zone of well-mixed air. Each zone is connected to the surrounding zones by airflow paths. These paths include doors, windows, cracks, holes, ventilation ducts, and other openings, such as those shown in Figure 8.1. Meanwhile, the space outside a building is treated as a separate ambient zone. The airflow inside a building with multiple rooms constitutes a flow network, and each room is deemed to be a well-mixed zone. The pollutant

Table 8.1 Overview of numerical models for solving indoor aerosol transport.

Model type	Pros	Cons	Application areas	Example software
Multizone model	Can handle large buildings with many rooms; high computational efficiency	Adopts well-mixed assumptions; no spatial information inside a zone	Indoor–outdoor and room–room ventilation and aerosol exchange, and average indoor aerosol concentrations	CONTAM, COMIS
Eulerian model	Provides particle concentration distribution directly; relatively low computational cost	Does not consider individual particle behavior	Spatial distribution of airflow and aerosol concentrations	ANSYS Fluent, OpenFOAM, STAR-CCM, COMSOL
Lagrangian model	Tracks the motion of each individual particle and accounts for force interactions	High computational cost when tracking a large number of particles	Spatial motion of aerosols, force interactions, and particle size variations	ANSYS Fluent, OpenFOAM, STAR-CCM, COMSOL
Coupled Euler–Lagrange model	Can model phase interfaces and droplet breakup/ coalescence	Computationally intensive	Spatial motion of liquids, gases, droplets, and droplet nuclei; particle phase changes and size evolution	ANSYS Fluent, OpenFOAM, STAR-CCM+, COMSOL

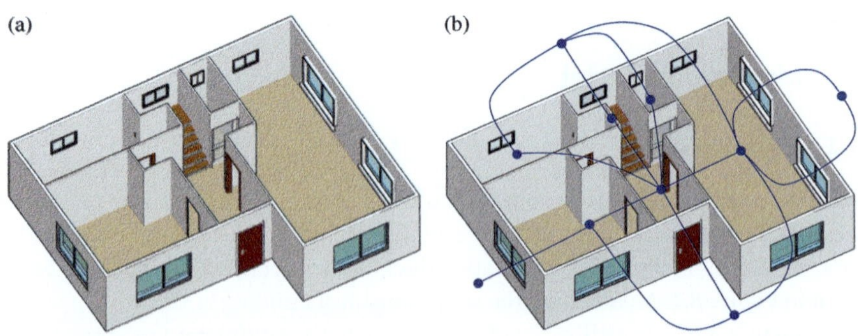

Figure 8.1 Illustration of multizone model in an apartment with multiple rooms: (a) schematic of the apartment; (b) the constituted flow network, where each room in the apartment, as well as the ambient space, is designated as an independent zone.

concentration in a zone obeys the mass conservation principle, whose governing equation is expressed as:

$$\frac{dm_i^\alpha}{dt} = \sum_j F_{j\to i}\left(1-\eta_j^\alpha\right)C_j^\alpha + G_i^\alpha + m_i\sum_\beta \kappa^{\alpha\beta}C_i^\beta - \sum_j F_{i\to j}C_i^\alpha - R_i^\alpha C_i^\alpha \quad (8.1)$$

where m_i^α is the mass of pollutant α in zone i, t is the time, $F_{j\to i}$ is the airflow rate from zone j to zone i, η_j^α is the filtration efficiency of pollutant α on the path from zone j to zone i, C_j^α is the average concentration of pollutant α in zone j, G_i^α is the release rate of pollutant α in zone i, m_i is the mass of the air in zone i, $\kappa^{\alpha\beta}$ is the kinetic reaction coefficient of substances α and β, and R_i^α is the removal rate of pollutant α. In Eq. (8.1), the term on the left-hand side represents the mass accumulation of the pollutant α in zone i, and the respective terms on the right-hand side represent the pollutant mass transported from zone j to zone i for all the neighboring zones, the generation or release of pollutant from a source in zone i, the generation or depletion of the pollutant by chemical reaction in zone i, the transport-out of the pollutant from zone i to zone j for all the neighboring zones, and the removal rate of the pollutant in zone i by filtration, deposition, or other mechanisms. Assuming that substance α is in excess, only the first order of chemical reaction is considered in Eq. (8.1).

Modeling of pollutant concentration by means of Eq. (8.1) is applicable to both gaseous pollutants and aerosols. The gravitational settling and deposition of aerosols onto surfaces can be accounted for by casting their contribution to the aerosol concentrations into the sink term, i.e., the last term in Eq. (8.1). However, the motion of aerosols inside a zone and the spatial variations in concentration within a zone are beyond the capability of the multizone model. Hence, the multizone model can only provide a uniform, well-mixed gaseous and/or aerosol concentration within a zone.

The mass flow rate F in the equation mentioned earlier must be solved from the airflow network that connects all the zones in the building. The governing equations of the multizone airflow include the mass conservation of the air and the simplified momentum equation through each opening, i.e., the pressure difference versus the flow rate at each opening. In each zone, the mass conservation of the air can be expressed as:

$$\sum_j F_{j\to i} + F_i = 0 \quad (8.2)$$

where the first term represents the sum of the mass flow rates from all the neighboring zones to zone i, and the second term stands for the non-flow processes that contribute to the mass of the air in zone i, such as production or depletion of the air due to chemical reaction. Note that $F_{j\to i}$ is negative if the actual net flow is from zone i to zone j. The value of F_i is zero in most cases when only physical processes are involved. Eq. (8.2) can then be simplified as:

$$\sum_j F_{j\to i} = 0 \tag{8.3}$$

The mass flow rate $F_{j\to i}$ is dependent on the pressure difference between zones j and i and can be written as:

$$F_{j\to i} = f(P_j - P_i) \tag{8.4}$$

where f is a coefficient and is non-dimensionless, P is the air pressure. Note that f can also be a function of pressure. Eq. (8.4) represents a simplified relationship between the pressure difference and the mass flow rate through the flow path. It does not imply that the mass flow rate depends linearly on the pressure difference. In fact, most infiltration models are cast into a power law relationship as:

$$F_{j\to i} = C(P_j - P_i)^n \tag{8.5}$$

where C is a coefficient and non-dimensionless, while n is an exponent. For most openings, n lies between 0.5 and 1. n is close to 0.5 for a large opening, while for the low-velocity flow through cracks or filters, n approaches 1. The readers can refer to Awbi (2003) and ASHRAE Fundamentals (2025) for the empirical models of the flow through various openings.

When Eq. (8.5) or Eq. (8.4) is substituted into Eq. (8.3), the final pressure-dependent equations can be established. In principle, the equations for one unknown variable are solvable. However, because Eqs. (8.4) and (8.5) are commonly nonlinear, and solving the pressures in each zone requires numerical iterations. Further details of the numerical solution can be found in National Institute of Standards and Technology (NIST) Technical Note 1887. Once $F_{j\to i}$ is known, Eq. (8.1) can be discretized for an easy solution of the concentration in each zone. The CONTAM (version 3.2) software package can be used to create the multizone geometry and carry out the corresponding numerical solution.

8.2.2 Demonstration of Multizone Modeling of Airflow and Transport of a Tracer Gas inside a House

Figure 8.2 presents an example of multizone modeling of the airflow and pollutant concentrations in a two-storey house (Zhao et al. 1998). The first floor consisted of a living room connected to a kitchen on the north side, a hall with a stairway to the second floor, a washroom on the north side, and a room labeled "J." on the south side. The second floor contained a bedroom, a hall, and two rooms labeled "W." The house was naturally infiltrated/exfiltrated without the opening of doors or windows, but the air in each room was well-mixed by fans. The details of the measurement procedure and the boundary conditions can be

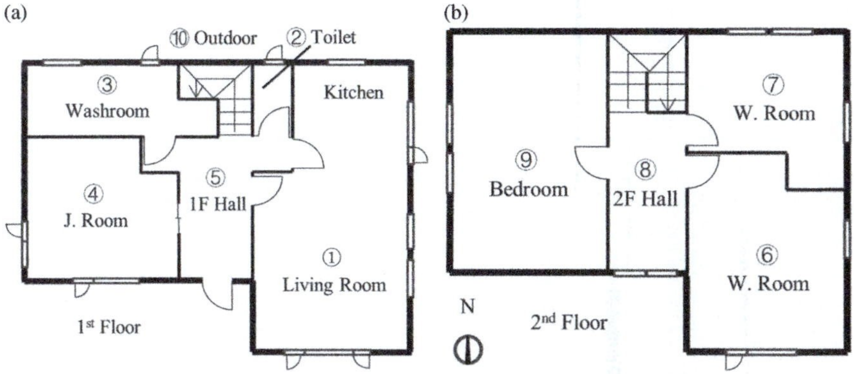

Figure 8.2 Planar view of a two-storey house with a gaseous pollutant released in the living room: (a) first floor; (b) second floor. *Source:* Zhao et al. (1998) / with permission of John Wiley & Sons.

found in Zhao et al. (1998). The sulfur hexafluoride (SF_6) tracer gas at a constant rate of 4 mg/s was released continually into the living room for 1 hour. Note that no aerosol source was released in this case, other than the SF_6 gas. Numerous studies have revealed that the transport of fine particles (with diameters less than 1.0 μm) follows the airflow closely. Hence, SF_6 gas transmission can also be used to represent cases involving fine particles.

Figure 8.3 displays the air change rates in the various zones. The maximum air change rates were found in the hallways of Zones 8 and 5 because these hallways contained the stairway connecting the two floors. This can also be seen in the airflow rates between the floors. The door and window cracks were well sealed, except for the east-facing windows in Zone 1 on the first floor. Therefore, most of the infiltration/exfiltration rates were quite small. The airflow results provided by the multizone model were generally in good agreement with the measurements, although some discrepancies existed.

Figure 8.4 compares the measured concentrations with the concentrations solved by the multizone model for the rooms on the first floor. The pollutant was released in Zone 1 (the living room) only for the first hour, and thus the pollutant concentration in Zone 1 was the highest. After one hour, the concentration in this zone started to decay. The rising and decay of the concentrations in the other rooms lagged behind those in Zone 1. In general, there was relatively good agreement between the measurement and the numerical solution, which implies that the multizone model performed well in this case.

It should be noted that the multizone model simply assumes that the air is well-mixed in each zone. The well-mixed assumption in this model makes the solution very efficient. Therefore, the multizone model can be used to roughly

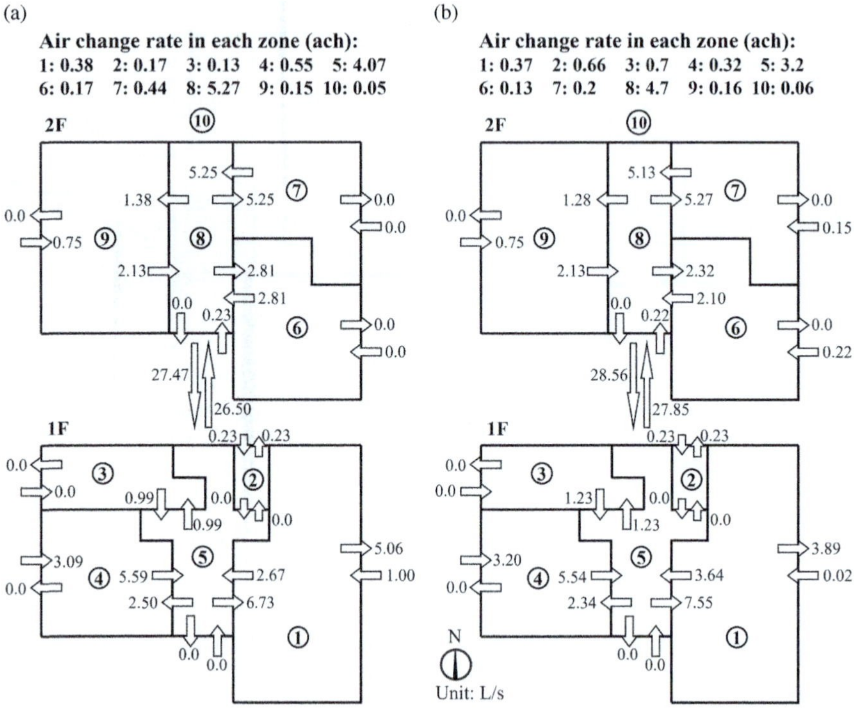

Figure 8.3 Comparison of the measured air change rates and mass flow rates with the rates that were numerically simulated by the multizone model, where the unit "ach" is air change per hour (h^{-1}): (a) measurement results. *Source:* Zhao et al. 1998 / with permission of John Wiley & Sons; (b) multizone modeling results.

determine the aerosol concentrations inside a room, especially when analysis of pollutant concentrations is needed over a long period for a building containing many rooms, and when very complicated meteorological conditions are encountered. In a large indoor space or in a space with weak air motion, attention should be taken that the well-mixed assumption may not hold. The readers should also be aware that the modeling of airflow rate versus pressure difference for the flow path between two neighboring zones may also be oversimplified.

8.2.3 Demonstration of Multizone Modeling of Particle Transport and CO_2 Concentration in a Test Room

For the sake of simplicity, an isolated room located in Dalian, a northern city in China, as shown in Figure 8.5, is analyzed in this case. The dimensions of the room were 5.0 m × 4.0 m × 2.9 m. The room was nearly empty, containing only a chair, an air purifier, a fan to mix the air, and an air conditioner for control of

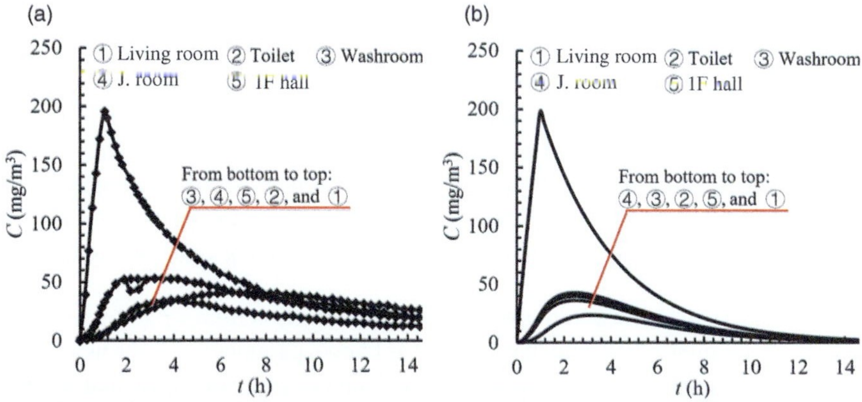

Figure 8.4 Comparison of the measured pollutant concentrations with the pollutant concentrations solved by the multizone model: (a) measured concentrations. *Source:* Zhao et al. (1998) / with permission of John Wiley & Sons; (b) solved concentrations.

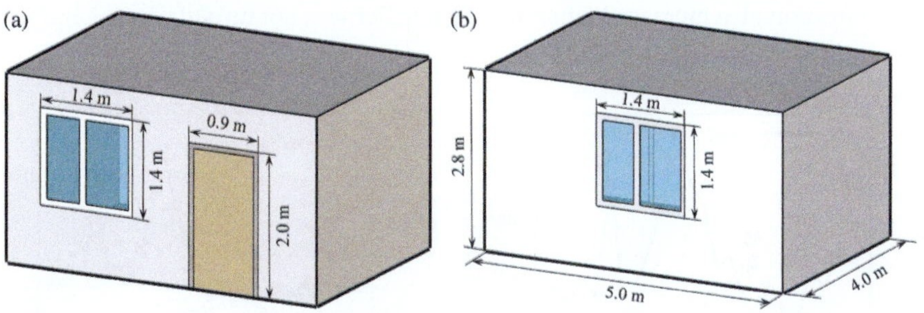

Figure 8.5 A test room for validation of the multizone model for solving particle concentrations: (a) view of the room from the northwest, with the window facing the north; (b) view of the room from the southeast, with the window facing the south.

the indoor temperature. Measurements were carried out from 0:00 to 22:00 on 3 January 2018. The atmospheric air was polluted with the occasionally high outdoor $PM_{2.5}$ concentrations. The air conditioner was operated from 9:00 to 18:30 with a temperature setting of 25 °C. During tests, all the windows were closed. The door was also closed, except when a student entered or exited the room twice to carry out certain tests. The student entered the room at 8:00 to turn on the portable air purifier for particle removal and stayed inside until 10:00. The student then turned off the air purifier and exited the room at 10:00. The student entered the room again at 12:00 and left at 14:00 without operating the air purifier. Indoor and outdoor environmental data were recorded during the experiment.

The geometry of the empty room, as shown in Figure 8.5 was created for the numerical solution. Airtightness test results provided the following information:

ELA (effective leakage area) = 176 cm^2, $C = 1$, and $n = 0.598$. The total leakage area was distributed across the envelope walls and ceiling, according to a leakage estimation conducted with an infrared camera. The measured outdoor wind speed, wind direction, air temperature, relative humidity, and solar irradiation intensity were specified as known input parameters. The transient indoor room ventilation rate and the indoor $PM_{2.5}$ and CO_2 concentrations were solved and compared with the measurement data. Further details of the experimental measurement procedure and numerical solution models can be found in Zhang et al. (2018) and Liu et al. (2021).

Figure 8.6a shows the outdoor wind speed and direction from a rooftop weather station on 3 January 2018, which were used as known parameters for the modeling. Figure 8.6b compares the measured air infiltration rates with the simulated results. The simulated airflow rates agreed well with the experimental data, with a mean relative deviation of less than 4.5%. The time-varying trend of the room infiltration rate resembled that of the outdoor air speed. Figure 8.6c compares the $PM_{2.5}$ concentrations. When the outdoor $PM_{2.5}$ concentration increased, the indoor $PM_{2.5}$ concentration also increased when the air purifier was not turned on. The indoor

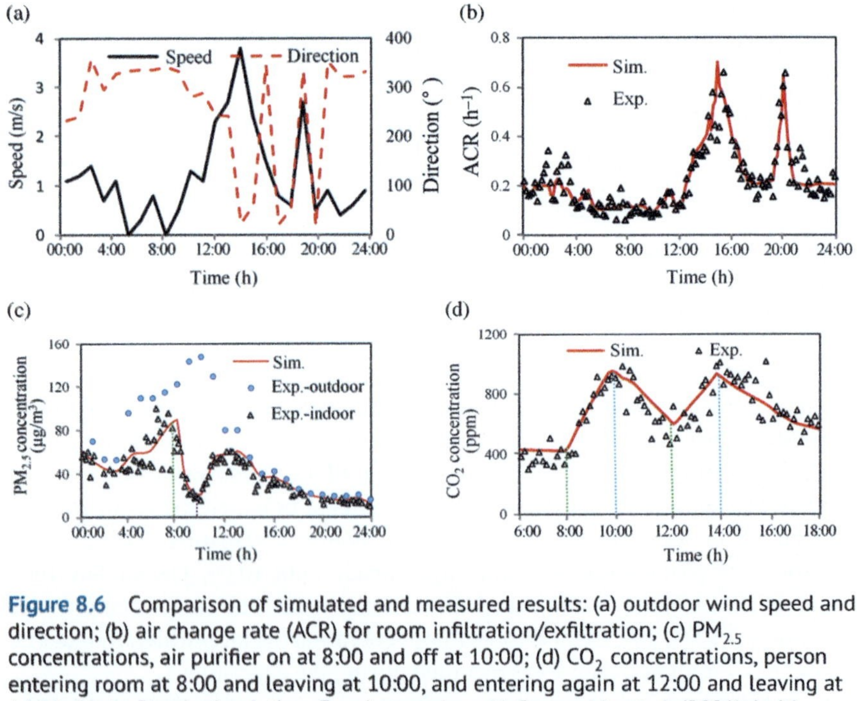

Figure 8.6 Comparison of simulated and measured results: (a) outdoor wind speed and direction; (b) air change rate (ACR) for room infiltration/exfiltration; (c) $PM_{2.5}$ concentrations, air purifier on at 8:00 and off at 10:00; (d) CO_2 concentrations, person entering room at 8:00 and leaving at 10:00, and entering again at 12:00 and leaving at 14:00. (Note: Sim. is simulation; Exp. is experiment.) *Source:* Liu et al. (2021) / with permission of Elsevier.

$PM_{2.5}$ concentration decreased quickly during operation of the purifier, from 8:00 to 10:00. When the air purifier was turned off, the indoor $PM_{2.5}$ concentration gradually rose again. The mean relative deviation between the simulated and measured $PM_{2.5}$ concentrations was 7%. Meanwhile, Figure 8.6d compares the simulated and measured CO_2 concentrations. The CO_2 concentration increased gradually when the student entered the room at 8:00 and again at 12:00. The concentration gradually decreased after the student left the room at 10:00 and again at 14:00. The mean relative deviation between the simulated and measured CO_2 concentrations was 4.2%.

In summary, the results of the numerical modeling agreed fairly well with the experimental data, which implies that the multizone model predicted the room airflow and the indoor particle and CO_2 concentrations with good accuracy.

8.3 CFD Modeling of Air Motion

CFD offers sophisticated modeling of the transport of mass, momentum, and energy, making it possible to resolve aerosol transport elaborately in space. Note that the transport of aerosols is mainly dependent on the airflow. This section describes the CFD modeling of airflow, while the modeling of particle motion is addressed in the subsequent sections.

It is known that the motion of a Newtonian fluid is governed by the Navier–Stokes (N–S) equations. Since air is a typical Newtonian fluid, the N–S equations should be solved for the spatial air motion. However, most indoor airflow is turbulent, with random and chaotic variations in motion that present challenges for a solution. In CFD modeling, there are various strategies for handling turbulence: direct numerical simulation (DNS), large-eddy simulation (LES), and solving the Reynolds-averaged Navier–Stokes (RANS) equations. Because both DNS and LES must resolve the time-variant turbulent flow, the solution burden is extremely high. DNS also requires grid meshes that are fine enough to capture the smallest turbulent vortices. Therefore, RANS modeling is the most frequently used affordable strategy both in the research community and in industry, and it is the approach taken in this textbook.

8.3.1 Governing Equations of RANS Modeling

The governing equations of Newtonian fluid flow, energy, and turbulence can be cast into the following generalized format:

$$\frac{\partial}{\partial t}(\rho\phi) + \frac{\partial}{\partial x_j}(\rho u_j \phi) = \frac{\partial}{\partial x_j}\left(\Gamma_{\phi,\text{eff}}\frac{\partial \phi}{\partial x_j}\right) + S_\phi \tag{8.6}$$

Table 8.2 Variables for the general scalar transport equation in CFD modeling of incompressible flow.

Equation	$\boldsymbol{\phi}$	$\boldsymbol{\Gamma_{\phi,\text{eff}}}$	$\boldsymbol{S_\phi}$
Continuity	1	0	0
Momentum	$\overline{u_i}$	$\mu + \mu_t$	$-\dfrac{\partial p}{\partial x_i} - \rho g_i \beta (T - T_0)$
Energy	i	$\dfrac{\mu}{\text{Pr}} + \dfrac{\mu_t}{\text{Pr}_t}$	S_i
Turbulent kinetic energy	k	$\dfrac{\mu + \mu_t}{\sigma_k}$	$G_k + G_b - \rho\varepsilon + S_k$
Dissipation rate of k	ε	$\dfrac{\mu + \mu_t}{\sigma_\varepsilon}$	$\dfrac{\varepsilon}{k}(C_{1\varepsilon} G_k + C_{3\varepsilon} G_b) - C_{2\varepsilon}\rho\dfrac{\varepsilon^2}{k}$ $-R_\varepsilon + S_\varepsilon$
Tracer gas concentration	C	$\dfrac{\mu + \mu_t}{\text{Sc}}$	S_C

where

$$\mu_t = \rho C_\mu \frac{k^2}{\varepsilon};\ G_k = -\rho\overline{u_i' u_j'}\frac{\partial u_j}{\partial x_i};\ G_b = \beta g_i \frac{\mu_t}{\text{Pr}_t}\frac{\partial T}{\partial x_i};$$

$$R_\varepsilon = \frac{C_\mu \rho \eta^3 (1 - \eta/\eta_0)}{1 + \beta\eta^3}\frac{\varepsilon^2}{k};\ \eta = Sk/\varepsilon;\ S = \sqrt{2S_{ij}S_{ij}};$$

$$S_{ij} = \frac{1}{2}\left(\frac{\partial u_j}{\partial x_i} + \frac{\partial u_i}{\partial x_j}\right);\ C_\mu = 0.0845;\ C_{1\varepsilon} = 1.42;\ C_{2\varepsilon} = 1.68;$$

$$C_{3\varepsilon} = 1.0;\ \sigma_k = 0.7194;\ \sigma_\varepsilon = 0.7194;\ \text{Pr} = 0.71;\ \text{Pr}_t = 0.9;$$

$$\text{Sc} = 1.0$$

where ρ is the fluid density, ϕ is a scalar variable, t is time, u_j is a velocity component in the three directions (x_j, $j = 1, 2, 3$) of a Cartesian coordinate system, $\Gamma_{\phi,\text{eff}}$ is the effective diffusion coefficient, and S_ϕ is the source term. As shown in Table 8.2, when different values are supplemented for ϕ, Eq. (8.6) can represent the continuity, momentum, energy, and turbulence (the renormalization group [RNG] k–ε model) equations.

8.3.2 Numerical Solution to the RANS Model

Eq. (8.6) is a partial differential equation over space and time. To solve Eq. (8.6) numerically on a computer, the generalized equation must be discretized over the grid meshes. Since it is difficult to represent complicated geometry with traditional structured quad meshes, discretizing Eq. (8.6) over unstructured meshes is more appropriate. Take the example of two unstructured grid cells in a two-dimensional

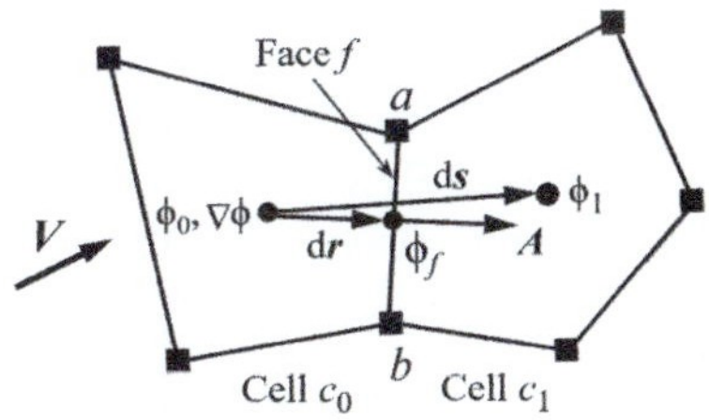

Figure 8.7 Control volume for unstructured grid cells c_0 and cell c_1 sharing a common face.

domain, as shown in Figure 8.7, where two adjacent cells share a common face f. The flow variables are stored at the cell centers. A similar derivation is applicable to the three-dimensional domain.

Integrating Eq. (8.6) over the control volume of cell c_0 yields:

$$\iiint \rho \frac{\partial[\phi(\tau)]}{\partial \tau} \mathrm{d}V + \iiint \frac{\partial}{\partial x_j}[\rho u_j \phi(\tau)]\mathrm{d}V = \iiint \frac{\partial}{\partial x_j}\left[\Gamma_\phi \frac{\partial \phi(\tau)}{\partial x_j}\right]\mathrm{d}V + \iiint S_\phi \mathrm{d}V \tag{8.7}$$

where V is the volume. Adopting the divergence theorem, Eq. (8.7) can be discretized into:

$$\rho \frac{\partial[\phi(\tau)]}{\partial \tau}\Delta V + \sum_f J_f \phi_f = \sum_f D_f + S_\phi \Delta V \tag{8.8}$$

where ΔV is the volume of cell c_0, J_f is the mass flow rate over face f, and D_f is the transport of ϕ due to diffusion. The first term on the left-hand side of Eq. (8.8) is the transient term, the second term on the left-hand side is the convection transport, the first term on the right-hand side is the diffusion transport, and the last term is the source contribution.

Integrating Eq. (8.8) over time from τ to $\tau + \Delta\tau$ gives:

$$\rho[\phi(\tau + \Delta\tau) - \phi(\tau)]\Delta V + \int_\tau^{\tau+\Delta\tau} \left(\sum_f J_f \phi_f\right)\mathrm{d}\tau = \int_\tau^{\tau+\Delta\tau} D_f \mathrm{d}\tau + \int_\tau^{\tau+\Delta\tau} S_\phi \Delta V \mathrm{d}\tau \tag{8.9}$$

With the time-implicit scheme to represent ϕ in the integral, the scalar variable at the next time step can be expressed as:

$$\phi(\tau + \Delta\tau) = \phi(\tau) - \frac{1}{\rho\Delta V}\left[\sum_f J_f \phi_f(\tau + \Delta\tau)\right]\Delta\tau + \frac{1}{\rho\Delta V}\left[\sum_f D_f(\tau + \Delta\tau)\right]\Delta\tau + \frac{1}{\rho}S_\phi \Delta\tau \tag{8.10}$$

For steady state cases, Eq. (8.10) can be simplified as:

$$\sum_f J_f \phi_f = \sum_f D_f + S_\phi \Delta V \tag{8.11}$$

Equations (8.10) and (8.11) are the discretized forms of Eq. (8.6) for the transient state and steady state, respectively. The convection term in the aforementioned equations involves interpolation of the face value ϕ_f. The first-order approximation is to use the value at the upwind cell to represent the face value:

$$\phi_f = \phi_{\text{upwind}} \tag{8.12}$$

For higher-order accuracy, the linear reconstruction of the face value leads to:

$$\phi_f = \phi_{\text{upwind}} + \nabla\phi_{\text{upwind}} \cdot \mathbf{dr} \tag{8.13}$$

where $\nabla\phi_{\text{upwind}}$ is the reconstruction gradient at the upwind cell, and $\mathbf{d}\boldsymbol{r}$ is the vector from the cell center to the face center. The reconstruction gradient can be calculated as:

$$\nabla\phi_{\text{upwind}} = \frac{\alpha}{\Delta V} \sum_f \overline{\phi_f} \cdot \boldsymbol{A_f} \tag{8.14}$$

where α is a factor to ensure that the reconstruction does not introduce local extrema. More details on α are described in the literature (Aftosmis et al. 1995; Barth and Jespersen 1989; Venkatakrishnan 1993). The face value of ϕ_f can be calculated as the average of the neighboring two cells, as follows:

$$\overline{\phi_f} = \frac{\phi_0 + \phi_1}{2} \tag{8.15}$$

where ϕ_0 is the variable value at cell c_0, and ϕ_1 is the variable value at cell c_1.

The diffusion transport can be calculated as (Mathur and Murthy 1997):

$$D_f = \Gamma_f \frac{(\phi_1 - \phi_0)}{\text{ds}} \frac{\boldsymbol{A_f} \cdot \boldsymbol{A_f}}{\boldsymbol{A_f} \cdot \boldsymbol{e_s}} + \Gamma_f \left(\overline{\nabla\phi} \cdot \boldsymbol{A_f} - \overline{\nabla\phi} \cdot \boldsymbol{e_s} \frac{\boldsymbol{A_f} \cdot \boldsymbol{A_f}}{\boldsymbol{A_f} \cdot \boldsymbol{e_s}} \right) \tag{8.16}$$

where Γ_f is the diffusion coefficient at the face, and $\boldsymbol{e_s}$ is the unit vector connecting two adjacent cell centers. Meanwhile, $\overline{\nabla\phi}$ is the average of the gradients at the two adjacent cells and can be calculated as:

$$\overline{\nabla\phi} = \frac{\nabla\phi_{c_0} + \nabla\phi_{c_1}}{2} \tag{8.17}$$

where $\nabla\phi_{c_0}$ and $\nabla\phi_{c1}$ can be computed by means of the divergence theorem using the averaged reconstructed values at the faces. For $\nabla\phi_{c_0}$, the expression is:

$$\nabla\phi_{c_0} = \frac{1}{\Delta V} \sum_f (\overline{\overline{\phi_f}} \boldsymbol{A}) \tag{8.18}$$

and $\overline{\overline{\phi_f}}$ is expressed as:

$$\overline{\overline{\phi_f}} = \frac{\phi_{f,0} + \phi_{f,1}}{2} \tag{8.19}$$

where $\phi_{f,0}$ and $\phi_{f,1}$ are calculated based on Eq. (8.13).

To facilitate the solution of the discretized equations, Eqs. (8.10) and (8.11) can be arranged into a linear system of equations:

$$a_P\phi_P = \sum_{nb} a_{nb}\phi_{nb} + S_P \tag{8.20}$$

where a_P is the coefficient of the variable ϕ_P at cell P, a_{nb} is the coefficient of the variable ϕ_{nb} for the cells neighboring P, and S_P is the source term. All the discretized equations can be solved with a general solver in a similar fashion.

The aforementioned numerical schemes have dealt only with the internal domain. The near-boundary regions should be handled in specific ways, such as by applying near-wall modeling. Walls are the main source of mean vorticity and turbulence. In the near-wall region, the variables have large gradients, and there may be strong transport of momentum, energy, and species. The airflow near walls cannot be accurately predicted without appropriate treatment of the near-wall effect.

The near-wall region can be roughly subdivided into three layers. In the region closest to the wall, there is a viscous sub-layer, where the flow is almost laminar, and the molecular viscosity plays a dominant role in the momentum, heat, and mass transfer. A fully turbulent layer is found in the outer region, where turbulence plays a major role. The interim region between the viscous sub-layer and the fully turbulent layer is a buffer layer where the molecular viscosity and turbulence are roughly equally important.

An economical method for handling the near-wall effect is to adopt wall functions. Wall functions yield semi-empirical formulas that link the variables at the near-wall CFD cells with the corresponding variables on the walls. Wall functions include laws of the wall for mean velocity, temperature, and other scalar variables. The standard wall functions proposed by Launder and Spalding (1974) have been widely used for resolving industrial flows. Wall functions are described in greater detail in the work of Launder and Spalding (1974) and in the Fluent documentation (ANSYS Inc., 2006).

In the solution of the fluid flow, the mass flow rate J_f in Eqs. (8.10) and (8.11) is still unknown. The continuity and momentum equations must be solved for the velocity fields and the mass flow rate J_f. The SIMPLE (semi-implicit method for pressure-linked equations) algorithm (Patankar 1980) provides a method to couple the continuity and momentum equations for resolving the flow. The SIMPLE algorithm is an iterative guess-and-correct procedure for the calculation of the pressure and velocity fields, as follows:

1) Initialize all field variables.
2) Guess or update the pressure field and solve the discretized momentum equations under the designated pressure field to obtain the intermediate velocity field.
3) Solve the pressure-correction equation (derived from the continuity equation).
4) Correct the mass flow rate, pressure, and velocity fields.
5) Solve all other discretized scalar transport equations, such as energy, turbulence, and contaminant concentrations.
6) Repeat from Step 2 until all the field variables have converged.

The SIMPLE algorithm has been successfully implemented in numerous commercial CFD codes. The pressure correction is good enough to correct the velocity field but not to correct the pressure, and thus the convergence may be slow. Hence, a number of improved schemes for pressure–velocity coupling have been developed, such as semi-implicit method for pressure linked equations revised (SIMPLER), semi-implicit method for pressure linked equations-consistent (SIMPLEC), and PISO (Versteeg and Malalasekera 2007).

8.3.3 Demonstration of CFD Modeling of Airflow, Heat Transfer, and Transport of a Tracer Gas in Indoor Environments

The indoor airflow, temperature, and the concentration of a tracer gas in two cases with displacement ventilation were solved. One case is that of an office, while the other is that of a workshop, as shown in Figure 8.8. Both cases have geometric dimensions of 4.92 m in length, 4.32 m in width, and 2.42 m in height because the same indoor environmental chamber was used to reproduce both cases.

For the office case, there was a file cabinet in one corner of the room, and two tables with traditional desk computers on them in the other two corners. Six fluorescent lamps were mounted on the ceiling to provide illumination for the office. Two occupants simulated by box-shaped manikins were seated sedentarily in front of the tables. Conditioned air was supplied through a corner-mounted, quarter-circular, perforated diffuser, and the internal air was exhausted at the center of the ceiling. Table 8.3 presents the major thermo-flow boundary conditions. Further details of the measurement test can be found in Zhang et al. (2009). The velocity, air temperature, and SF_6 tracer gas concentrations were measured along vertical poles indexed by number, as shown in the planar view of the geometric setup. A gas mixture (1% SF_6 + 99% N_2) was continuously released at a constant rate from a small tube at the head of one manikin but not from the other manikin. The gas discharge momentum was very weak and therefore could be omitted.

In the workshop case, it was assumed that four persons were doing assembly work. Four heated boxes on the tables were used to mimic heat generation from

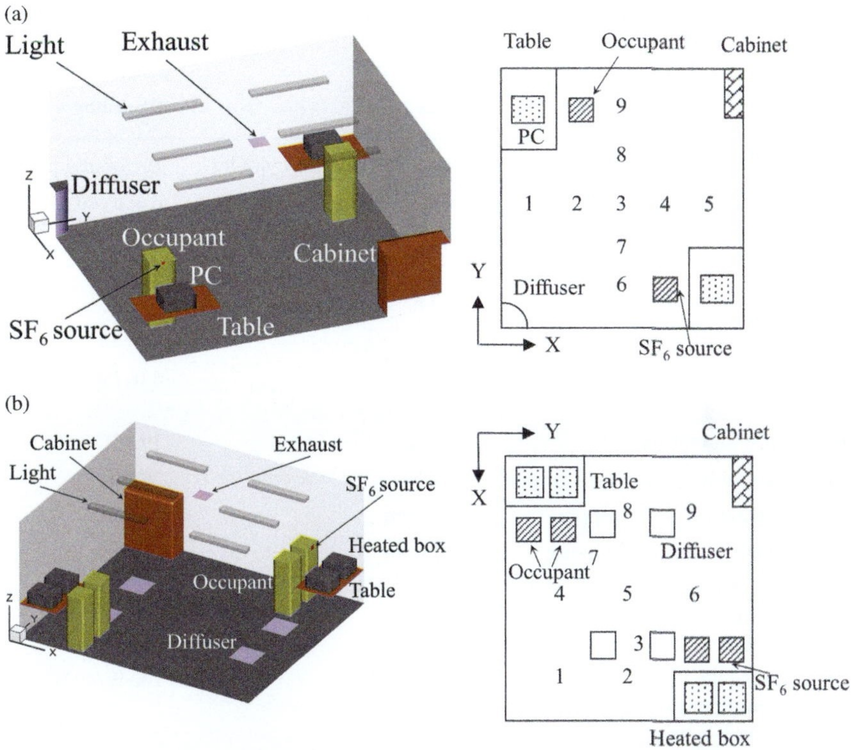

Figure 8.8 Two indoor environment cases with displacement ventilation for CFD modeling, where the numbers indicate vertical poles along which the velocity, temperature, and SF_6 tracer gas concentrations were measured, with an overview on the left and a planar view on the right: (a) office case; (b) workshop case. *Source:* Zhang et al. (2009) / with permission of John Wiley & Sons.

the assembly lines. The heat generation rate from the manikins was much higher than that in the office case because the manikins represented workers engaged in heavy labor. Conditioned air was supplied through four perforated, square panels, and the internal air was exhausted at the center of the ceiling. The same SF_6 tracer gas was introduced at the head level of one of the manikins. The major boundary conditions can be viewed in Table 8.3, and further details of the case setup and measurement can be found in the literature (Zhang et al. 2009).

The geometric domains of the two cases were created with the GAMBIT software (version 2.1). Both combined structured (hexahedral grids) and unstructured (tetrahedral grids) meshes were created using the Tet/Hybrid scheme with Hex core. The grid cell size near the diffuser was approximately 0.5 cm, and then was gradually shifted to 8 cm in the core region of the domain. The grid cell numbers

Table 8.3 Boundary conditions for the office and workshop environments with displacement ventilation.

Case	Office	Workshop
Air supply diffuser type	Perforated quarter-cylinder	Perforated panel
Diffuser dimensions	$r = 0.28$ m $h = 0.8$ m	$a = b = 0.5$ m
Diffuser's effective opening area ratio	0.67	0.67
Room air change rate	4.57 h^{-1}	8.62 h^{-1}
Air supply turbulence intensity	8%	8%
Hydraulic diameter of air supply openings	0.005 m	0.007 m
Supply air temperature	19.0 °C	19.0 °C
Exhaust air temperature	26.2 °C	26.7 °C
SF_6 concentration at the supply	0.18 ppm	0.20 ppm
SF_6 concentration at the exhaust	0.95 ppm	0.88 ppm
Total sensible heat production rate by each occupant/measured average surface temperature	75 W/30 °C	200 W/35 °C
Total sensible heat production rate by each computer or heated box/measured average surface temperature	100 W/40 °C	100 W/40 °C
Total sensible heat production rate by each lamp/measured average surface temperature	64 W/40 °C	64 W/40 °C

Source: Zhang et al. (2009) / with permission of John Wiley & Sons.

were 247,000 for the office case and 383,000 for the workshop case. The RNG k-ε model together with the standard wall functions (Yakhot et al. 1992) was used due to its stable performance (Zhang et al. 2007). Notably, the wall Prandtl number was tuned to accurately represent the near-wall heat transfer according to the recommendations by Zhang et al. (2013). The differential equations were solved with the finite volume method by dividing the domain into many spatial cells.

Besides, to ensure that both the air supply velocity out of the diffuser and the total air supply rate were correctly specified as the boundary conditions, the effective opening area ratio of each diffuser was appropriately assigned by blocking a certain ratio of CFD grid cells. This means some of the grid cells on the diffuser surface are allowed for discharging air, while others are not, similar to the realistic situations. More details of the method can be found in the literature

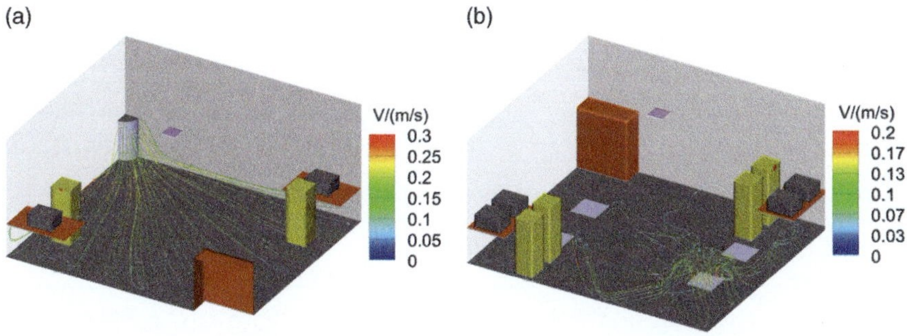

Figure 8.9 CFD-predicted airflow streamlines from the air supply openings: (a) in the office case; (b) in the workshop case. *Source:* Zhang et al. (2009) / with permission of John Wiley & Sons.

(Zhang et al. 2009). Interested readers can also refer to the numerical solution cases by accessing the electronic resources of this book's webpage.

Figure 8.9a presents the solved flow trajectories from the quarter-circular diffuser in the office case. Because the supplied conditioned air was cool, the air jets dropped down to the floor and then moved forward. The velocities of the air jets decayed as the jets proceeded. The numerically solved flow pathlines were very smooth and clearly showed how the air moved inside the office.

Figure 8.10 displays the vertical profiles of air velocity, temperature, and SF_6 concentration along representative vertical poles and a comparison with the measurement data. The results are presented at only four poles (P3, P5, P6, and P9), as shown in Figure 8.8a. Air velocities were generally low (less than 0.15 m/s) with displacement ventilation. The computed results were in good agreement with the measured data. The velocities along the lower portions of P3, P5, and P6 were higher than along the upper portions due to the jet from the diffuser, as can be seen in Figure 8.9a. The computed air temperature profiles were in excellent agreement with the measured data, as shown in Figure 8.10b. The jet impact from the diffuser was also reflected in the air temperature profiles. Because the floor was warmer than the adjacent air, a decrease in the air temperature from the floor to ankle height could be seen. The SF_6 concentration profiles computed by CFD were in reasonably good agreement with the measured profiles, as could be seen in Figure 8.10c. Since P6 was in close vicinity to the SF_6 source, the concentration in the upper region was higher. The concentration along P9 was uniform because this location was far away from the source. Similar to the air temperature stratification, the SF_6 concentration was also stratified.

The workshop had a higher cooling demand and thus higher airflow rates than the office. However, the discharge velocity from the perforated panels was quite low because the total air supply area of the perforated panels was large. The normal

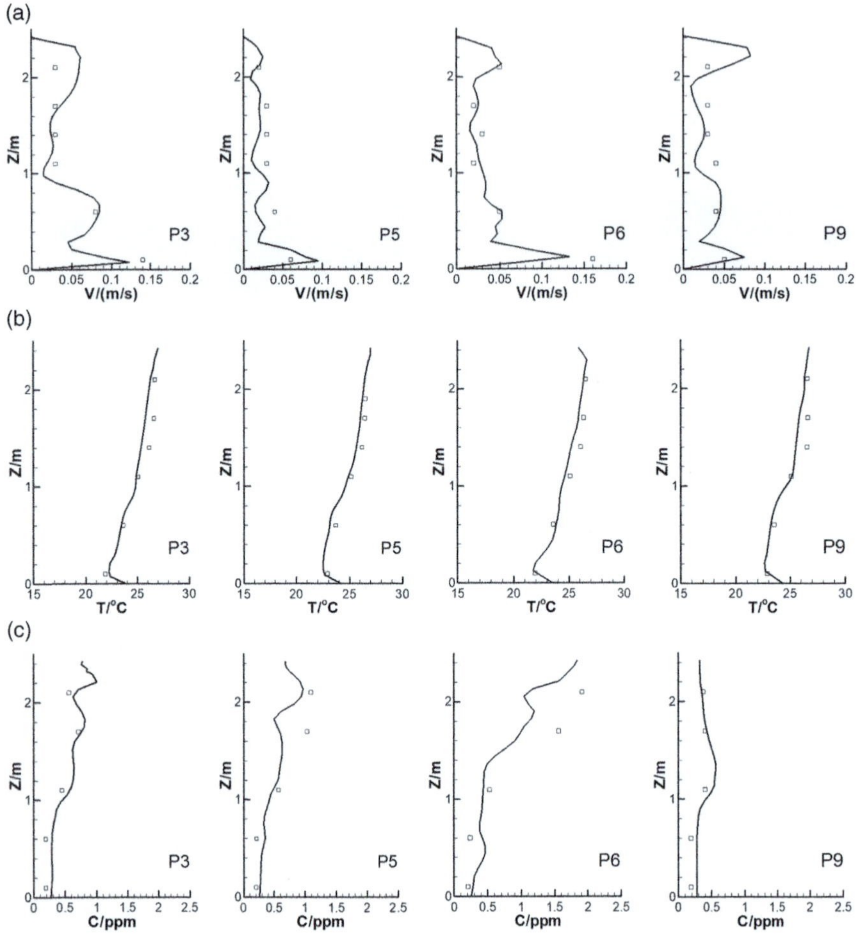

Figure 8.10 Comparison of the vertical profiles in four representative locations as provided by CFD and by experiment for the office with the quarter-circular-perforated diffuser, where lines represent CFD and symbols indicate the experiment: (a) air velocity; (b) air temperature; (c) SF_6 concentration. *Source:* Zhang et al. (2009) / with permission of John Wiley & Sons.

velocity was 0.19 m/s on the panel, and the volumetric flow rate was 0.12 m^3/s. Figure 8.9b displays the jet flow path lines, where the jet was clearly weak. The jets traveled up to approximately 0.5 m high and then dropped due to the cold buoyancy of the supplied cool air. After that, the air was lifted up again by the heat sources inside the room, but at a much lower velocity.

As shown in Figure 8.11a, the air velocity in the workshop was lower than 0.1 m/s at the four representative positions (P1, P3, P4, and P8, as shown in

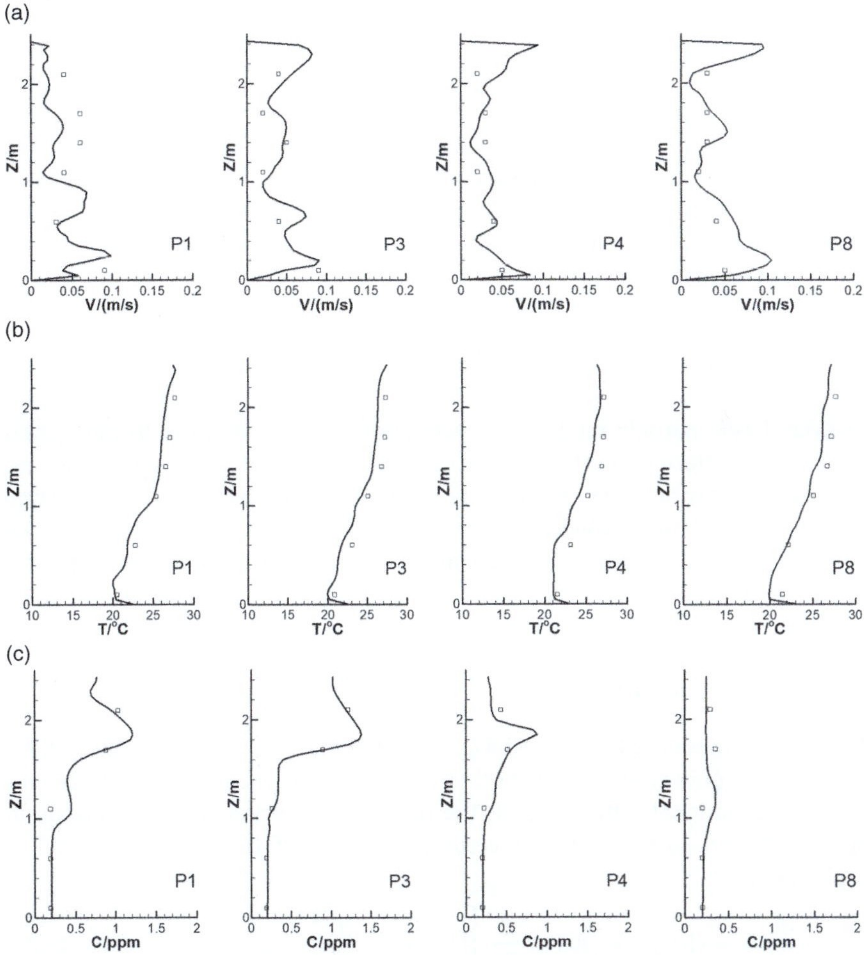

Figure 8.11 Comparison of the vertical profiles in four representative locations by CFD and by experiment for the workshop with four floor-perforated-panel diffusers, where lines represent CFD and symbols indicate the experiment: (a) air velocity; (b) air temperature; (c) SF_6 concentration. *Source:* Adapted from Zhang et al. (2009).

Figure 8.8b) since the air was supplied to the workshop with low momentum. The computed velocities agreed reasonably well with the measured values. Air temperature stratification, with 20 °C at the floor of the workshop and 27 °C near the ceiling, as shown in Figure 8.11b, was more evident here than in the office. Again, the agreement was very good between the computed and measured air temperatures. The computed SF_6 concentration profiles were in excellent agreement with the measurement data, as shown in Figure 8.11c. The SF_6

concentration stratification could also be observed. A higher SF_6 concentration was found in the upper part of the room ($Z > 1.6$ m) because the weak jet flow did not mix the room air well. Since P3 was in close proximity to the SF_6 source, the concentration along the upper portion of this pole was very high.

The two cases mentioned earlier demonstrate that the airflow, air temperature, and tracer gas concentration provided by CFD modeling are in good agreement with the measurement data. Once the indoor airflow has been solved, the motion of air-laden particles and the particle concentrations can be modeled and numerically solved accordingly.

8.4 Eulerian-Based Particle Modeling

Eulerian-based particle modeling is very similar to the solution to the spatial transport of a tracer gas. However, aerosols are a discrete phase within the air. The aerosols may not trace the air motion very well, especially in the case of large particles with significant inertia. In contrast, ultrafine particles can closely trace the airflow, but their diffusion effect may be significant. Consequently, particle settling or deposition on surfaces due to gravity or Brownian diffusion may need to be taken into account when carrying out the Eulerian-based modeling.

8.4.1 Basic Principles

The Eulerian-based drift-flux model (Lai and Cheng 2007) is widely used to solve the particle concentration distribution in space. This model can well account for the gravitational settling effect, the turbulent diffusion, and the deposition of particles on surfaces. The governing equation of the drift-flux model is expressed as:

$$\frac{\partial C}{\partial t} + \frac{\partial}{\partial x_j}[(U_j + V_{s,j})C] = \frac{\partial}{\partial x_j}\left[(D + \varepsilon_P) \cdot \frac{\partial C}{\partial x_j}\right] + S_C \tag{8.21}$$

where C is the particulate mass or number concentration, such as mg/m^3 or #/m^3; t is time, s; $x_j (j = 1, 2, 3)$ is the position in Cartesian coordinates, m; $U_j (j = 1, 2, 3)$ is the velocity component of air, m/s; $V_{s,j}$ is the particle settling velocity in the x_j direction, m/s; D is the Brownian diffusivity, m^2/s; ε_P is the particle eddy diffusivity, m^2/s; and S_C is the source term, m^{-3}/s. The terminal settling velocity can be calculated by Stokes's law, further details of which are provided in Chapter 2. For airborne particles with diameters less than 10 μm (PM_{10}), the relaxation time is very short (typically less than the magnitude of 10^{-3} seconds), as shown in Table 3.1. In this context, the assumption that the particle eddy diffusivity, ε_P, is equal to the carrier fluid turbulent viscosity yields generally good accuracy (Hinze and Uberoi 1975).

To account for the particle deposition effect, the nominal particle concentration on a wall (Gao and Niu 2007) can be simplified as:

$$C_{\mathrm{wall},i} = C_{\mathrm{cell}} - C_{\mathrm{cell}}(V_{\mathrm{Dep}} - iV_{\mathrm{s}})\frac{\delta y}{D + \varepsilon_{\mathrm{p}}} \tag{8.22}$$

where $C_{\mathrm{wall},i}$ is the nominal particle concentration on a wall, which serves as the boundary condition; i is the wall orientation index ($i = 0$ for a vertical wall, $i = 1$ for an upward-facing horizontal wall, and $i = -1$ for a downward-facing horizontal wall); C_{cell} is the concentration in the near-wall grid cell, which is beyond the concentration boundary layer; V_{Dep} is the particle deposition velocity; V_{s} is the particle settling velocity; and δy is the distance between the center of the grid cell and the wall. Lai and Nazaroff (2000) proposed a semi-empirical model to calculate the deposition velocity, V_{Dep}. They simplified the deposition flux as one-dimensional and constant within the concentration boundary layer. Equation (8.22) can account for the effects of the Brownian and eddy diffusion as well as the gravitational settling. With governing Eq. (8.21) and boundary Eq. (8.22), the evolution of particle concentrations over time throughout the domain can be solved once the flow information has been obtained and the particle source information has been specified.

8.4.2 Demonstration of Eulerian-Based Particle Modeling in a Ventilated Enclosure

An empty rectangular enclosure with mechanical ventilation is investigated here, as shown in Figure 8.12a. Conditioned air was supplied to the enclosure at a speed of 0.225 m/s through a square opening at the center of the left-hand wall, next to the ceiling. The internal air was exhausted out of the enclosure through a square outlet at the center of the right-hand wall, near the floor. The dimensions of the air supply opening and the air exhaust opening were identical, i.e., 0.04 m × 0.04 m. The whole enclosure was maintained under isothermal conditions. Spherical particles of silver-coated hollow glass with a diameter of 10 μm were continuously injected into the air supply at a constant rate. The particles had a density of 1,400 kg/m^3. Both airflow velocity and particle concentration were measured by a phase Doppler anemometry (PDA) system. The evolution of the particle plume was recorded for 30 minutes. Further details of the experimental test can be found in the literature (Chen et al. 2006).

The enclosure geometry was created by the GAMBIT software (version 2.3.16), and then rectangular grid cells of 40 × 20 × 20 were generated by means of the "Submap" scheme. The CFD meshes were imported into the FLUENT software, and all the boundary settings were then appropriately specified. The RNG k–ε model (Yakhot et al. 1992) was adopted to resolve the turbulence, and the

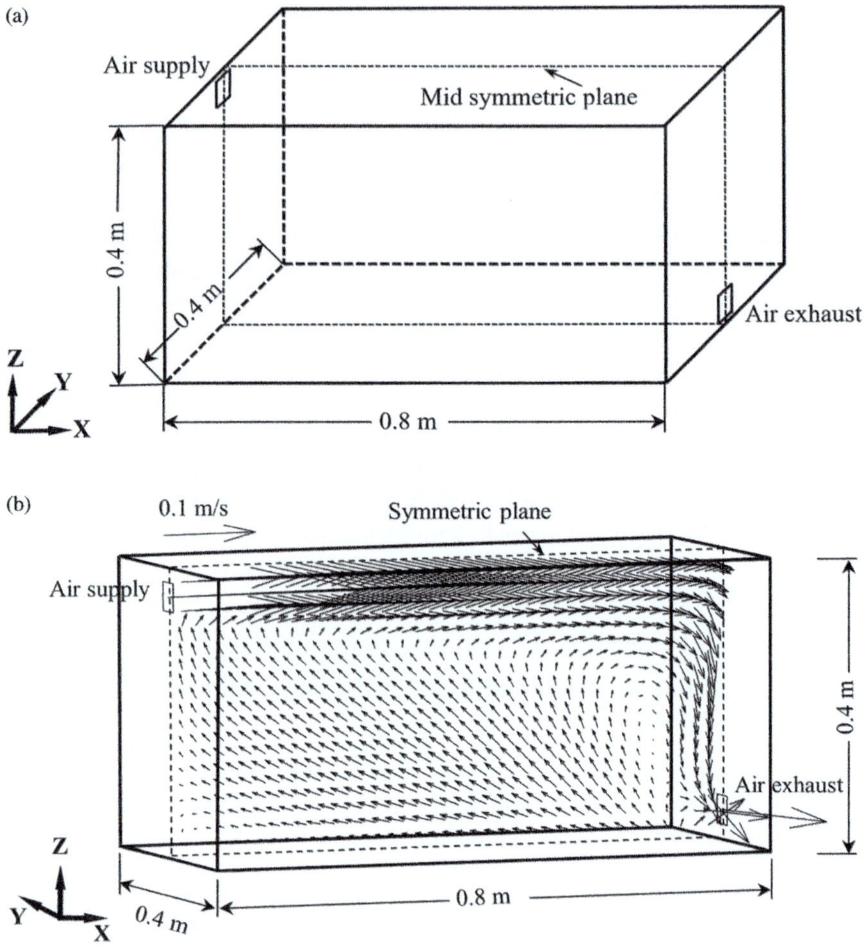

Figure 8.12 A ventilation enclosure in which particle transport was modeled: (a) geometric dimensions. *Source:* Adapted from Chen et al. (2006); (b) CFD-solved airflow pattern on the symmetric Y plane.

standard wall function was employed for the near-wall treatment. The discretization scheme for pressure and momentum was the second-order upwind type.

Figure 8.12b presents the solved airflow pattern on the middle symmetric plane. The air jets from the air supply opening traveled horizontally to the right and formed a generally clockwise circulating flow. Some of the flow was extracted by the air exhaust. Figure 8.13 compares the solved horizontal velocity component, i.e., the X-velocity, with the measurement along three vertical poles on the symmetric plane. A relatively high speed can be seen in the upper region due to the

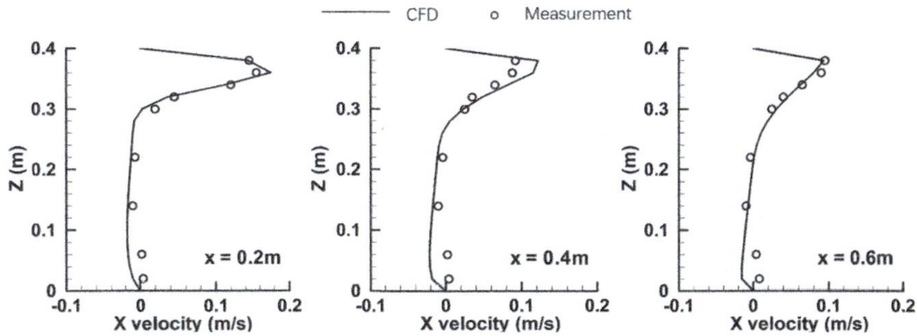

Figure 8.13 Comparison of the predicted horizontal velocity component with the measurement at three vertical poles on the symmetric Y plane.

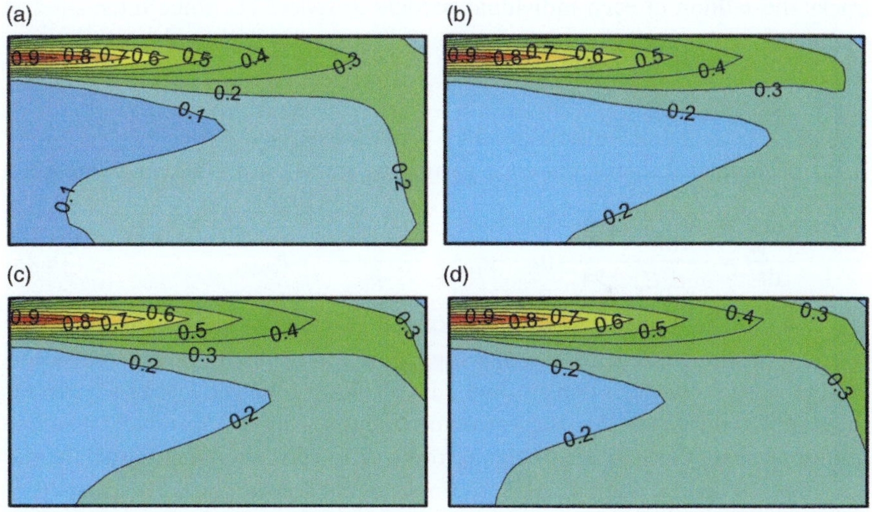

Figure 8.14 Normalized particle concentration distribution on the vertically symmetric plane at different times solved by the Eulerian-based drift-flux model: (a) at $t = 60$ seconds; (b) at $t = 180$ seconds; (c) at $t = 300$ seconds; (d) at $t = 1{,}800$ seconds.

horizontal jet. The CFD simulation provided a velocity distribution that agreed very well with the measurement.

Figure 8.14 presents the normalized particle concentration on the vertical symmetric plane at different times. The particle concentration at the air supply was deemed to be unity. At $t = 60$ seconds, as shown in Figure 8.14a, the decay of the particle concentration with the jet could be seen. The particle concentration in the bottom region was lower due to the incomplete spreading of the particles within the first minute. With the elapse of time, the concentrations in the lower left region increased, as could be seen in Figure 8.14b and c. However, even at $t = 1{,}800$ seconds, as shown in Figure 8.14d, the particle concentration in the lower left region was still lower because of the gravitational settling of the particles. The concentrations of the studied 10 μm particles never became as uniform as the concentration of 1 μm particles (Chen et al. 2006) or that of the tracer gas. This implies that the particle settling effect is quite evident for the studied 10 μm particles. A quantitative comparison of the predicted particle concentrations with the measurement data is provided in the next section.

8.5 Lagrangian-Based Particle Modeling

Unlike the Eulerian-based particle modeling, the Lagrangian-based modeling tracks the motion of each individual particle in space. The force interaction is examined for the studied particles on the basis of Newton's second law.

8.5.1 Basic Principles

In the Lagrangian frame of reference, particle acceleration in space can be modeled as:

$$\frac{du_{\mathrm{p},i}}{dt} = \frac{18\mu}{\rho_\mathrm{p} d_\mathrm{p}^2 C_\mathrm{c}} \frac{C_\mathrm{D}\mathrm{Re}_\mathrm{p}}{24}(u_i - u_{\mathrm{p},i}) + \frac{g_i(\rho_\mathrm{p} - \rho)}{\rho_\mathrm{p}} + F_{\mathrm{Add},i} \tag{8.23}$$

where $u_{\mathrm{p},i}$ is the particle velocity component, C_c is the Cunningham correction factor, C_D is the drag coefficient, u_i is the air velocity, g_i is the gravitational acceleration, ρ_p is the particle density, and ρ is the air density. The first two terms on the right-hand side are the drag force and buoyancy force, respectively. The last term represents the sum of additional forces acting on the particles, such as the Brownian diffusion, Saffman's lift, thermophoretic forces, and so on. Note that Eq. (8.23) applies to particles with a unit mass, and the particles' Reynolds number is defined as:

$$\mathrm{Re}_\mathrm{p} \equiv \frac{\rho d_\mathrm{p}|u_\mathrm{p} - u|}{\mu} \tag{8.24}$$

The drag coefficient can be estimated as:

$$C_\mathrm{D} = \frac{24}{\mathrm{Re}_\mathrm{p}}(1 + 0.27\mathrm{Re}_\mathrm{p})^{0.43} + 0.47\left[1 - \exp\left(-0.04\mathrm{Re}_\mathrm{p}^{0.38}\right)\right], \text{ for } \mathrm{Re}_\mathrm{p} < 2 \times 10^5 \tag{8.25}$$

or

$$C_\mathrm{D} = a_1 + \frac{a_2}{\mathrm{Re}_\mathrm{p}} + \frac{a_3}{\mathrm{Re}_\mathrm{p}^2} \tag{8.26}$$

where a_1, a_2, and a_3 are constants (Morsi and Alexander 1972) for smooth spherical particles. The Cunningham correction can be activated for submicron particles using the following equation (ANSYS Inc. 2006):

$$C_c = 1 + \frac{2\lambda}{d_p}\left[1.257 + 0.4\exp\left(-\frac{1.1d_p}{2\lambda}\right)\right] \tag{8.27}$$

Commonly, the additional forces may be neglected because they are smaller than the drag or buoyancy force by two orders of magnitude in an indoor space (Zhang and Chen 2006). Only within the turbulent boundary layer may some of these forces occasionally be comparable to the fluid-exerted drag force (Li and Ahmadi 1992). In those cases in which the additional forces must be considered, the readers are advised to refer to the Fluent user guide (ANSYS Inc. 2006) for the basic principles of modeling the "virtual mass" force, Brownian diffusion force, thermophoretic force, and Saffman's lift force, as well as some additional forces in the rotating reference frames. Both the Brownian diffusion force and the thermophoretic force are also addressed in Chapter 5.

Before a particle can be tracked, the air velocity, u_i, must be provided to Eq. (8.23). It is supposed that the airflow and particle motion are under one-way coupling, which means that the airflow contributes to the particle motion, but not vice versa. The one-way coupling principle is generally acceptable if the volumetric particle numbers are not very large. In such a situation, the airflow can be solved separately. In the RANS modeling, u_i consists of a time-averaged component, $\overline{u}_i$, plus an instantaneous component, u_i'. There are a large number of eddy-viscosity turbulence models (Wilcox 1993) that can be applied to calculate u_i'. If the most widely used k–ε model is employed and the turbulence is assumed to be isotropic, the instantaneous velocity can be provided by the discrete random walk (DRW) model as:

$$u_i' = \xi\sqrt{2k/3} \tag{8.28}$$

where ξ is a Gaussian random number with zero mean and unit variance between −1 and 1, and k is the turbulent kinetic energy. However, in the vicinity of walls, the turbulence can be anisotropic. The instantaneous velocity components parallel with and perpendicular to a wall can be significantly different, so each component of velocity should be modeled separately (Zhang and Chen 2009). The isotropic DRW model may not be sufficiently accurate in the near-wall region.

When particles strike a solid surface, they may become trapped, or they may rebound and then resuspend. However, only when the striking velocity is greater than a certain critical threshold can a particle rebound, as introduced in Section 4.5. For indoor particles that strike surfaces because of gravitational settling, the chance of particle resuspension is minimal (Riley et al. 2002).

The Eulerian model can output the particle concentration directly, but the Lagrangian model cannot. The Lagrangian model contains only the particle trajectory information, so it is necessary to convert the trajectories into the particle number concentration. The particle source in-cell (PSI-C) scheme (Crowe et al. 1977) can be applied in the post-processing. This scheme correlates the particle concentration with the residence time of each trajectory in a CFD cell. The particle trajectory is obtained by integrating the ordinary differential Eq. (8.23) twice over time.

8.5.2 Demonstration of Lagrangian-Based Particle Modeling in a Ventilated Enclosure

The case of particle transport addressed in Section 8.4.2 was also resolved by the Lagrangian model. Lagrangian modeling was carried out on the basis of the same airflow as shown in that section. For individual tracking of particles after their release through the air supply opening, the automated tracking scheme was activated in the ANSYS Fluent, which then automatically switched between the lower-order integration scheme and the higher-order scheme. The lower-order scheme employed the analytical exponential solution, so it was numerically stable and efficient, but it might be inaccurate when a large time step was used. Meanwhile, the higher-order scheme employed the Runge–Kutta numerical scheme, which was stable only in a limited time range. To ensure accuracy and numerical stability, a constant time step of 0.01 seconds was used.

The number of particle trajectories should be sufficiently large to obtain a concentration that is statistically stable. This investigation evaluated two different trajectory rates, 400/s and 4,000/s, and found no meaningful difference between the two. It was concluded that 400 trajectories per second were sufficient for the studied case. When a particle struck the bottom surface, it was set to adhere to the surface, and the particle tracking was terminated. When striking other surfaces, the particle was assumed to rebound as in an elastic collision.

Figure 8.15 presents the solved trajectories of particles for a period of 60 seconds after their release at the air supply inlet. The trajectories near the ceiling were very dense, and the particles had relatively high velocity. As the particles approached the vertical wall, their trajectories generally bent downward. The trajectories then became more dispersed, and the velocity further decreased. Some of the particles were entrained in the region surrounding the air supply inlet. Meanwhile, a very small number of particles struck the bottom surface of the enclosure. Figure 8.15 presents only a relatively short duration of the particle trajectories, for the sake of clear viewing.

Figure 8.16 displays the particle concentrations converted from the solved particle trajectories on the vertical symmetric plane. The highest particle concentration

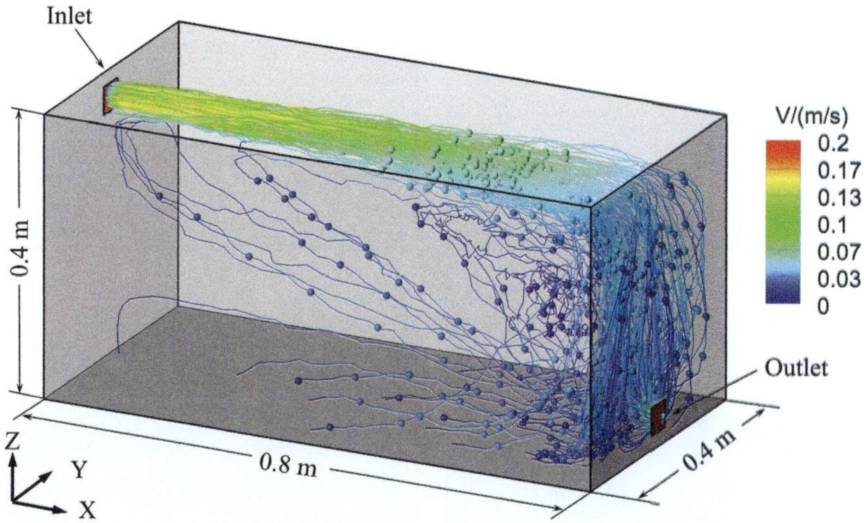

Figure 8.15 Particle trajectories solved by the Lagrangian model, with velocity magnitude represented by a color scale.

was in the upper part of the enclosure, while in the lower part and in the left-hand region below the air supply jets, the concentration was much lower. When compared with the concentration distribution solved by the Eulerian model, shown in Figure 8.14, the overall particle distribution pattern provided by the Lagrangian model was very similar. However, the concentration contours from the Lagrangian model were not as smooth as those from the Eulerian model. This implied that the results obtained from the Lagrangian model were more random. The concentration gradients near the walls were also more evident in the Lagrangian model's predictions than in those of the Eulerian model.

Figure 8.17 presents the dimensionless particle concentration profiles along the height of the enclosure after an elapsed time of 1,800 seconds from the initial release. The dimensionless concentration was normalized by the particle concentration at the air supply. The concentration reached a peak at around $Z = 0.35$ m, which was close to the center of the air supply jets. Due to gravitational settling, there was a noticeable concentration gradient within the domain. Note that the time constant of the enclosure was 356 seconds. Without the gravitational settling, the particle distribution would have approached a uniform state after a period of 1,800 seconds. This proves that the gravitational settling cannot be omitted for the studied 10 μm particles. Below $Z = 0.2$ m, the concentration variation was not very evident. There were slight differences in the solved particle concentrations between the Eulerian and Lagrangian models. Fortunately, both models more or less captured the major particle distribution profiles. The Eulerian model

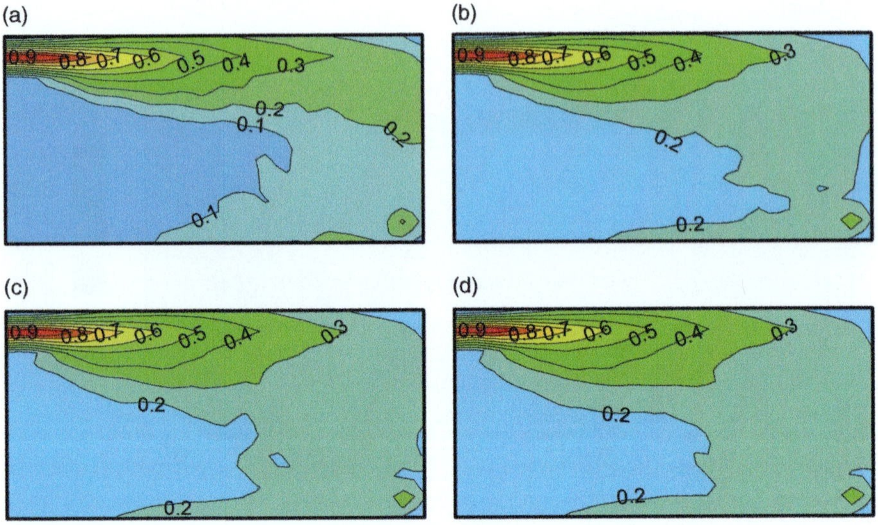

Figure 8.16 Particle concentration distribution on the vertically symmetric plane at different times solved by the Lagrangian model: (a) at $t = 60$ seconds; (b) at $t = 180$ seconds; (c) at $t = 300$ seconds; (d) at $t = 1{,}800$ seconds.

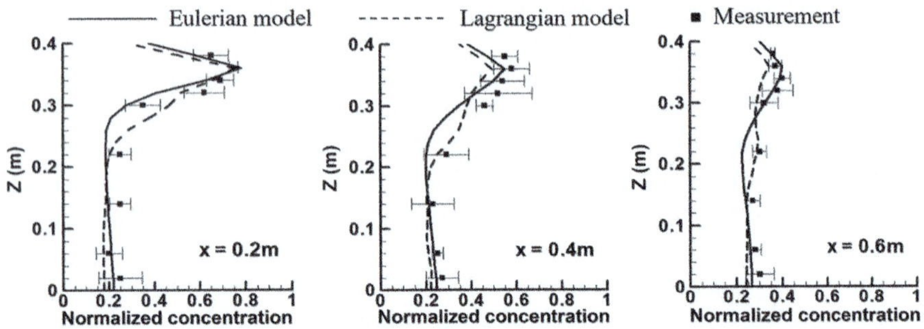

Figure 8.17 Comparison of the concentrations predicted by the Eulerian-based drift-flux model and the Lagrangian model with the measured concentrations (together with uncertainty illustrated by error bars) at three different locations on the symmetric Y plane.

seemed to compute a sharper concentration peak at the top. Nevertheless, the concentration profiles predicted by both models were close to the measurement.

The solutions presented in Sections 8.4.2 and 8.5.2 underscore the capability of RANS-based CFD to provide the flow and particle distribution in reasonably good agreement with the measurement. Both the Eulerian and Lagrangian models were able to predict the particle concentrations and their distribution with generally good accuracy.

8.6 Multiphase Particle Modeling

In addition to pure solid or liquid aerosols, combined solid and liquid aerosols are very common, and they are known as multiphase aerosols. These aerosols may contain a mix of soluble and insoluble solids, such as inorganic and organic salts and dust, within liquid droplets. For example, in cough droplets, soluble salts dissolve in the liquid phase, together with some insoluble substances inside. As the droplets evaporate, the dissolved salts precipitate, forming a solid residue together with insoluble substances. This process significantly changes the aerosol's physical and dynamic properties, affecting its behavior and interactions with the air and the solid boundaries. Due to the complexity of multiphase aerosols and the intricate interactions among phases, precise modeling of their transport is more challenging. This section addresses the modeling of these multiphase aerosols, including their formation, evolution in space, and spatial distribution.

8.6.1 Basic Principles

The breakup of liquids is a primary mechanism in the generation of multiphase aerosols. Liquid breakup may emit numerous fine droplets, with diameters ranging from a few microns to several hundred microns. Larger droplets tend to deposit, while small droplets can evaporate rapidly. Once the liquid has completely evaporated or has reached an equilibrium state, the residual solids, or a mixture of solids and unevaporated liquid, form droplet nuclei with smaller diameters. This process involves the dynamics of gas–liquid breakup as well as heat and mass transfer.

8.6.1.1 Multiphase Flow Modeling

If the liquid moves significantly while emitting liquid aerosols, it is necessary to model the flow of the bulk liquid. Here, multiphase flow modeling based on the Eulerian method, known as the volume of fluid (VOF) method, is presented. It is particularly suited to tracking the interface between two immiscible fluids, such as water and air.

In the VOF method, the volume fraction, α, is used to index the type of fluid in multiphase flow and track the interface between phases. The volume fraction equation of the ith phase fluid is as follows:

$$\frac{1}{\rho_i}\left[\frac{\partial}{\partial t}(\alpha_i\rho_i) + \nabla\cdot(\alpha_i\rho_i u_i) = S_i\right] \tag{8.29}$$

where S_i is the source term of the ith phase fluid. Taking the air–water flow as an example, the volume fraction in each CFD cell has three possible states: (i) there is no air phase in the cell; (ii) the cell is fully occupied by the air; or (iii) the cell contains the interface between the air and water phases. These states correspond to the following values of α_i: $\alpha_i = 0$, $\alpha_i = 1$, or $0 < \alpha_i < 1$, respectively, supposing α_i represents the air phase. The sum of the volume fractions of different phases is equal to unity:

$$\sum_{i=1}^{n} \alpha_i = 1 \tag{8.30}$$

In the VOF method, the governing equation applies to all phases. The governing equations for multiphase flow are identical to Eq. (8.6), except that ρ and μ must be calculated as volume fraction-based averages. These equations are not listed in this section to prevent repetition.

8.6.1.2 Coupled Euler–Lagrange Modeling

To track the interfaces of numerous tiny droplets using the VOF method, the mesh resolution must be at the submicron scale. Consequently, billions of CFD cells would be required for accurate tracking, and the computational cost would be prohibitively high. Alternatively, tracking each individual fine droplet using the Lagrangian method can significantly reduce the computational burden. This approach constitutes coupled Euler–Lagrange modeling. The coupled method adopts the Eulerian approach for modeling the motion of large droplets while utilizing the Lagrangian approach to resolve fine droplets. Once droplets meet a specified criterion of size, they are converted into particles for Lagrangian-based tracking.

Coupled Euler–Lagrange modeling consists of three main steps: (i) identification of detached droplets or liquid fragments in the VOF field under the Eulerian framework; (ii) calculation of their physical properties, including velocity, temperature, and diameter; and (iii) conversion of droplets that meet the size conversion criterion into particles for Lagrangian tracking, and removal of these droplets from the Eulerian framework (Heinrich and Schwarze 2020).

A schematic diagram of the droplet identification method is provided in Figure 8.18. The identification of droplets is achieved using the 3D connected-component labeling (CCL) algorithm (Heinrich and Schwarze, 2020). First, the grid cells with a liquid phase fraction α in the Eulerian framework are labeled based on the threshold value of $\alpha_t = 0.2$, as shown in Figure 8.18a and b. All grid cells with α values greater than the threshold are marked, while the remaining cells are treated as gas. Next, if a marked droplet is to be transformed into a particle within the Lagrangian framework, it requires a unique identification (ID) for the purpose of particle tracking and to avoid ambiguity in parallel computation. The CCL algorithm (Heinrich and Schwarze, 2020) allows the propagation of numerical values from a given grid to the neighboring grids. In this context, all liquid grids are sequentially labeled in ascending order, and the smallest ID is

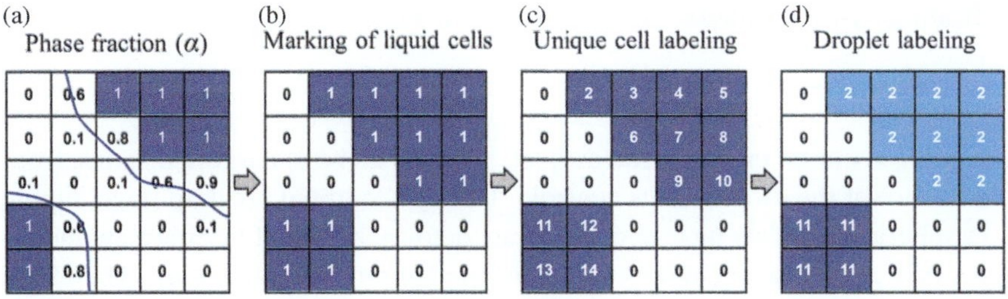

Figure 8.18 Schematic diagram of the droplet identification approach: (a) phase fraction distribution; (b) marking of liquid cells; (c) labeling of liquid cells in ascending order; (d) droplet identification. *Source:* Heinrich and Schwarze, 2020 / Elsevier / CC BY 4.0.

propagated to all neighboring liquid grid cells. Therefore, each droplet, which is a collection of cells with the same ID, is identified and labeled, as shown in Figure 8.18c and d.

The next step is to calculate the physical properties of the identified droplets. In the first step, all the marked droplet grid cells are looped through so that their physical information can be accessed. Next, the physical properties of the entire quantity of droplets are computed using the following formulas:

$$V_{\mathrm{p}} = \sum_i \alpha_i V_i \tag{8.31}$$

$$x_{\mathrm{p}} = \frac{1}{V_{\mathrm{p}}} \sum_i \alpha_i V_i x_i \tag{8.32}$$

$$\boldsymbol{u}_{\mathrm{p}} = \frac{1}{V_{\mathrm{p}}} \sum_i \alpha_i V_i \boldsymbol{u}_i \tag{8.33}$$

$$d_{\mathrm{p}} = \sqrt[3]{\frac{6V_{\mathrm{p}}}{\pi}} \tag{8.34}$$

where V_{p} is the volume of the identified detached droplet, V_i is the volume of the ith grid cell, x_{p} is the position of the droplet, x_i is the coordinate of the ith grid cell, $\boldsymbol{u}_{\mathrm{p}}$ is the velocity vector of the droplet, $\boldsymbol{u}_i$ is the velocity vector of the ith grid, and d_{p} is the diameter of the droplet.

The conversion criteria for each droplet are checked. Due to the complexity and variability of the breakup process, these criteria are difficult to define on the basis of physical properties alone. For example, many instabilities, such as Kelvin–Helmholtz and Rayleigh–Taylor instabilities (Zuzio et al. 2017), occur during liquid breakup. Therefore, geometric characteristics are used as conversion criteria: (i) the droplet diameter must be smaller than a specified threshold, and (ii) the sphericity Ψ_{p} of the droplet must be less than 2. The degree of similarity of a geometric shape to a sphere is characterized by the sphericity, with $\Psi_{\mathrm{p}} = 1$

indicating a perfect sphere. The sphericity is defined as:

$$\Psi_p = \frac{2r_{max}}{d_p} \tag{8.35}$$

where Ψ_p is the sphericity of the droplet, and r_{max} is the maximum distance from the surface of the droplet to its centroid. During the initial breakup stage, highly non-spherical droplets have complex drag, deformation, and interface characteristics that cannot be accurately solved in the Lagrangian framework. Therefore, highly non-spherical droplets must continue to be solved within the VOF framework and will not be converted into particles for Lagrangian tracking.

If the aforementioned criteria are met, the droplets in the Eulerian framework are converted into particles in the Lagrangian framework for spatial tracking. Simultaneously, the corresponding cells of the converted droplets are set to $\alpha = 0$ and are therefore removed from the VOF volume fraction field. If the tracked droplets impact a surface and subsequently splash, the so-called wall-film model (O'Rourke and Amsden 2000) can be employed.

8.6.1.3 Heat and Mass Transfer Modeling

Multiphase aerosols often exist as droplets. If volatile components are present, they may evaporate under certain conditions, exchanging heat and mass with the bulk gases. Evaporation is triggered when the temperature of a droplet reaches the vaporization temperature. The evaporation process will persist until the volatile fraction of the droplet is fully exhausted or until the temperature of the droplet reaches the boiling point (ANSYS Inc. 2021):

$$T_{vap} \leq T_p < T_{bp} \tag{8.36}$$

$$m_p > (1 - f_v)m_{p,0} \tag{8.37}$$

where T_{vap} is the vaporization temperature, K; T_p is the temperature of the droplet, K; T_{bp} is the boiling point, K; m_p is the current mass of the droplet, kg; f_v is the volatile fraction inside the droplet; and $m_{p,0}$ is the initial mass of the droplet.

The vaporization rate of a multiphase aerosol is the sum of the vaporization rates of all individual components. In instances of slow vaporization, the process is subject to gradient diffusion. The flux of vapor transfer from droplet to gas phase is dependent on the concentration difference between the droplet's surface and the bulk gas, as follows:

$$N_v = k_c(C_{v,s} - C_{v,\infty}) \tag{8.38}$$

where N_v is the flux of vapor transfer, kmol/(m^2·s); k_c is the convection mass transfer coefficient, m/s; $C_{v,s}$ is the vapor concentration at the droplet surface, kmol/m^3; and $C_{v,\infty}$ is the vapor concentration in the bulk gas, kmol/m^3. It is assumed that N_v is positive for evaporation and negative for condensation.

The vapor concentration at the surface of the droplet is determined by assuming that the partial pressure of vapor at the surface is equal to the saturated vapor pressure:

$$C_{\mathrm{v,s}} = \frac{p_{\mathrm{sat}}(T_{\mathrm{p}})}{\mathrm{R_u} T_{\mathrm{p}}} \tag{8.39}$$

where $p_{\mathrm{sat}}(T_{\mathrm{p}})$ is the saturated vapor pressure at temperature T_{p}, and $\mathrm{R_u}$ is the universal gas constant. The vapor concentration in the bulk gas is calculated as:

$$C_{\mathrm{v},\infty} = X_{\mathrm{v}} \frac{p}{\mathrm{R_u} T_{\infty}} \tag{8.40}$$

where X_{v} is the local bulk mole fraction of vapor, p is the local pressure, and T_{∞} is the local bulk temperature of the gas.

The mass change of the droplet is calculated as:

$$m_{\mathrm{p}}(t + \Delta t) = m_{\mathrm{p}}\left(t\right) - N_{\mathrm{v}} A_{\mathrm{p}} M_{\mathrm{w,v}} \Delta t \tag{8.41}$$

where $M_{\mathrm{w,v}}$ is the molecular weight of vapor, and A_{p} is the surface area of the droplet, m^2.

For high vaporization rates, convection transfer of the evaporated vapor from the droplet surface to the surrounding gas (Stefan flow) is critical. The contribution of the convection mass transfer to the droplet mass can be modeled as (Miller et al. 1998; Sazhin 2006):

$$\frac{dm_{\mathrm{p}}}{dt} = k_{\mathrm{c}} A_{\mathrm{p}} \rho \ln(1 + B_{\mathrm{m}}) \tag{8.42}$$

Here B_{m} is the Spalding mass number, given by:

$$B_{\mathrm{m}} = \frac{Y_{\mathrm{v,s}} - Y_{\mathrm{v},\infty}}{1 - Y_{\mathrm{v,s}}} \tag{8.43}$$

where $Y_{\mathrm{v,s}}$ is the vapor mass fraction at the droplet surface, and $Y_{\mathrm{v},\infty}$ is the vapor mass fraction in the bulk gas.

The temperature of the droplet can then be updated based on the heat balance. The heat balance correlates the droplet's sensible heat transfer rate with the convection and latent heat transfer between the droplet and the bulk gas phase, and also accounts for the radiation heat transfer as:

$$m_{\mathrm{p}} c_p \frac{\mathrm{d}T_{\mathrm{p}}}{\mathrm{d}t} = hA_{\mathrm{p}}(T_{\infty} - T_{\mathrm{p}}) - \frac{\mathrm{d}m_{\mathrm{p}}}{\mathrm{d}t} h_{\mathrm{fg}} + A_{\mathrm{p}} \varepsilon_{\mathrm{p}} \sigma \left(\theta_{\mathrm{R}}^4 - T_{\mathrm{p}}^4\right) \tag{8.44}$$

where c_p is the specific heat capacity of the droplet, (J/(kg·K)); h is the convection heat transfer coefficient, $\mathrm{W/(m^2 \cdot K)}$; h_{fg} is the latent heat of the liquid, J/kg; ε_{p} is the particle emissivity; σ is the Stefan–Boltzmann constant; and θ_{R} is the surrounding mean radiation temperature, K. Here, $\mathrm{d}m_{\mathrm{p}}/\mathrm{d}t$ represents the rate of evaporation, which can be obtained directly from Eq. (8.42) or indirectly from Eq. (8.41) but enforcing a negative sign. The reason for such enforcing of a negative sign is that Eq. (8.41) describes the actual mass balance of a droplet during change of phases, in which the actual mass loss corresponds to the evaporation.

8.6.2 Demonstration of Modeling of Cough Droplets in Space

This section demonstrates the multiphase modeling of cough droplets from a healthy adult male. The individual coughed once in a calm indoor environment with an ambient temperature of 22 °C and a relative humidity of 50% (Li et al. 2022). Particle tracking velocimetry (PTV) was used to capture the trajectories of the cough droplets for comparison with the multiphase modeling result. The droplets were illuminated by an LED light source and photographed with a high-speed camera. The expelled airflow rates were measured as a function of time, which was specified as the boundary conditions for the CFD modeling. As shown in Figure 8.19a, the expelled airflow rate rose sharply to a peak within 0.05 seconds during a cough and then gradually decayed to zero within 1 second. The flow rate profiles measured by Li et al. (2022) were comparable to those

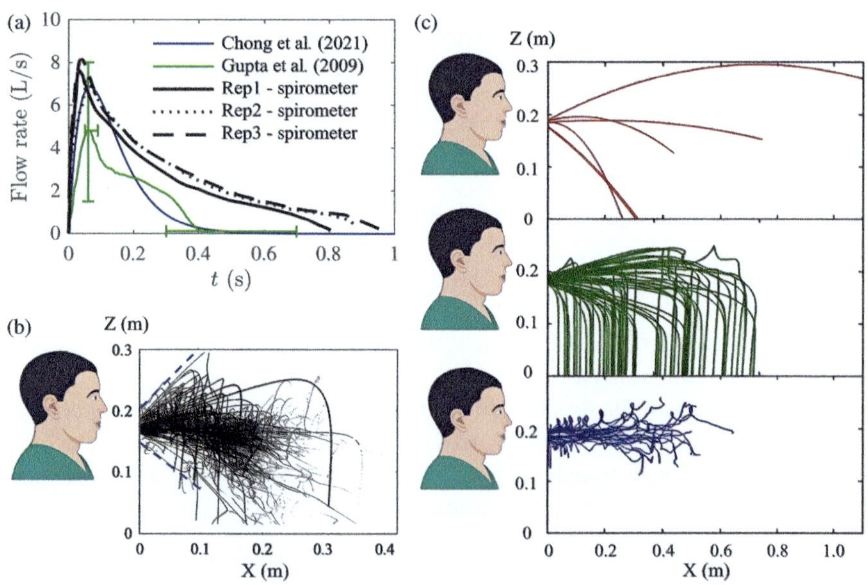

Figure 8.19 Comparison of the numerically solved coughing-induced droplet trajectories with the measurement (Li et al. 2022): (a) measured time-varying flow rate profile in a single cough as the boundary condition; (b) measured droplet trajectories of a single cough by PTV; (c) CFD-solved droplet trajectories, where the red, green, and blue lines correspond to large ($d_0 > 500\,\mu m$), medium ($d_0 \approx 90\,\mu m$), and small ($d_0 < 40\,\mu m$) droplets, respectively. *Source:* Li et al. (2022) / Elsevier / CC BY 4.0.

conducted by other researchers (Gupta et al. 2009; Chong et al. 2021), but the duration of the cough was slightly longer.

A fixed circular opening with a diameter of 23 mm was used to represent the mouth in the CFD modeling. The mouth opening was set as a flow inlet with the flow rate profile shown in Figure 8.19a. The temperature of the coughing jet out of the mouth was 34 °C. A total of 5,000 droplets were emitted through the mouth opening at a constant rate for 0.4 seconds. The initial size distribution of the released droplets was the same as that used by Xie et al. (2009). Since the droplets were sparsely distributed in space, droplet-to-droplet interaction was omitted.

In the numerical solution, the gas phase was treated as an incompressible fluid, with the thermal buoyancy modeled by means of the Boussinesq approximation. The gas phase equations were solved by DNS with a staggered, second-order finite difference scheme. Time integration was carried out using a fractional-step, third-order Runge–Kutta approach. For the droplets, the spherical point-particle model (Li et al. 2022) was employed. As the droplets evaporated, the vapor mass fraction and temperature of the bulk air were updated, which in turn altered the density of the local gas phase. Assumed that the momentum equation for the gas phase remained unchanged in the presence of droplets, while the updating of the gas density might have affected the momentum transfer.

Figure 8.19c presents the numerically solved droplet trajectories between $t = 0$–0.2 seconds. The red, green, and blue lines represent large, medium, and small droplets, respectively. The large droplets moved along ballistic trajectories over longer distances than the medium droplets. The velocities of the medium droplets decayed quickly, dropping to nearly zero in the horizontal direction before the droplets reached the ground. Meanwhile, the small droplets did not exhibit pronounced deposition owing to their smaller diameters. The small droplets had a larger specific surface area and evaporated rapidly. This led to a further reduction in their sizes and finally the formation of droplet nuclei. Consequently, small droplets traced the coughing airflow closely and remained suspended in the air for extended periods.

The numerically solved droplet trajectories can be compared with those measured by the PTV, as shown in Figure 8.19b. The measured droplet trajectories lasted approximately 0.4 seconds. The thicker lines in the figure represent the trajectories of droplets with larger diameters, while the fine lines indicate smaller droplets or droplet nuclei. The general motion trajectories predicted by the CFD model for droplets of different sizes were similar to the measured trajectories. However, the CFD model overestimated the distances traveled by droplets from the mouth opening. The reason may have been the underestimation of flow turbulence by the model, which resulted in smaller drag forces on the moving droplets. Nevertheless, multiphase particle modeling can predict droplet ejection, motion, evaporation, and particle size evolution in reasonably good agreement with the measurements.

8.6.3 Demonstration of Modeling of Liquid Aerosol Generation and Transport during the Flushing of a Squat Toilet

When a squat toilet is flushed with water, droplets may be emitted. The process involves the flow of water from a tank to a squat pan, collision of the water flow with the pan, emission of droplets into the air, and motion and evaporation of the droplets in space. The modeling and numerical solution of flushing-induced droplet emission and transport are described in this section.

A cubicle with a squat pan inside, as shown in Figure 8.20, was studied. The cubicle had a length of 1.2 m, a width of 1.0 m, and a height of 2.3 m. The planar dimensions of the squat pan were 0.59 m × 0.46 m, as depicted in Figure 8.20a.

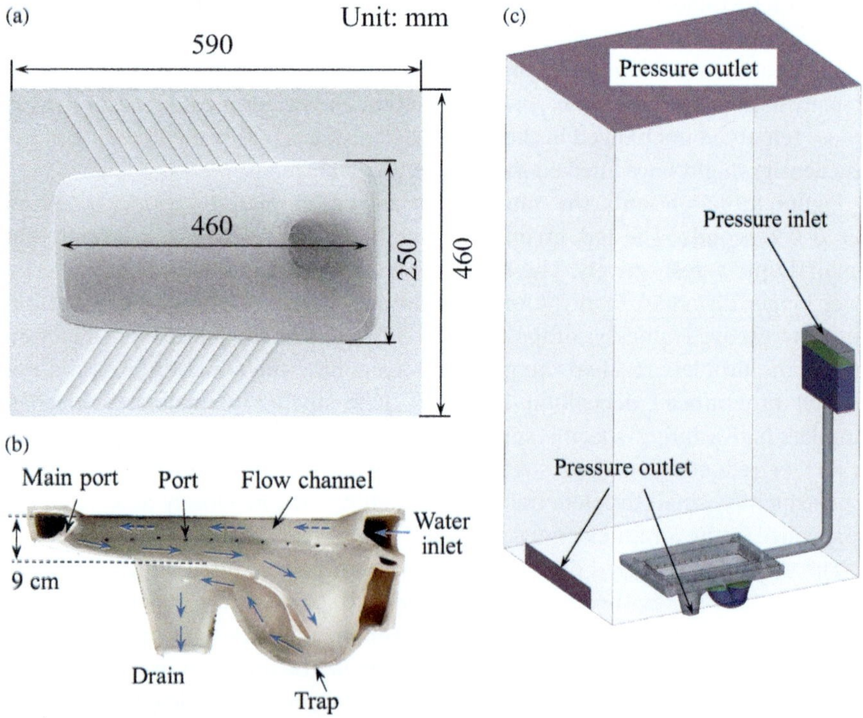

Figure 8.20 Schematic diagram of a squat toilet inside a lavatory cubicle for CFD modeling: (a) planar view of the squat pan; (b) sectional view of the squat pan and the path of flushing water flow, *Source:* Zhang et al. (2022)/Springer Nature/CC BY 4.0; (c) geometry of the lavatory cubicle and the boundary conditions used in modeling.

When the toilet was flushed, the water in the tank flowed into internal channels along the periphery of the toilet. The water inside the internal channels was then discharged into the pan through small ports evenly distributed around the pan. The water jets flushed the entire surface of the pan and forced the waste into the trap. Through the U-bend, all the water and waste were dumped into the drain. The water tank was 0.9 m high and was located outside the cubicle to minimize obstruction of the space inside the cubicle.

The whole geometry, including the cubicle, water tank, water duct, and squat pan, as shown in Figure 8.20, was created in CFD. The upper interface of the water tank was set as an inlet with zero pressure, while the gap at the bottom of the cubicle door, the opening connected to the drainage duct, and the upper imaginary surface of the cubicle were set as outlets at zero pressure. When tracking particles, the bowl surfaces and the internal channel surfaces were set as a wall-film boundary, while the remaining surfaces were set as an escape boundary. The volume of the flushing water was 6 l. Once the flush button was pressed, the water flowed into the duct and the squat pan under the force of gravity. The ambient temperature and relative humidity were set at 24 °C and 50%, respectively. The temperature of the water in the tank was 21.5 °C.

Polyhedral-hexahedral core meshes were created in the solution domain. The total number of grid cells was 1.193 million, with a cell size of 1 mm at the water jet ports of the squat pan and 5 mm in the internal flow channels. The maximum grid cell size in space was 60 mm. These grid meshes were verified as sufficient to reach grid independence. The numerical simulation utilized the CFD software ANSYS Fluent (version: 2021 R2) as the tool. The realizable k-ε model was chosen to resolve the turbulence. The PISO algorithm was adopted to couple the velocity and pressure. The diffusion term was discretized using the least squares cell-based scheme, while the pressure was discretized using the PRESTO! scheme (ANSYS Inc. 2021). The remaining convective terms were discretized by the second-order upwind scheme. The Geo-Reconstruct method was used for phase interface reconstruction. An adaptive time step with the initial value of 0.5 ms was employed with the global Courant number less than 2.

Upon complete evaporation, the fine droplets formed droplet nuclei. These nuclei were assumed to originate from the solid content present in the tap water used for toilet flushing. The solids included both soluble and insoluble components. According to data from the water utility company, the soluble solids were primarily sodium chloride, at a concentration of 100 mg/l, while the concentration of the insoluble solids was assumed zero for simplicity. Considering that a few droplets were generated during the toilet flushing, only the heat and mass transfer within the droplets themselves were considered, and their influence on the ambient temperature and humidity was neglected.

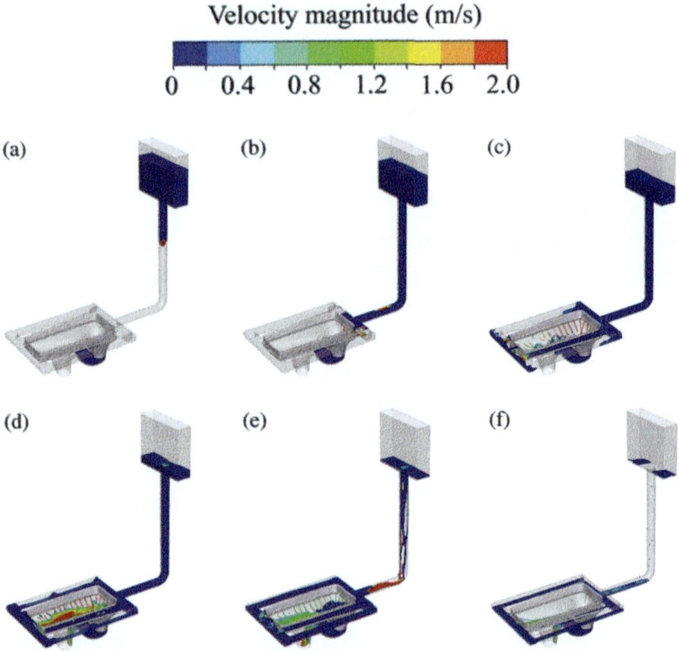

Figure 8.21 CFD-predicted water flow since the flush button of the squat toilet was pressed: (a) at $t = 0.2$ seconds; (b) at $t = 0.5$ seconds; (c) at $t = 1.0$ seconds; (d) at $t = 1.5$ seconds; (e) at $t = 2.5$ seconds; (f) at $t = 5.0$ seconds.

Figure 8.21 presents the numerically solved water flow at different times during the flushing of the toilet. Prior to flushing, there was water only in the water tank and in the trap of the squat pan. There was no water inside the water duct or the internal channels of the squat toilet. At $t = 0.2$ seconds, as shown in Figure 8.21a, the water stored in the tank began to flow down into the vertical duct under the force of gravity. At $t = 0.5$ seconds, the water entered the horizontal internal channels of the squat toilet. The water started to exit the ports around the squat pan at $t = 1.0$ seconds, and the water flow velocities reached 2.0 m/s. At $t = 1.5$ seconds, the outer flow channels on both sides were also filled with water. The water flow toward the trap had a velocity of approximately 1.5 m/s. In the central region, the velocities were even higher, reaching around 2.0 m/s. The water flow velocities started to decrease at $t = 2.5$ seconds, since the tank was nearly empty, and the flushing process ended at $t = 5$ seconds.

Figure 8.22 illustrates the airflow induced by the movement of water during the toilet flushing. It can be seen that the streamlines with high velocities emanate from the rear part of the pan. High-speed jets were directed toward the water tank. In contrast, in the front part of the bowl, the surrounding air was drawn into the

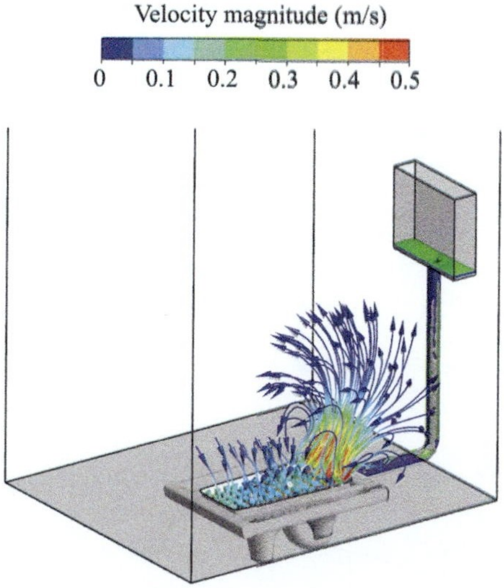

Figure 8.22 Streamlines of the induced airflow 2.0 seconds after the flush button of the squat toilet was pressed.

bowl due to the flow entrainment. As a result, the high-speed jet flows tended to move up and toward the cubicle door. The upward airflow might carry the generated liquid aerosols to a certain height inside the cubicle.

Figure 8.23 illustrates the emission of water droplets, their possible breakup into smaller droplets and evaporation into droplet nuclei, and their motion inside the cubicle. At t = 0.7 seconds, the flushing water flow impacted the internal channel walls inside the squat toilet. The impact resulted in the splashing of water and produced millimeter-sized droplets. The droplet conversion criterion was set to 5 mm, so that droplets smaller than 5 mm with a sphericity less than 2 were converted into particles for Lagrangian tracking. These tracked droplets had certain velocities and, upon colliding with other surfaces, they underwent secondary breakup to generate smaller droplets with diameters up to 0.05 mm (Motzkus et al. 2008). At t = 1.0 seconds, the fine droplets were expelled through some of the ports into the squat pan. Some broken-up droplets with small sizes gradually evaporated into droplet nuclei. At t = 2 seconds, the droplet nuclei were driven out of the squat pan by the water-induced airflow. At t = 5 seconds, the flushing had already ended, and the droplet nuclei began to disperse across a wider region. At t = 30 seconds, the droplet nuclei had dispersed to a height of 1 m above the squat pan. These droplet nuclei dispersed across a wider region with the passage of time. At t = 120 seconds, the droplet nuclei had reached a height of 1.6 m above the squat pan. Notably, the droplet nuclei mainly resided at the rear of the cubicle, far from the cubicle door. This was attributed to the driving of airflow upward toward the water tank by the flushing water. Even at t = 300 seconds, there were a large number of suspended droplet nuclei that had

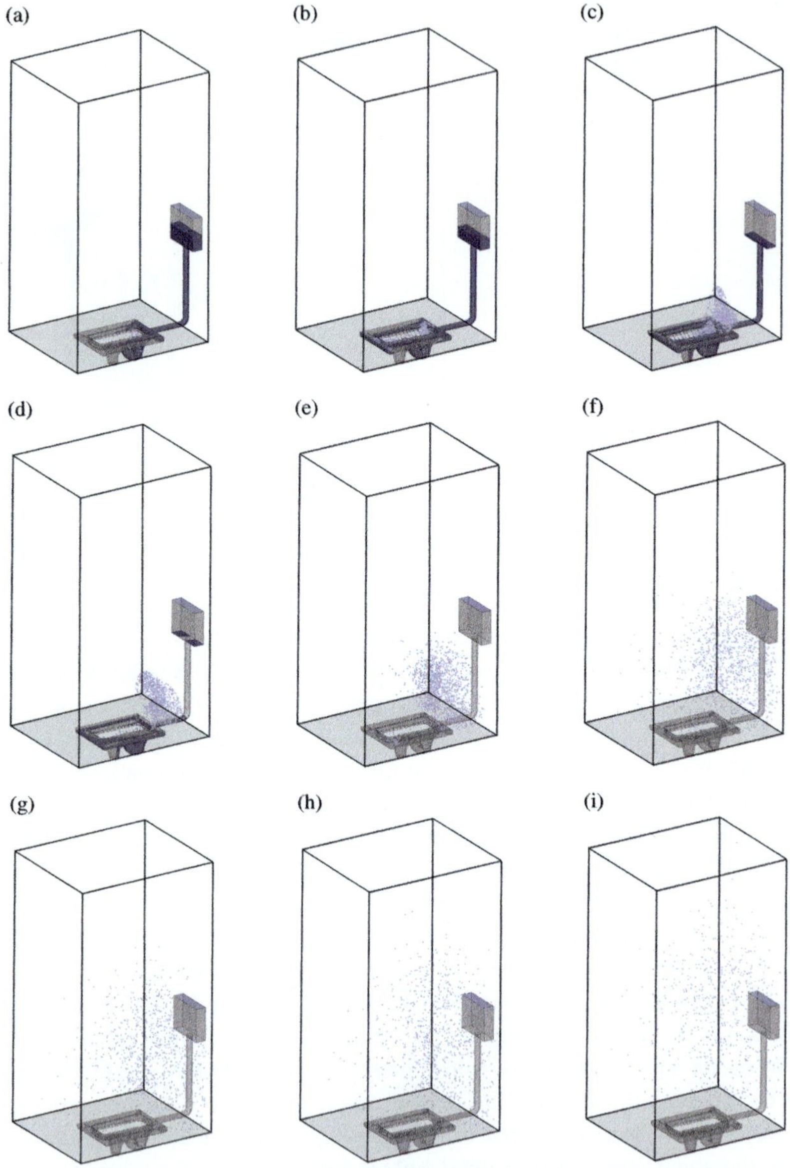

Figure 8.23 Emission of liquid droplets and their evolution in a cubicle since the flush button of the squat toilet was pressed: (a) at $t = 0.7$ seconds; (b) at $t = 1.0$ seconds; (c) at $t = 2.0$ seconds; (d) at $t = 5.0$ seconds; (e) at $t = 10.0$ seconds; (f) at $t = 30.0$ seconds; (g) at $t = 60.0$ seconds; (h) at $t = 120.0$ seconds; (i) at $t = 300.0$ seconds.

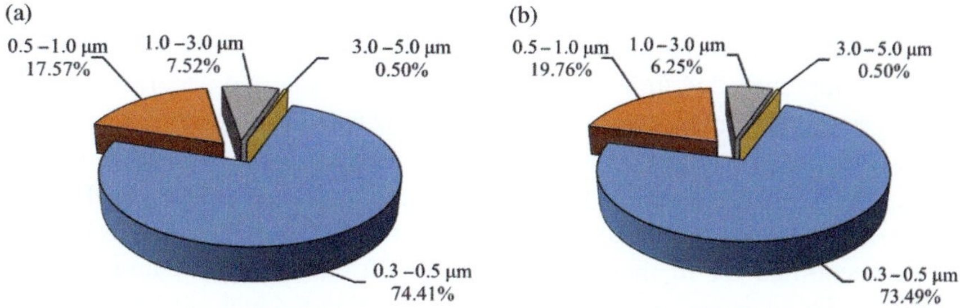

Figure 8.24 Comparison of the size distribution of droplet nuclei after the flushing of a squat toilet, as obtained by: (a) experimental measurement. *Source:* Adapted from Zhang et al. (2022); (b) CFD modeling.

not settled onto surfaces. Consequently, significant attention should be paid to the probable inhalation risk encountered by a toilet user not wearing a face mask.

Figure 8.24 compares the size distribution of the droplet nuclei provided by the measurement and by CFD modeling. The measured particle size distribution was obtained with the use of an optical particle counter, through the extraction of airborne particles from the cubicle for analysis. The background particles that were present prior to toilet flushing were excluded. It can be seen in Figure 8.24 that most of the droplet nuclei fell within the size range of 0.3–0.5 μm, with a number ratio of approximately 74%. Aerosols larger than 3 μm were quite few in number, and their ratio was less than 1%. Droplet nuclei with a particle size smaller than 3 μm constituted over 99%. The size distribution of droplet nuclei predicted by CFD was in excellent agreement with the measurement.

8.7 Summary

This chapter addresses the basic principles of indoor aerosol transport modeling, encompassing multizone and CFD modeling approaches. For a building with multiple rooms, the multizone model treats each room as a well-mixed zone, and the rooms are interconnected by airflow paths. The mass conservation of air and airborne pollutants and the relationship between flow rate and pressure difference across each airflow path are solved. The multizone model is applicable to solving both gaseous pollutants and aerosols transport over a long period in a building with many rooms. The model is very efficient, but it cannot provide spatial information on flow, gaseous pollutants, or aerosols. Moreover, it should be noted that there are many situations in which the well-mixed assumption may not be valid, especially when the room airflow is very weak and the room space is large.

In contrast, CFD modeling is sophisticated and much more accurate. CFD is able to resolve the very detailed spatial transport of gaseous pollutants and aerosols, accounting for airflow, heat and mass transfer, turbulence, and so on. The Navier–Stokes equations form the basis of these models, with a focus on the RANS approach due to its good balance between solution accuracy and the required computational cost. Modeling of the particle transport can be based on the Eulerian or Lagrangian principles. Eulerian-based particle modeling is very similar to the solution to the spatial transport of a tracer gas. Particle settling or deposition onto surfaces due to gravity or Brownian diffusion can also be accounted for in the Eulerian-based drift-flux model. Meanwhile, the Lagrangian-based model tracks the motion of each individual particle in space. The force interaction is examined for each particle based on Newton's second law.

For liquid droplets or aerosols with phase changes, multiphase particle modeling is required. Flow motion of both gas and liquid phases, as well as the phase conversion and the associated heat and mass transfer, can be modeled. The VOF method is particularly suited to tracking the interface between two immiscible fluids. Coupled Euler–Lagrange modeling can handle droplets of different sizes, where the Eulerian approach examines the motion of pure fluids and large droplets, while the Lagrangian approach tracks small droplets. Once droplets meet the specified size criterion, they are converted into particles for Lagrangian-based tracking. The evaporation of droplets from the liquid phase into the gas phase, or, vice versa, their condensation, involves intricate heat and mass transfer processes. Consequently, the heat and mass convection, the heat conduction or mass diffusion, the latent heat transfer, and possibly the thermal radiation must be accounted for in an examination of the energy balance.

In the cases presented in this chapter, the CFD models exhibited relatively good accuracy and powerful capabilities in resolving the emission, spatial transport, and evolution of particles, including or excluding phase changes in indoor space.

Problems

8.1 What are the fundamental assumptions of the multizone model? Explain how airflow paths between zones are characterized in this model, and discuss the limitations of this approach.

8.2 The multizone model assumes well-mixed conditions in each zone. Discuss scenarios where this assumption might fail, and propose potential solutions to address these limitations.

8.3 What are the key differences among DNS, LES, and RANS turbulence modeling approaches in CFD? Why is RANS commonly used for indoor airflow simulations?

8.4 Explain the role of wall functions in RANS modeling. How do they simplify near-wall turbulence calculations, and what limitations might arise from their use?

8.5 Compare the Eulerian and Lagrangian approaches for modeling particle transport. What are the advantages and disadvantages of each method in predicting aerosol distribution?

8.6 How does the Eulerian-based drift-flux model (Eq. 8.21) account for gravitational settling and turbulent diffusion of particles? Define the term "deposition velocity" in this context.

8.7 The concentration contours provided by the Lagrangian model in Figure 8.16 are less smooth than those from the Eulerian model. What factors contribute to this discrepancy, and how might statistical averaging improve the results?

8.8 Describe the coupled Euler–Lagrange method for multiphase aerosol modeling. Why is this approach necessary for simulating droplet breakup and evaporation?

8.9 In multiphase modeling, droplet identification relies on geometric criteria such as sphericity (Eq. 8.35). Discuss potential errors in this method and propose alternative strategies for handling complex droplet shapes.

8.10 This chapter emphasizes CFD validation with experimental data. Propose a framework for standardizing validation protocols in indoor aerosol transport studies, and also discuss the uncertainties associated with experimental measurements.

8.11 Design a case study involving a building with multiple rooms (e.g., a dormitory, office building, or laboratory). Specify the weather and ventilation conditions, and use the multizone modeling approach to solve and analyze the airflow and particle transport within the building.

8.12 Design a case study of an indoor environment (e.g., an office room, laboratory, or train carriage). Clearly specify the boundary conditions for ventilation and particle transport, and use a CFD software package to examine the airflow, temperature, and particle concentration distribution in this environment.

References

Aftosmis, M., Gaitone, D., and Tavares, T. (1995). Behavior of linear reconstruction techniques on unstructured meshes. *AIAA Journal* 33 (11): 2038–2049.

ANSYS Inc. (2006). *ANSYS Fluent User's Guide Release 2006*. Canonsburg, PA: ANSYS Inc.

ANSYS Inc. (2021). *ANSYS Fluent Theory Guide Release 2021 R2*. Canonsburg, PA: ANSYS Inc.

ASHRAE (American Society of Heating, Refrigerating and Air-conditioning Engineers) (2025). *ASHRAE Handbook – Fundamentals*. Atlanta: GA.

Awbi, H.B. (2003). *Ventilation of Buildings*. London: Routledge.

Barth, T. and Jespersen, D. (1989). The design and application of upwind schemes on unstructured meshes. In: *27th Aerospace Sciences Meeting 0366*, 1–12. Reno, Nevada, USA: *American* Institute of Aeronautics and Astronautics (AIAA).

Chen, F., Yu, S.C.M., and Lai, A.C.K. (2006). Modeling particle distribution and deposition in indoor environments with a new drift-flux model. *Atmospheric Environment* 40 (2): 357–367.

Chong, K.L., Ng, C.S., Hori, N. et al. (2021). Extended lifetime of respiratory droplets in a turbulent vapor puff and its implications on airborne disease transmission. *Physical Review Letters* 126 (3): 034502.

Crowe, C.T., Sharma, M.P., and Stock, D.E. (1977). The particle-source-in cell (PSI-CELL) model for gas-droplet flows. *Journal of Fluids Engineering* 99 (2): 325–332.

Gao, N.P. and Niu, J.L. (2007). Modeling particle dispersion and deposition in indoor environments. *Atmospheric Environment* 41 (18): 3862–3876.

Gupta, J.K., Lin, C.H., and Chen, Q. (2009). Flow dynamics and characterization of a cough. *Indoor Air* 19 (6): 517–525.

Heinrich, M. and Schwarze, R. (2020). 3D-coupling of volume-of-fluid and Lagrangian particle tracking for spray atomization simulation in OpenFOAM. *SoftwareX* 11: 100483.

Hinze, J.O. and Uberoi, M.S. (1975). Turbulence. *Journal of Applied Mechanics* 27 (3): 256–275.

Lai, A.C.K. and Cheng, Y.C. (2007). Study of expiratory droplet dispersion and transport using a new Eulerian modeling approach. *Atmospheric Environment* 41 (35): 7473–7484.

Lai, A.C.K. and Nazaroff, W.W. (2000). Modeling indoor particle deposition from turbulent flow onto smooth surfaces. *Journal of Aerosol Science* 31 (4): 463–476.

Launder, B. and Spalding, D. (1974). The numerical computation of turbulent flow. *Computer Methods in Applied Mechanics and Engineering* 3: 269–289.

Li, A. and Ahmadi, G. (1992). Dispersion and deposition of spherical particles from point sources in a turbulent channel flow. *Aerosol Science and Technology* 16: 209–226.

Li, M., Chong, K.L., Ng, C.S. et al. (2022). Towards realistic simulations of human cough, effect of droplet emission duration and spread angle. *International Journal of Multiphase Flow* 147: 103883.

Liu, S., Song, R., and Zhang, T. (2021). Residential building ventilation in situations with outdoor PM2.5 pollution. *Building and Environment* 202: 108040.

Mathur, S.R. and Murthy, J.Y. (1997). A pressure-based method for unstructured meshes. *Numerical Heat Transfer* 31 (2): 195–215.

Miller, R.S., Harstad, K., and Bellan, J. (1998). Evaluation of equilibrium and non-equilibrium evaporation models for many droplet gas-liquid flow simulations. *International Journal of Multiphase Flow* 24 (6): 1025–1055.

Morsi, S.A. and Alexander, A.J. (1972). An investigation of particle trajectories in two-phase flow systems. *Journal of Fluid Mechanics* 55 (2): 193–208.

Motzkus, C., Géhin, E., and Gensdarmes, F. (2008). Study of airborne particles produced by normal impact of millimetric droplets onto a liquid film. *Experiments in Fluids* 45 (5): 797–812.

O'Rourke, P.J. and Amsden, A.A. (2000). A spray/wall interaction submodel for the KIVA-3 wall film model. *SAE Transactions* 1: 281–298.

Patankar, S. (1980). *Numerical Heat Transfer and Fluid Flow*. New York: Hemisphere Publishing Corporation.

Riley, W.J., McKone, T.E., Lai, A.C.K., and Nazaroff, W.W. (2002). Indoor particulate matter of outdoor origin, importance of size-dependent removal mechanisms. *Environmental Science & Technology* 36 (2): 200–207.

Sazhin, S.S. (2006). Advanced models of fuel droplet heating and evaporation. *Progress in Energy and Combustion Science* 32: 162–214.

Venkatakrishnan, V. (1993). On the accuracy of limiters and convergence to steady state solutions. In: *31st Aerospace Sciences Meeting 0880*, 1–10. Reno, Nevada, USA: American Institute of Aeronautics and Astronautics (AIAA).

Versteeg, H. and Malalasekera, W. (2007). *An Introduction to Computational Fluid Dynamics*, 2ndе. Essex: Pearson Prentice Hall.

Wilcox, D.C. (1993). Comparison of two-equation turbulence models for boundary layers with pressure gradient. *AIAA Journal* 31 (8): 1414–1421.

Xie, X., Li, Y., Sun, H., and Liu, L. (2009). Exhaled droplets due to talking and coughing. *Journal of the Royal Society Interface* 6 (Suppl_6): S703–S714.

Yakhot, V., Orzag, S., Thangam, S., and Gatski, T. (1992). Development of turbulence models for shear flows by a double expansion technique. *Physics of Fluids A Fluid Dynamics* 4 (7): 1510–1520.

Zhang, Z. and Chen, Q. (2006). Experimental measurements and numerical simulations of particle transport and distribution in ventilated rooms. *Atmospheric Environment* 40 (18): 3396–3408.

Zhang, Z. and Chen, Q. (2009). Prediction of particle deposition onto indoor surfaces by CFD with a modified Lagrangian method. *Atmospheric Environment* 43 (2): 319–328.

Zhang, Z., Zhang, W., Zhai, Z., and Chen, Q.Y. (2007). Evaluation of various turbulence models in predicting airflow and turbulence in enclosed environments by CFD, part 2 – comparison with experimental data from literature. *HVAC&R Research* 13 (6): 871–886.

Zhang, T., Lee, K., and Chen, Q. (2009). A simplified approach to describe complex diffusers in displacement ventilation for CFD simulations. *Indoor Air* 19 (3): 255–267.

Zhang, T., Zhou, H., and Wang, S. (2013). An adjustment to the standard temperature wall function for CFD modeling of indoor convective heat transfer. *Building and Environment* 68: 159–169.

Zhang, T., Su, Z., Wang, J., and Wang, S. (2018). Ventilation, indoor particle filtration, and energy consumption of an apartment in northern China. *Building and Environment* 143: 280–292.

Zhang, T., Yao, L., Gao, Z., and Wang, F. (2022). Particle exposure risk to a lavatory user after flushing a squat toilet. *Scientific Reports* 12 (1): 21088.

Zhao, Y., Yoshino, H., and Okuama, H. (1998). Evaluation of the COMIS model by comparing simulation and measurement of airflow and pollutant concentration. *Indoor Air* 8 (2): 123–130.

Zuzio, D., Estivalezes, J.-L., and DiPierro, B. (2017). An improved multiscale Eulerian-Lagrangian method for simulation of atomization process. *Computers & Fluids* 176: 285–301.

9

Indoor Aerosol Exposure and Health Effects

9.1 Introduction

Excessive exposure to aerosols can lead to numerous health effects, such as asthma, rhinitis, cancer, and cardiovascular diseases. Airborne aerosols of certain sizes can be inhaled into the respiratory tract and lungs and deposit inside. The resulting adverse impacts are mainly due to the oxidative stress, inflammation, or damage to the immune system induced by the deposited aerosols. This chapter addresses the mechanisms of aerosol deposition in the respiratory system, the mechanisms of the toxicity of aerosol components, and the associated health outcomes for both nonbiological aerosols and noninfectious bioaerosols.

9.2 Mechanisms of Respiratory Exposure

The respiratory system can be divided into three regions: the extrathoracic region (also known as the head airways or the nasopharyngeal region), the tracheobronchial region (also referred to as the lung airways), and the alveolar region (also known as the pulmonary region), as shown in Figure 9.1a. These regions differ greatly in terms of geometric structure, airflow patterns, function, and the deposition rate of inhaled particles. The extrathoracic region includes the nose, mouth, pharynx, and larynx. The inhaled air is warmed and humidified in the extrathoracic region, and the air velocities are relatively high. The tracheobronchial region includes the airways from the trachea to the terminal bronchioles and looks like an inverted tree, as shown in Figure 9.1b from generation 0 to generation 14. The air velocities progressively drop as the air approaches the terminal bronchioles. The alveolar region is the place where gas exchange occurs, such as the exchange of oxygen and carbon dioxide. This region includes generations 15–23, as shown in Figure 9.1b. There are an average

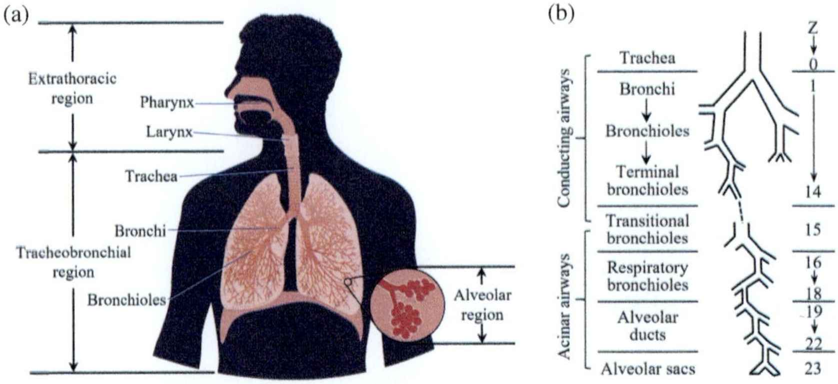

Figure 9.1 The respiratory system: (a) overview. *Source:* Stadnytskyi et al. (2021) / with permission of John Wiley & Sons; (b) schematics of the conducting airways and acinar airways.

of approximately 480 million alveoli, each of which has an average diameter of 200 μm (Ochs et al. 2004), and they constitute a total surface area of 50–100 m^2 for gas exchange.

A typical adult inhales or exhales 0.5 l of tidal air in each breath during rest, at a rate of approximately 12 breaths per minute. Both the tidal volume and the breathing rate during heavy work may be triple those in the resting state. During normal breathing, approximately 2.4 l of reserve air is not exhaled, and this amount is known as the thoracic volume. Under forced exhalation, around 1.2 l of reserve air is expelled. Because the flow during inhalation and exhalation is time-varying, the final portion of inhaled fresh air remains inside the conducting airways until it is expired; this portion of air does not reach the alveoli. Similarly, the final portion of air exhaled from the alveoli remains in the conducting airways; this portion will be drawn into the alveoli as the next inspiration begins. The volume of the conducting airways for a typical adult is approximately 0.15 l and is known as the anatomic dead space. The conducting airways constitute a portion of the lung that transports air but does not exchange gases.

The deposition of particles in the lung follows mechanisms that are similar to those of particle collection by filters. However, the relative contributions of the mechanisms may be very different. The respiratory flow is transient; the airway geometry, flow direction, and velocities change with time and the breathing cycle. The main mechanisms for particle deposition inside the respiratory system are impaction, settling, and diffusion. The particle deposition rate and location are dependent on the density, size, and shape of the particles, the geometry of the airway, and the breathing pattern. Particle deposition can differ greatly among healthy adults, children, and people with respiratory diseases. The breathing frequency, the volume of inhaled air, and the length of the pause between inhalation and exhalation can all affect the deposition of inhaled particles.

The air inhaled through the nose/mouth into the alveolar region undergoes complicated changes in flow direction. Due to inertia, some of the large particles near the airway surfaces may deposit there when the flow direction changes. Particle deposition by impaction is concentrated at or near the first carina and the tracheal bifurcation because of the sharp bend in the airway. In the smaller airways, for example, in the alveolar region, settling plays the most critical role in particle deposition. The flow velocities are quite low in the alveolar region, and the airway dimensions are very small. Brownian diffusion can also cause fine particles to deposit on airway surfaces, especially in the smaller airways, where the particles can stay for a relatively long duration. In smaller airways where particles have a long residence time, both settling and Brownian diffusion may become major mechanisms of particle deposition.

Particle deposition in any respiratory region is strongly dependent on the deposition in the preceding regions and the local deposition efficiency. Figure 9.2 presents the predicted total and regional deposition ratios for particles with

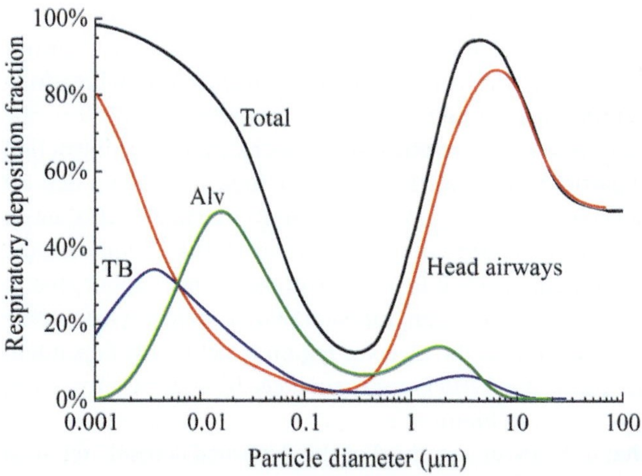

Figure 9.2 Predicted average total and regional deposition ratio of aerosols in the respiratory system for an adult engaged in light exercise with nasal breathing, based on the ICRP deposition model, where Alv represents the alveolar region, and TB represents the tracheobronchial region. Only the average data for males and females are displayed. *Source:* Hinds and Zhu (2022) / with permission of John Wiley & Sons.

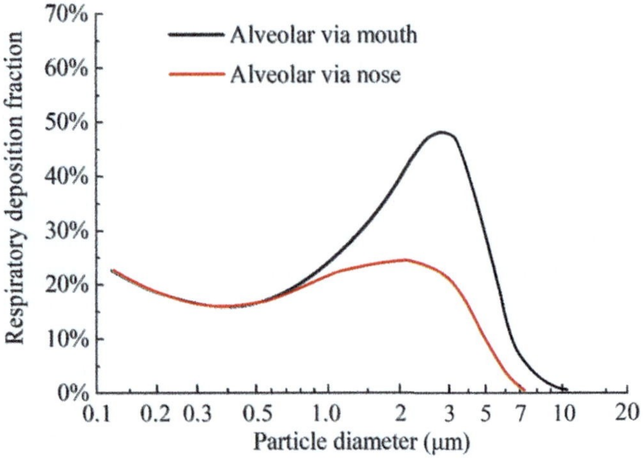

Figure 9.3 Deposition ratio of inhaled aerosols in the alveolar region. *Source:* Hinds and Zhu (2022) / with permission of John Wiley & Sons.

different aerodynamic sizes. Coarse particles generally deposit in the head airways. The nasal hairs and airway bends remove the majority of coarse particles by impaction. The particles deposited on the ciliated cavity surfaces of the nasal passage are cleared to the pharynx and are finally swallowed into the digestive system. The particle deposition in the head airways is less during mouth breathing than during nasal breathing. Meanwhile, relatively few particles deposit in the tracheobronchial region, where the contributing mechanisms are impaction and settling. Ultrafine particles, such as those smaller than 10 nm, can deposit significantly in the tracheobronchial region as a result of Brownian diffusion. In general, it is difficult for particles larger than 10 μm to reach the alveolar region, and only a few particles ranging from 2 to 10 μm can reach this region. Due to low air speeds and small airways, Brownian diffusion plays a more important role than flow. The particle deposition in the alveolar region is not highly dependent on the particle size. The deposition ratio is approximately 10–20% for particles in the range of 0.1–1 μm, as shown in Figure 9.3.

It should be noted that only about one-third of the deposited aerosols retained in the lungs may have adverse health impacts. This is because approximately one-third of the deposited aerosols are removed by mucociliary clearance, and another third are depleted by the alveolar macrophages, as shown in Figure 9.4. The ultimately retained particles are known as biopersistent particles. Generally, coarse and fine particles affect only the local airway deposition sites, whereas ultrafine particles can penetrate the alveolar epithelial barrier into the blood circulation and then act on the cardiovascular system.

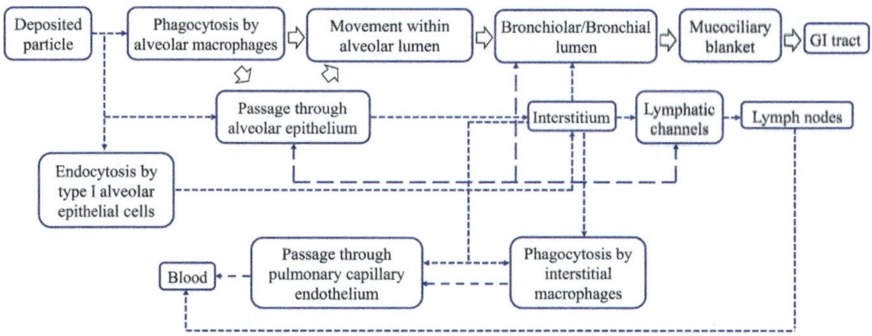

Figure 9.4 Transport pathways of particles deposited in the respiratory tract. *Source:* Ruzer and Harley (2012) / with permission of Taylor & Francis.

9.3 Mechanisms of Toxicity of Aerosol Components

The adverse health effects of the inhaled particulate matter (PM) have been identified mainly through studies in epidemiology and toxicology, along with some clinical studies. According to the available epidemiological studies, PM exposure can be correlated with a number of health effects. However, the correlations claimed in these studies should not be misunderstood as cause-and-effect relationships. In contrast, clinical studies can provide clear, firm cause-and-effect analyses between PM exposure and a number of diseases. Unfortunately, for ethical reasons, the test subjects can only be exposed to very low concentrations of PM, and the number of test subjects is very limited. Hence, evidence from clinical studies is still scarce.

To evaluate the adverse health impacts of high-dose, long-term PM exposure, animal models are commonly used. Numerous studies have addressed animal-based PM exposure, using rodents such as mice and rats. The particle surface area per lung weight is the most relevant dose metric in toxicological studies, and this metric can be projected to assess the adverse health effects for humans. An exception is made for exposure to long (>5 μm) fiber-like particles, where particle number has been found to be a more relevant dose metric (Shvedova et al. 2014). In practice, however, only the particle mass and number concentrations are currently monitored. Therefore, the gap between the particle surface area per lung weight and the monitored particle mass or number concentrations may require filling.

The deposited particles can trigger inflammatory and/or immunological responses, either directly or indirectly. In response to the deposition/contact/ingestion of particles, the neutrophils, macrophages, and eosinophils produce reactive oxygen and nitrogen species (Ward et al. 1988; Poli 2002). Typical reactive oxygen metabolites (ROMs) include superoxide radical ($O_2^{\bullet-}$), hydrogen peroxide (H_2O_2), hydroxyl radical (•OH), and singlet O_2 and hypochlorous acid.

The ROMs induce inflammation (Chang and Crapo 2002), lipid peroxidation, and the breakage of DNA strands (McCord 1993). Oxidant stress is claimed to be the primary pathway for negative health effects. *In vitro* studies (Quay et al. 1998) have indicated that transition metals, endotoxins, and organic carbon in the PM can trigger inflammation. The aforementioned agents on the pulmonary macrophages (Soukup and Becker 2001; Monn et al. 2003), lung epithelial cells (Fujii et al. 2001), and peripheral blood monocytes (Monn and Becker 1999) activate molecular signaling pathways for inflammation.

Meanwhile, the crustal dusts and residual oil fly ash from combustion may contain transition metals (the first-row transition metals include titanium, vanadium, chromium, manganese, iron, cobalt, nickel, and copper) in valence states. These metals can cycle electrons, leading to the production of the hydroxy radical (•OH) following the Haber–Weiss reaction, as shown in Table 9.1 (Ward et al. 1988). The hydroxy radical is highly reactive and causes injury to the respiratory system. The metals in dusts from many sources other than the residual oil fly ash are also able to generate ROMs (Ghio et al. 1996; Pritchard et al. 1996; Frampton et al. 1999).

Some organic components of PM can also produce ROMs. For example, the polycyclic aromatic hydrocarbons (PAHs) and PAH-derived quinones in diesel exhaust particles have been found to produce oxidant stress *in vitro* (Li et al. 2000). Further details of the basic mechanisms can be found in the literature (Li et al. 2002).

Endotoxins can also lead to inflammation, similarly to the transition metals and organic components (Becker et al. 1996; Monn et al. 2003; Soukup and Becker 2001). Endotoxins are lipopolysaccharides (LPS) in the walls of the outer cells of Gram-negative bacteria. These endotoxins can activate many signaling pathways, as do the organic components, by means of interactions with specific lymphocyte receptors (Monick and Hunninghake 2002). This gives rise to the inflammatory potential of endotoxins.

Meanwhile, aeroallergens from plants, animals, and insects may stimulate Immunoglobulin E- (IgE-) mediated pathways of the immune responses, as

Table 9.1 Haber–Weiss reaction for the production of hydroxyl radicals.

Generation of hydroxyl radicals catalyzed by ferrous iron[a]	
$Fe^{3+} + O_2^{\bullet -} \rightarrow Fe^{2+} + O_2$	(9.1)
$Fe^{2+} + H_2O_2 \rightarrow Fe^{3+} + OH^- + \bullet OH$	(9.2)
$O_2^{\bullet -} + H_2O_2 \rightarrow O_2 + OH^- + \bullet OH$	(9.3)

a) Note, all first-row transition metals may participate in similar chemical reactions.
Source: Ward et al. (1988) / with permission of Elsevier.

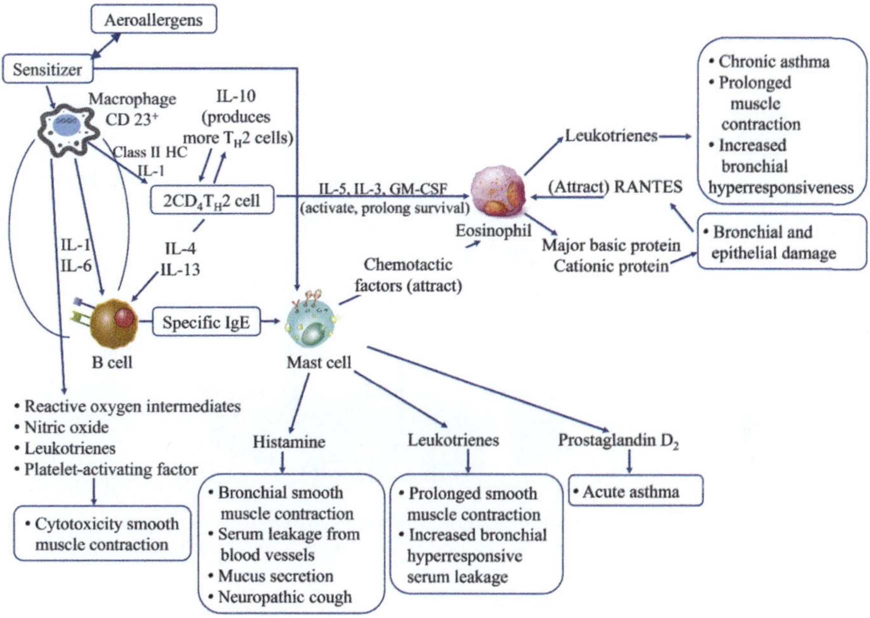

Figure 9.5 Major immunologic pathways related to IgE-mediated immune responses. *Source:* Ruzer and Harley (2012) / with permission of Taylor & Francis.

shown in Figure 9.5. Like the organic components of PM, aeroallergens enhance response mechanisms through these pathways and restrain negative feedback controls by specific cytokines.

9.4 Health Effects of Nonbiological Aerosol Exposure

PM is thought to be the most lethal airborne pollutant in the atmospheric environment, accounting for approximately 51% of premature deaths across the globe (Landrigan et al. 2018). Aerosols that are permanently deposited in the lungs may have short-term, acute effects on health as well as long-term, chronic health impacts, as shown in Figure 9.6. These impacts can occur not only in the lungs but also in the cardiovascular system, brain, or reproductive system. Lung cancer, rhinitis/pneumonia, decreased lung function, and reduced lung growth are some of the typical pulmonary diseases. Meanwhile, the cardiovascular diseases induced by aerosol exposure include myocardial infarction, elevated blood pressure, atherosclerosis, and heart stroke. Some of the major health effects are outlined below.

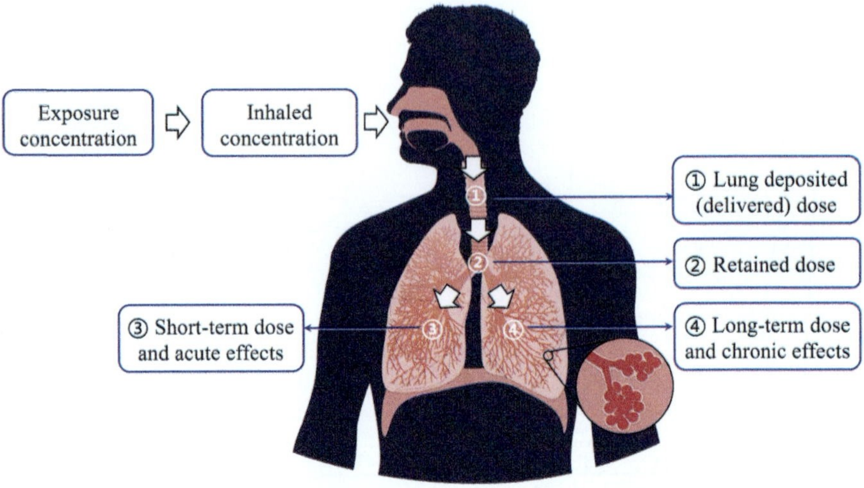

Figure 9.6 Schematic of aerosol inhalation and deposition in the respiratory system, and the associated health effects. *Source:* Stadnytskyi et al. (2021) / with permission of John Wiley & Sons.

9.4.1 Birth Outcomes

Exposure to ambient aerosols ($PM_{2.5}$ and PM_{10}) can significantly affect birth and neonatal outcomes. Relevant gaseous species that also contribute to birth outcomes include SO_2, NO_x, CO, O_3, and PAHs. Adverse outcomes encompass neonatal mortality, intrauterine growth retardation, low birth weight, small size for gestational age, preterm delivery, and stillbirth (Ruzer and Harley 2012).

9.4.2 Mortality

Based on cohort studies (Ruzer and Harley 2012), there is a strong association between the risk of death and the mean concentrations of $PM_{2.5}$ and PM_{10}. It has been speculated that a 10 μg/m^3 rise in average $PM_{2.5}$ may increase mortality from all causes by 5% (95% CI: 2–11%), from cardiopulmonary disease by 9% (3–16%), and from lung cancer by 14% (4–23%) (Pope III et al. 2002). Figure 9.7 presents the mortality due to exposure to household air pollution in 2016; the highest mortality was found in the least developed countries.

9.4.3 Atopic Allergy and Asthma

The reported rates of atopic allergy and asthma have steadily increased in recent decades, and the affected population's age continues to decrease (CDC 2000).

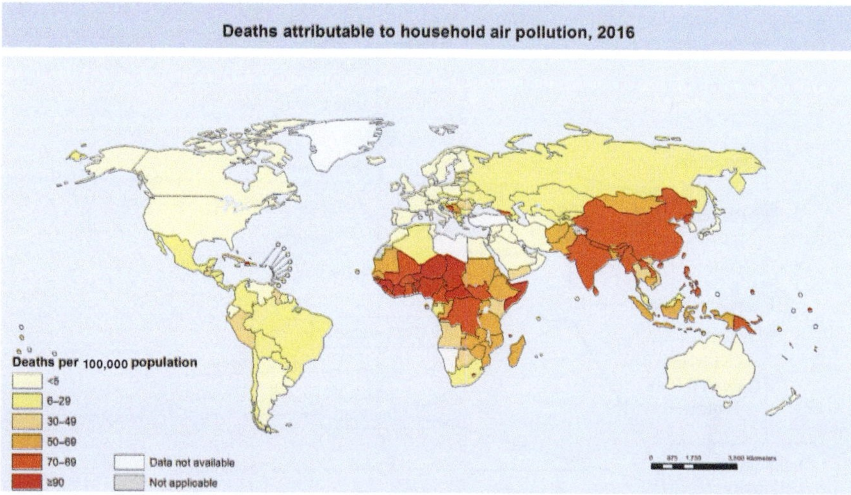

Figure 9.7 Mortality attributed to household air pollution. *Source:* World Health Organization (2018) / with permission of World Health Organization.

Although it has been claimed that asthma symptoms worsen with increased atmospheric aerosol concentrations, the contribution of the aerosol and/or other ambient or indoor pollutants to the asthma etiology should be quite small. In addition, atmospheric aerosols may contain biological allergens such as fungal spores, filaments, or their fragments.

9.4.4 Cancer

Atmospheric aerosol concentrations are strongly correlated with lung cancer risks. The exact aerosol components that lead to lung cancer are not well known. The US EPA has published a model of the possible mechanisms leading from exposure to diesel exhaust aerosols to the induction of carcinogenesis, as shown in Figure 9.8.

9.4.5 Cardiovascular Problems

The inflammation induced by aerosol exposure can worsen existing cardiovascular diseases. Additionally, excessive aerosol exposure can affect heart rate variability, increase cardiac arrhythmias, induce a procoagulant state, alter the endothelial cell function, and so on. All these may increase mortality in older adults (Ruzer and Harley 2012).

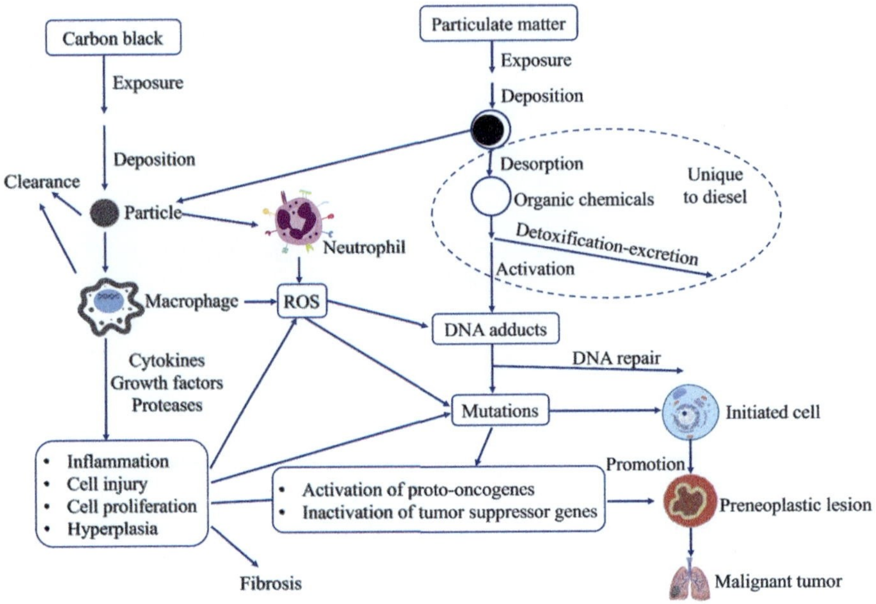

Figure 9.8 Speculated mechanisms of carcinogenesis in rats exposed to high-concentration diesel exhaust gases. *Source:* Ruzer and Harley (2012) / with permission of Taylor & Francis.

9.5 Health Effects of Noninfectious Bioaerosol Exposure

Respiratory airway inflammation can also be triggered by toxins, pro-inflammatory agents, or allergens. Exposure to specific bioaerosols can cause various symptoms or conditions. However, it should be noted that some nonbiological agents, such as cigarette smoke, chemical substances, or atmospheric pollutants, can result in similar symptoms. The following summarizes some of the respiratory diseases/symptoms arising from specific bioaerosol exposure.

9.5.1 Asthma

The symptoms of asthma include wheezing, breathlessness, cough, and chest tightness. Exposure to bacterial endotoxins (Douwes et al. 2002), dust mites, cockroaches, pets (IOM 2000), and fungi (IOM 2004; Heseltine and Rosen 2009), is reportedly associated with asthma in the general population. Viral infections may exacerbate asthma (Johnston et al. 1995), possibly due to the impairment of the innate immune system after infection. Mold exposure is estimated to account for an approximately 30–50% increase in respiratory and asthma-related health outcomes (Fisk et al. 2006).

9.5.2 Rhinitis

The typical symptoms of rhinitis are nasal congestion (runny or stuffy nose), rhinorrhea, sneezing, itchy eyes, conjunctivitis, and so on. Rhinitis is essentially an allergic or nonallergic inflammation of the mucosa in the upper respiratory tract. The percentages of allergic and nonallergic rhinitis are nearly equal. Allergic rhinitis can be caused by pollen or fungi. The types of pollen that may induce hay fever include ragweed, Russian thistle, English plantain, timothy, rye, and other grasses, as well as birch, maple, oak, and willow trees (IOM 2000).

9.5.3 Chronic Airflow Obstruction

Chronic airflow obstruction may lead to greater variation in or a reduction in the peak expiratory flow. Exposure to endotoxins or bioaerosol exposure resulting in elevated $(1 \rightarrow 3)$-β-D-glucan was found to be related to airflow obstruction (Douwes et al. 2000). People in damp, moldy environments have a relatively high chance of suffering from chronic airflow obstruction.

9.5.4 Hypersensitivity Pneumonitis

Hypersensitivity pneumonitis (HP) is a restrictive/obstructive pulmonary condition with delayed, febrile, systemic symptoms. Acute HP involves symptoms of fever, chills, dyspnea, chest tightness, cough, etc. Sufferers from subacute HP may have dyspnea and cough for several weeks or months. Chronic HP is commonly afebrile but has symptoms of dyspnea, cough, malaise, weakness, and weight loss. Exposure to thermophilic microorganisms, such as *Saccharopolyspora rectivirgula, Saccharomonospora viridis, Thermoactinomyces sacchari, Thermoactinomyces vulgaris, Mycobacterium immunogenum*, and fungi, may induce HP (Ruzer and Harley 2012).

9.5.5 Chronic Bronchitis

Chronic bronchitis may involve airway obstruction, but not necessarily. Some chronic bronchitis sufferers have symptoms like dyspnea, chest tightness, and wheezing. Both smoking and bioaerosol exposure can lead to chronic bronchitis. Farmers, cotton processing workers, and compost workers are reportedly at higher risk of developing chronic bronchitis. Chronic bronchitis is characterized by hyperplasia of mucus glands and an increase in mucus production.

9.5.6 Pulmonary Hemorrhage

Pulmonary hemorrhage can involve diffuse bleeding or hemorrhage of the alveoli. If the cause is unknown, it is called idiopathic pulmonary hemorrhage (IPH). Exposure to *Stachybotrys chartarum* (*S. chartarum*) may lead to acute IPH (CDC 1994, 2000; Etzel et al. 1998), since a hemolytic protein (stachylysin) has been found in *S. chartarum*. However, some controversy exists around the association between exposure to *S. chartarum* and acute IPH (Heseltine and Rosen 2009; IOM 2004).

9.5.7 Organic Dust Toxic Syndrome (ODTS)

ODTS is also known as "toxic alveolitis" or "inhalation fever." The symptoms of ODTS are similar to those of influenza but commonly last for only one day. The typical symptoms include dry cough, chest tightness, fever, shivering, headache, muscle and joint pain, dyspnea, fatigue, nausea, and general malaise (Douwes et al. 2003). ODTS is a nonallergic, acute, febrile illness. It is commonly experienced by farmers and textile fiber processing workers, and by people in internal spaces in which the humidifier and/or air-conditioning systems are severely contaminated by microorganisms. ODTS is triggered by exposure to microorganisms and bacterial endotoxins.

9.5.8 Cancer

Exposure to zoonotic viral or bacterial agents, mycotoxins, or pesticides has been postulated to lead to cancer in the lungs, brain, lips, stomach, prostate, connective tissues, and hematological system (Douwes et al. 2003). Livestock workers, farmers, abattoir workers, and butchers are at increased risk for these types of cancer.

9.5.9 Building-related Illnesses and Symptoms

Building-related illnesses (BRIs) are diagnosable diseases with well-known etiologies. BRIs include but are not limited to acute viral infections, tuberculosis, and legionellosis; carbon monoxide poisoning; irritant-induced or exacerbated asthma or rhinitis; and HP. In contrast, building-related symptoms (BRSs) or sick building syndrome (SBS) is a combination of nonspecific symptoms in the absence of diagnosable diseases, and is observed among people who spend long periods inside buildings. Typical symptoms of SBS include but are not limited to, headache, sore throat, hoarseness, dry cough, chest discomfort, fatigue, skin and eye irritation, and nasal itching and dryness. Occasional vomiting, nausea, concentration difficulty, joint pain, low-grade fever, frequent respiratory infections, and olfactory failure or displacement have also been reported (Ruzer and Harley 2012). SBS may arise from the inflammatory effects of exposure to bacteria, fungi, and certain organic agents. Excessive exposure to ozone and some chemical contaminants may also lead to SBS.

9.6 Summary

Exposure to PM reportedly accounts for about 51% of premature deaths across the globe. Nearly one-third of inhaled aerosols remain in the lungs and may adversely impact people's health, while the other two-thirds of the inhaled aerosols are digested or destroyed by macrophages. In general, particles larger than 10 μm are unlikely to reach the alveolar region, and only a few particles ranging from 2 to 10 μm reach this region. Meanwhile, ultrafine particles, such as those smaller than 10 nm, can deposit significantly in the tracheobronchial region by means of Brownian diffusion. The deposited particles can trigger inflammatory and/or immunological responses by producing reactive oxygen and nitrogen species. Negative health effects may be observed not only in the lungs but also in the cardiovascular system, the brain, and the reproductive system. Lung cancer, rhinitis/pneumonia, decreased lung function, and reduced lung growth are pulmonary diseases typically associated with nonbiological aerosol exposure. Respiratory airway inflammation and symptoms can also be triggered by toxins, pro-inflammatory agents, or allergens after exposure to bioaerosols. Exposure to specific bioaerosols can cause various symptoms or conditions.

Problems

9.1 What are the three main regions of the respiratory system? Please briefly describe the role of each region in aerosol deposition and the dominant deposition mechanism.

9.2 How does aerosol particle size influence the deposition ratio in the alveolar region? Please explain the differences in deposition patterns between ultrafine particles (<10 nm) and coarse particles (>10 μm) in the respiratory tract. You may refer to Figures 9.2 and 9.3.

9.3 How do transition metals and organic compounds in aerosols induce inflammatory responses via oxidative stress?

9.4 Please list at least three cardiovascular diseases associated with nonbiological aerosol exposure and briefly describe their pathological mechanisms.

9.5 Please compare the molecular mechanisms by which nonbiological aerosols (e.g., $PM_{2.5}$) and noninfectious bioaerosols (e.g., endotoxins) trigger inflammation. Which signaling pathways would commonly be activated?

9.6 What are the typical symptoms and triggering factors of organic dust toxic syndrome (ODTS)? Which occupational groups are at higher risk of ODTS?

9.7 According to Figure 9.7, mortality from household air pollution is significantly higher in developing countries than in developed countries. Discuss the ways in which socioeconomic and technological interventions could reduce this health disparity.

References

Becker, S., Soukup, J.M., Gilmour, M.I. et al. (1996). Stimulation of human and rat alveolar macrophages by urban air particulates: effects of oxidant radical generation and cytokine production. *Toxicology and Applied Pharmacology* 141: 637–648.

Centers for Disease Control and Prevention (CDC) (2000). Update: pulmonary hemorrhage/hemosiderosis among infants – Cleveland, Ohio, 1993–1996. *Morbidity and Mortality Weekly Report* 49 (9): 180–184.

Centers for Disease Control and Prevention (CDC) (1994). Acute idiopathic pulmonary hemorrhage among infants. *Morbidity and Mortality Weekly Report* 53: 1.

Chang, L.Y. and Crapo, J.D. (2002). Inhibition of airway inflammation and hyperreactivity by an antioxidant mimetic. *Free Radical Biology and Medicine* 33: 379–386.

Douwes, J., Thorne, P., Pearce, N. et al. (2003). Bioaerosol health effects and exposure assessment: progress and prospects. *Annals of Occupational Hygiene* 47 (3): 187–200.

Douwes, J., Pearce, N., and Heederik, D. (2002). Does environmental endotoxin exposure prevent asthma? *Thorax* 57 (1): 86–90.

Douwes, J., Zuidhof, A., Doekes, G. et al. (2000). (1→3)-β-glucan and endotoxin in house dust and peak flow variability in children. *American Journal of Respiratory and Critical Care Medicine* 162: 1348–1354.

Etzel, R.A., Montana, E., Sorenson, W.G. et al. (1998). Acute pulmonary hemorrhage in infants associated with exposure to *Stachybotrys atra* and other fungi. *Archives of Pediatrics & Adolescent Medicine* 152 (8): 757–762.

Fisk, W.J., Lei-Gomez, Q., and Mendell, M.J. (2006). Meta-analyses of the associations of respiratory health effects with dampness and mold in homes. *Indoor Air* 17: LBNL-59363.

Fujii, T., Hayashi, S., Hogg, J.C. et al. (2001). Particulate matter induces cytokine expression in human bronchial epithelial cells. *American Journal of Respiratory Cell and Molecular Biology* 25 (3): 265–271.

Frampton, M.W., Ghio, A.J., Samet, J.M. et al. (1999). Effects of aqueous extracts of PM_{10} filters from the Utah valley on human airway epithelial cells. *American Journal of Physiology* 277 (5 Part 1): L960–L967.

Ghio, A.J., Stonehuerner, J., Pritchard, R.J. et al. (1996). Humic-like substances in air pollution particulates correlate with concentrations of transition metals and oxidant generation. *Inhalation Toxicology* 8: 479–494.

Heseltine, E. and Rosen, J. (2009). *WHO Guidelines for Indoor Air Quality: Dampness and Mould*. Geneva: World Health Organization.

Hinds, W.C. and Zhu, Y. (2022). *Aerosol Technology: Properties, Behavior, and Measurement of Airborne Particles*, 3e. Hoboken, NJ: Wiley.

Institute of Medicine (IOM) (2000). *Clearing the Air: Asthma and Indoor Air Exposures*. Washington, DC: National Academy Press.

Institute of Medicine (IOM) (2004). *Damp Indoor Spaces and Health*. Washington, DC: National Academy Press.

Johnston, S.L., Pattemore, P.K., Sanderson, G. et al. (1995). Community study of role of viral infections in exacerbations of asthma in 9–11 year old children. *BMJ* 310 (6989): 1225–1229.

Landrigan, P.J., Fuller, R., Acosta, N.J. et al. (2018). The lancet commission on pollution and health. *The Lancet* 391: 462–512.

Li, N., Kim, S., Wang, M. et al. (2002). Use of a stratified oxidative stress model to study the biological effects of ambient concentrated and diesel exhaust particulate matter. *Inhalation Toxicology* 14: 459–486.

Li, N., Venkatesan, M.I., Miguel, A. et al. (2000). Induction of heme oxygenase-1 expression in macrophages by diesel exhaust particle chemicals and quinones via the antioxidant-responsive element. *The Journal of Immunology* 165 (6): 3393–3401.

McCord, J.M. (1993). Human disease, free radicals, and the oxidant/antioxidant balance. *Clinical Biochemistry* 26 (5): 351–357.

Monick, M.M. and Hunninghake, G.W. (2002). Activation of second messenger pathways in alveolar macrophages by endotoxin. *European Respiratory Journal* 20 (1): 210–222.

Monn, C., Naef, R., and Koller, T. (2003). Reactions of macrophages exposed to particles $<10\,\mu m$. *Environmental Research* 91 (1): 35–44.

Monn, C. and Becker, S. (1999). Cytotoxicity and induction of proinflammatory cytokines from human monocytes exposed to fine (PM2.5) and coarse particles

(PM10–2.5) in outdoor and indoor air. *Toxicology and Applied Pharmacology* 155 (3): 245–252.

Ochs, M., Nyengaard, J.R., Jung, A. et al. (2004). The number of alveoli in the human lung. *American Journal of Respiratory and Critical Care Medicine* 169 (1): 120–124.

Poli, G. (2002). Introduction – serial review: reactive oxygen and nitrogen in inflammation. *Free Radical Biology and Medicine* 33 (3): 301–302.

Pope, C.A. III, Burnett, R.T., Thun, M.J. et al. (2002). Lung cancer, cardiopulmonary mortality, and long-term exposure to fine particulate air pollution. *JAMA* 287 (9): 1132–1141.

Pritchard, R.J., Ghio, A.J., Lehmann, J.R. et al. (1996). Oxidant generation and lung injury after particulate air pollutant exposure increase with concentration of associated metals. *Inhalation Toxicology* 8: 457–477.

Quay, J.L., Reed, W., Samet, J., and Devlin, R.B. (1998). Air pollution particles induce IL-6 gene expression in human airway epithelial cells via NF-κB activation. *American Journal of Respiratory Cell and Molecular Biology* 19 (1): 98–106.

Ruzer, L.S. and Harley, N.H. (2012). *Aerosols Handbook: Measurement, Dosimetry, and Health Effects*. Boca Raton, FL: CRC Press.

Shvedova, A.A., Yanamala, N., Kisin, E.R. et al. (2014). Long-term effects of carbon containing engineered nanomaterials and asbestos in the lung: one year postexposure comparisons. *American Journal of Physiology-Lung Cellular and Molecular Physiology* 306 (2): 170–182.

Soukup, J.M. and Becker, S. (2001). Human alveolar macrophage responses to air pollution particulates are associated with insoluble components of coarse material, including particulate endotoxin. *Toxicology and Applied Pharmacology* 171 (1): 20–26.

Stadnytskyi, V., Anfinrud, P., and Bax, A. (2021). Breathing, speaking, coughing or sneezing: what drives transmission of SARS-CoV-2? *Journal of Internal Medicine* 290 (5): 1010–1027.

Ward, P.A., Warren, J.S., and Johnson, K.J. (1988). Oxygen radicals, inflammation, and tissue injury. *Free Radical Biology and Medicine* 5 (5–6): 403–408.

World Health Organization (2018). Deaths attributable to household air pollution, 2016. https://www.who.int/images/default-source/maps/global_hap_deaths_2016.png?sfvrsn=21545575_0 (accessed 23 July 2025).

10

Indoor Pathogenic Aerosols and Transmission of Infection

10.1 Introduction

In recent decades, recurrent outbreaks of respiratory infectious diseases have underscored the necessity of investigating airborne transmission of pathogenic aerosols. The COVID-19 pandemic has resulted in immense losses on a worldwide scale. To identify the mechanisms of airborne infectious disease transmission, researchers have explored how humans emit respiratory droplets and how these droplets carrying pathogens are transported in space. Understanding the respiratory infectious disease transmission mechanisms and accurately predicting the resulting infection risks can aid in appropriately mitigating the risk of cross-infection in indoor environments. This chapter addresses the shedding of pathogenic aerosols from the respiratory tract, the transmission of these aerosols in indoor environments, the infectious dose, and the infection risk after susceptible individuals inhale pathogenic aerosols.

10.2 Shedding of Pathogenic Aerosols

10.2.1 Brief Introduction

Droplet emission during respiratory activity involves the atomization of liquid into liquid parcels or particles (Morawska 2006). This intricate process is strongly affected by the velocity of the respiratory airflow, which varies significantly from one type of respiratory activity to another. Different velocities lead to different levels of turbulence within the airways, influencing the atomization of exhaled droplets. This, in turn, gives rise to droplets of various sizes. After exhalation, these droplets undergo evaporation and are transformed into droplet nuclei that may harbor and transmit pathogens.

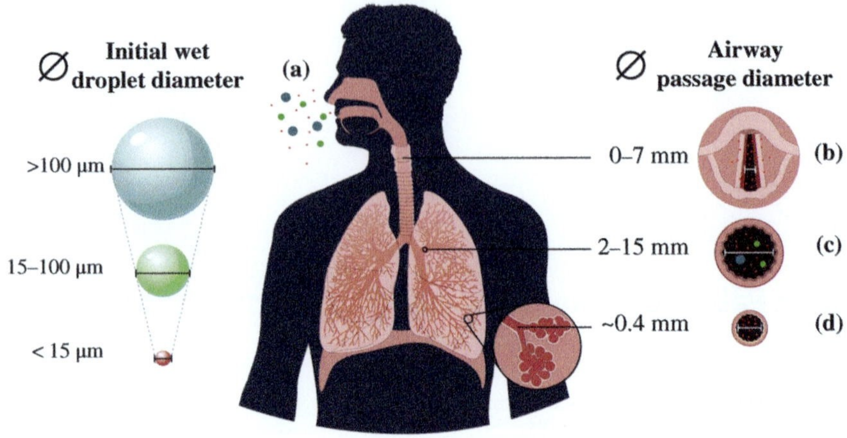

Figure 10.1 Respiratory droplets emitted during exhalation, vocalization, coughing, or sneezing, spanning a wide range of sizes in accordance with the site of origin: (a) large droplets formed near the front of the oral cavity during vocalization; (b) small droplets emitted by vocal folds during vocalization; (c) droplets of varying sizes generated by coughing, sneezing, or sudden exhalation; (d) transient closure of the distal airways for generation of very small breath droplets. *Source:* Stadnytskyi et al. (2021) / with permission of John Wiley & Sons.

Droplets can generally be categorized into two groups based on their size: those smaller than 5 μm in diameter are designated as small droplets, while those exceeding 5 μm are considered as large droplets (Fennelly 2020). Notably, some studies (e.g., Chen et al. 2020) adopt 50 μm as the critical diameter to distinguish droplets of small and large sizes. Because most droplets evaporate rapidly after emission, the resulting final sizes of droplet nuclei may not differ much. Large droplets tend to deposit rapidly, making them highly prone to deposition and direct contact by susceptible individuals in close proximity. In contrast, small droplets may evaporate into droplet nuclei that remain suspended in the air for extended periods. As shown in Figure 10.1, the largest droplets are generated during vocalization near the front of the oral cavity, where airflow is modulated by varying the gaps between the lips, tongue, and teeth. Small droplets are generated by the vocal folds during vocalization. Transient closure of the distal airways generates very small breath droplets. This section addresses the shedding of pathogen-laden droplets during various respiratory activities, and the droplets' size distribution and transmission distances.

10.2.2 Shedding of Liquid Aerosols by Breathing

During breathing, the generation of aerosols is mainly due to the reopening of small airways, especially those located in the lower respiratory tract. This process

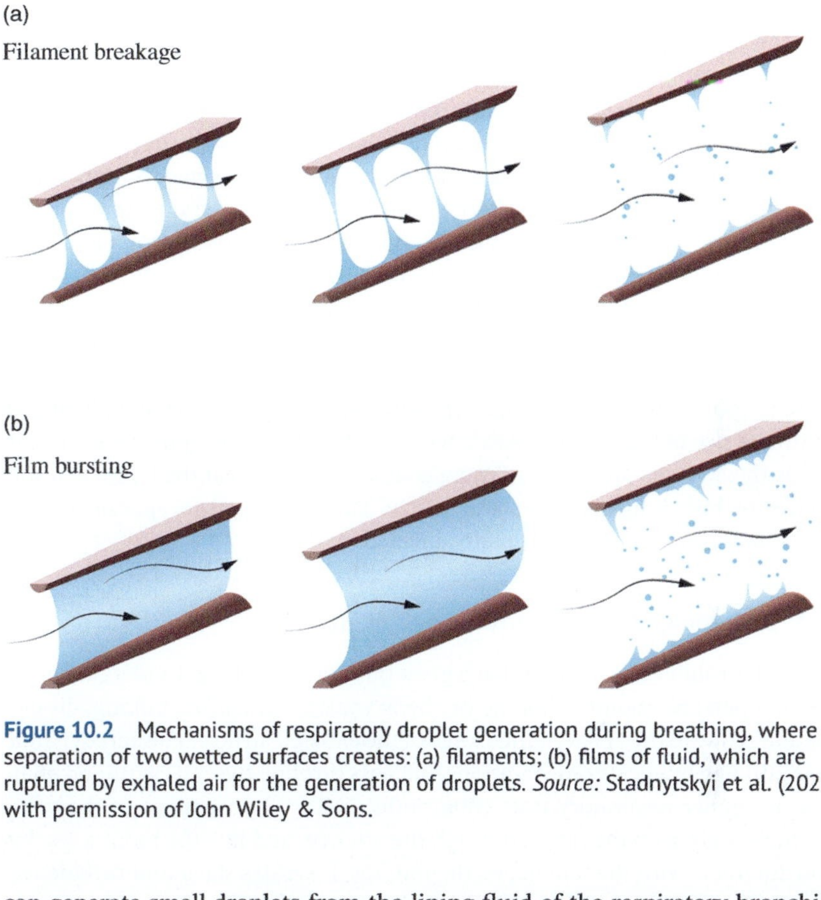

Figure 10.2 Mechanisms of respiratory droplet generation during breathing, where separation of two wetted surfaces creates: (a) filaments; (b) films of fluid, which are ruptured by exhaled air for the generation of droplets. *Source:* Stadnytskyi et al. (2021) / with permission of John Wiley & Sons.

can generate small droplets from the lining fluid of the respiratory bronchioles. The separation of two wetted surfaces creates filaments (Figure 10.2a) or films (Figure 10.2b) of fluid that become thinner as the surfaces separate; these filaments and films are ruptured by exhaled air, creating droplets that travel with the exhalation airflow (Stadnytskyi et al. 2021). Subsequently, during exhalation, these droplets are expelled and dispersed into the air, thereby undergoing the process of aerosolization (Johnson and Morawska 2009). The shear forces created by the airflow over the mucosal surfaces in the upper respiratory tract may also contribute to the production of small droplets.

The airflow rate produced by breathing and the quantity of exhaled droplets are mainly impacted by the metabolic rate (Xu et al. 2017). Furthermore, the exhalation velocity is influenced by the mode of breathing (nasal or oral), with an average velocity of 1.4 m/s for nasal breathing and 1.3 m/s for mouth breathing (Tang et al. 2013).

The aerosols produced during breathing are generally small, often in the submicron range (less than 1 μm in diameter). The study by Fabian et al. (2008) showed that the sizes of aerosols during breathing ranged from 0.3 to 5 μm, and more than 87% had diameters less than 1 μm. The concentration of the aerosols emitted during breathing is generally lower than that during coughing or sneezing. It has been reported that breathing releases approximately 18 particles per minute (Asadi et al. 2020). Although the aerosols emitted during breathing may not be large in number, they can remain suspended in the air for a long time due to their small sizes and the negligible gravitational settling effect.

When liquid aerosols are expelled from the respiratory tract into the air, they undergo a rapid evaporation process because they are moving from the high relative humidity (RH) inside the respiratory tract to the lower RH of the surrounding air. The evaporation process, which typically lasts a few seconds, greatly affects the ultimate size of the aerosols. At a lower RH, the evaporation process is shorter, and thus the aerosols are smaller (Wang et al. 2021). Note that the liquid aerosols generated by breathing are very small in size. They commonly evaporate quickly (<0.1 seconds) and are thus not greatly affected by RH.

10.2.3 Shedding of Liquid Aerosols by Talking

A study (Bourouiba 2020) found that a great quantity of talking droplets are emitted from a person's mouth when he or she is vocalizing. During talking, droplet formation is due to the aerodynamic force of the exhaled airflow. The aerodynamic force is initiated by vocal cord vibrations that pass through the oral cavity and pharynx in the upper respiratory tract (Bourouiba 2020). The vibration of the vocal cords propels air from the lungs, through the trachea, and into the mouth. As this airflow interacts with the tongue, teeth, and lips, it creates shear and turbulence. The turbulence disrupts the liquid films of saliva and mucus along the oral and pharyngeal mucosa, resulting in the generation of droplets. This droplet generation is particularly pronounced during the articulation of plosive sounds, such as "p" and "b," where the abrupt acceleration and shear forces by the airflow lead to greater droplet production (Abkarian and Stone 2020).

The velocity of exhaled airflow during talking varies with the talking content and the surrounding environment. Consequently, the reported airflow velocities during talking differ among the published studies. One study measured an average exhaled airflow velocity of 4.07 m/s for men and 2.31 m/s for women using particle image velocimetry (PIV) (Kwon et al. 2012). In another study, the average exhaled airflow velocity was measured at 4.3 m/s for males and 3 m/s for females, also using PIV (Chao et al. 2009).

Talking-generated droplets range widely in size. One study showed that approximately 82% of the droplets are smaller than 100 μm (Dai and Zhao 2021).

The talking strength is a key factor affecting the droplet size distribution and transmission distance. Louder talking tends to generate more droplets, many of which are smaller in size, as compared to quieter talking (Asadi et al. 2019). Conversely, quieter talking tends to generate fewer and larger droplets, which fall to the ground more rapidly due to gravitational settling.

10.2.4 Shedding of Liquid Aerosols by Coughing and Sneezing

Coughing and sneezing are natural physiological responses that are typically triggered by respiratory irritation, infection, or allergic reactions. Droplets generated during coughing and sneezing serve as the primary vehicle for transmission of viruses, bacteria, and other pathogens. Although coughing and sneezing are infrequent, the airflow velocity and droplet concentration during coughing and sneezing are significantly higher than during breathing or talking. Notably, coughing is a key symptom of COVID-19 and influenza.

When an individual coughs or sneezes, the air in the respiratory tract is strongly compressed. The air is then forced to move quickly through the trachea, throat, and mouth, forming a high-speed jet. During this process, the liquid on the surface of the respiratory tract is strongly sheared by the airflow, which produces droplets. These droplets contain water, mucus, and probably pathogens. The initial velocity of exhaled air during coughing is around 22 m/s (Zhu et al. 2006), and the exhaled air velocity during sneezing ranges from 30 to 100 m/s (Dai and Zhao 2021). Both sneezing and coughing create a turbulent multiphase gas and particle cloud (Abkarian and Stone 2020), which slows down evaporation of the droplets. Large droplets of saliva and mucus shoot out of the mouth after a sneeze but fall relatively quickly. A turbulent cloud composed of small droplets can be transmitted farther. As shown in Figure 10.3, the mucosal salivary fluid ejected by sneezing creates elongated ligaments with fluid beads. These beads are then elongated and broken up, which produces droplets.

Interestingly, studies have shown that there is no big difference in particle size distribution between coughing and talking (Dai and Zhao 2021). Approximately 64% of coughing droplets are smaller than 100 μm, and the remaining 36% are larger than 100 μm. Droplets larger than 500 μm are present at a higher frequency during coughing than during talking. The size of coughing droplets is slightly larger than that of talking droplets due to the different mechanisms of droplet generation.

The initial velocity of the exhaled airflow is much higher during sneezing than during talking or coughing. Consequently, the droplets generated by sneezing are much greater in number and larger in particle size than those generated by talking or coughing. The size distribution of droplets generated by sneezing can be either unimodal or bimodal (Figure 10.4). Most of the droplets range from 1 to

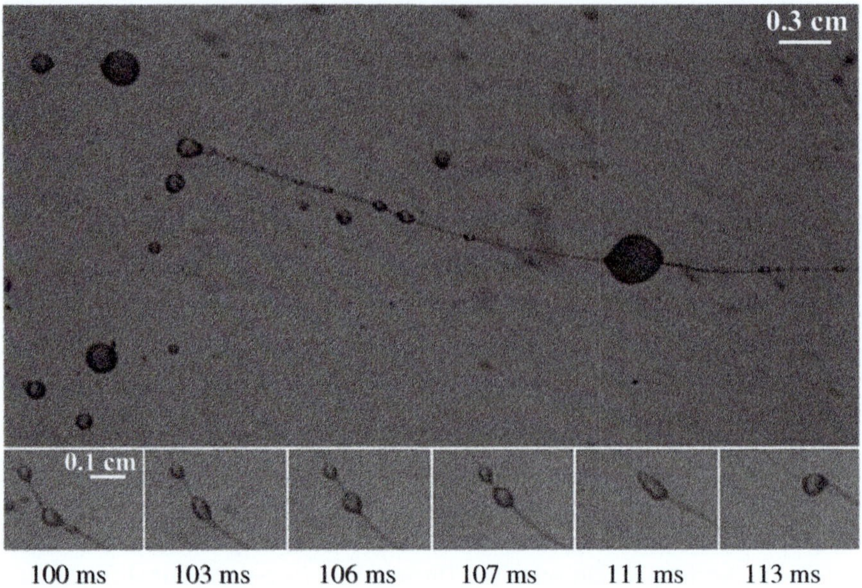

Figure 10.3 Sneezing-generated droplets captured by a high-speed camera: elongated ligaments of mucosal salivary fluid ejected by the sneeze and formation of droplets after breakup. *Source:* Scharfman et al. (2016) / with permission of Springer Nature.

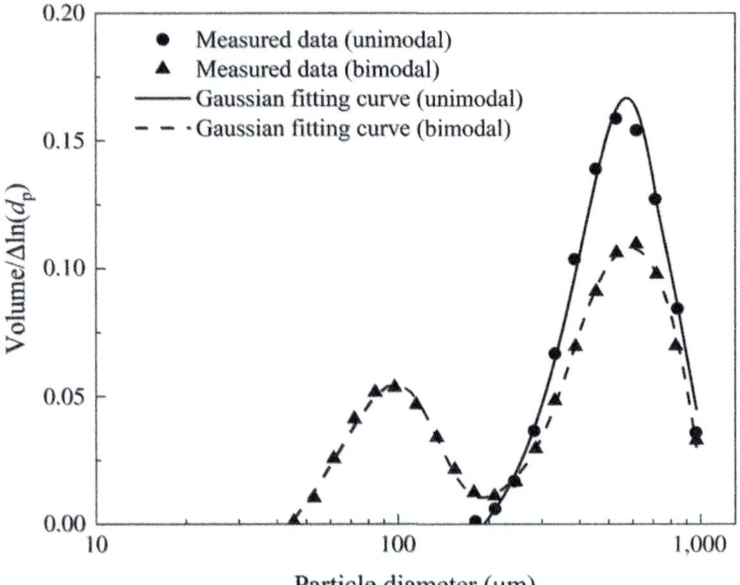

Figure 10.4 Measured data and fitting curves of sneezing-generated droplet sizes (unimodal and bimodal distributions). *Source:* Han et al. (2013) / with permission of The Royal Society.

Table 10.1 The emitted droplet sizes, rates, and release velocity by different respiratory activities.

Respiratory activity	Droplet diameter (μm)	Emission rate	Release velocity (m/s)
Breathing	0.3–5 (Fabian et al. 2008)	0.31 droplets/s (Asadi et al. 2020)	1.29–1.6 (Xu et al. 2017)
Talking	1–200 (Chao et al. 2009)	1–50 droplets/s (Asadi et al. 2019)	2.31–4.07 (Kwon et al. 2012)
Coughing	1–500 (Duguid 1945; Dai and Zhao 2021)	360–5,800 droplets/cough (Duguid 1945)	22 (Zhu et al. 2006)
Sneezing	1–1,000 (Duguid 1945; Dai and Zhao 2021)	3,700–46,000 droplets/sneeze (Duguid 1945)	30–100 (Dai and Zhao 2021)

1,000 μm in diameter (Duguid 1945). Droplets from sneezing may be transmitted 6–7 m, while droplets from coughing are usually transmitted 2–3 m (Jayaweera et al. 2020).

Table 10.1 summarizes the emitted droplet sizes, release rates, and release velocities by different respiratory activities. The quantification of droplet numbers in the table varies with the respiratory activities. The emission rate of breathing and talking is measured in droplets per second. Coughing and sneezing are quantified as droplets per single event. Different respiratory activities produce droplets with varying sizes and quantities. Breathing generates small droplets (0.3–5 μm) at a low rate (0.31 droplets/s) and a low velocity (1.29–1.6 m/s). Talking emits larger (1–200 μm) and more (1–50 droplets/s) droplets at a higher velocity (2.31–4.07 m/s). Coughing releases 360–5,800 droplets (1–500 μm) at 22 m/s, while sneezing produces the largest number (3,700–46,000) and widest size range (1–1,000 μm), with velocities up to 100 m/s. The more intensive are the respiratory activities, the more and larger droplets are released at higher speeds.

10.3 Transmission of Pathogenic Aerosols

10.3.1 Brief Introduction

Respiratory infectious pathogens are encapsulated in exhaled droplets of various sizes and transported by external forces, leading to spatial transmission. There are three possible transmission mechanisms for respiratory pathogenic aerosols: airborne aerosol transmission, direct droplet transmission, and surface contact transmission. Droplets smaller than 50 μm evaporate rapidly after emission, forming droplet nuclei. These droplet nuclei can remain suspended in the air for long periods and are transmitted over long distances. In contrast, large droplets (>50 μm) are quickly deposited on nearby surfaces. If the droplets are deposited on the mucous membranes of the mouth, nose, and eyes, or are directly inhaled or swallowed, there is a risk of droplet transmission. Additionally, frequent touching of the mouth, nose, or eyes with contaminated hands carrying pathogens can lead to contact transmission. This section provides an overview of the three transmission mechanisms of pathogens, including the modes of transmission and some influencing factors.

10.3.2 Droplet Transmission

Droplet transmission refers to the spread of infectious agents, such as viruses or bacteria, by liquid droplets expelled from the respiratory tract of an infected individual during coughing, sneezing, talking, or even breathing. These expelled droplets can vary in size. Some large droplets may be deposited directly on the mucous membranes of another individual's mouth, nose, or eyes, or on surfaces that others may subsequently touch.

Droplet transmission involves two key parameters: critical distance and critical size. The critical distance for droplet transmission refers to the maximum distance that respiratory droplets can travel in the air before they settle to the ground or lose their infectiousness. This distance is not a fixed number; rather, it can vary with the size of the droplets, the velocity at which they are expelled, and the humidity and temperature of the surrounding air. Droplets larger than 50 μm tend to fall to the ground relatively quickly, within a few feet (approximately 1–2 m). Coughing at 10 m/s can propel droplets over 2 m, whereas breathing at 1 m/s limits travel to less than 1 m (Xie et al. 2007). Consequently, an average critical distance of 1.5 m has been suggested (Zhao et al. 2022). However, for the purpose of understanding droplet transmission specifically, the critical distance is often designated within close proximity (e.g., within 2 m) of an infected person (Liu et al. 2017a).

Meanwhile, the critical size, when used to describe droplet transmission, refers to a threshold dimension that distinguishes droplets from aerosols in terms of their potential for transmitting infectious agents in the air. Droplets larger than the critical diameter are deposited on the mucous membranes of the face before they evaporate completely, constituting droplet transmission. Droplets smaller than the critical diameter evaporate completely into droplet nuclei, which enter the airborne transmission mode. The World Health Organization (WHO 2014) has specified 5 μm as the critical droplet diameter. This is because particles smaller than 5 μm are more likely to enter the alveoli (Wells 1934). However, one study (Chen et al. 2020) reported 50 μm as the size below which droplets are less likely to be deposited on the faces of susceptible individuals, implying that infectious respiratory droplets must be larger than 50 μm for droplet transmission. In fact, the critical diameter of droplets also depends on the ambient RH and the exhaled airflow. Droplets evaporate more slowly in moist air than in dry air, and thus they fall more quickly in moist air. It has been determined that the critical sizes are approximately 125, 100, 85, and 60 μm for RH levels of 0%, 50%, 70%, and 90%, respectively, in indoor environments (Xie et al. 2007).

Wells (1934) first discovered that droplets smaller than 100 μm evaporate faster than they are deposited. The evaporation-falling curve reveals the relationship among droplet size, evaporation time, and deposition time, as shown in Figure 10.5. If the air is not saturated with water vapor, all droplets evaporate as they fall. Sufficiently large droplets reach the ground, where they continue to evaporate (deposition curve on the right side of the figure). In contrast, small droplets evaporate so rapidly that their evaporation is complete before they reach the ground (left-hand evaporation curve). There is a turning point between the evaporation and deposition curves. The droplet size at this turning point is the critical droplet size. Droplets larger than the critical size are deposited on the ground before complete evaporation. Droplets smaller than the critical size evaporate quickly and settle slowly; consequently, they evaporate completely in the air before reaching the ground. Many researchers have investigated the impacts of

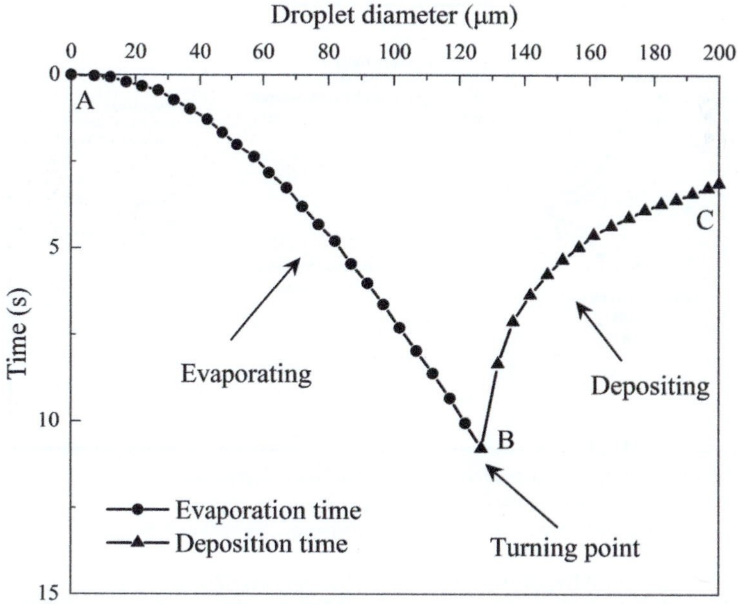

Figure 10.5 Evaporation and deposition times of droplets with varying diameters ($T_{p0} = 33\,°C$, $T_{\infty} = 18\,°C$, $RH = 0$). *Source:* Xie et al. (2007) / with permission of John Wiley & Sons.

respiratory activity (Xie et al. 2007), initial salt solution concentration (Liu et al. 2017b), and turbulence (Wei and Li 2015) on the evaporation-falling curve.

10.3.3 Airborne Transmission

Airborne transmission refers to the process by which droplets containing pathogens lose their moisture in the air and form smaller particles known as droplet nuclei or aerosols. Droplets with an initial diameter less than 50 μm can evaporate into droplet nuclei. These tiny particles, consisting of proteins and pathogens, can remain suspended in the air for significant periods and travel over long distances. When these droplet nuclei are inhaled by susceptible individuals, respiratory infection may result (Li and Feng 2023).

Both short-distance and long-distance airborne transmission can occur, as shown in Figure 10.6. Short-distance airborne transmission refers to the spread of pathogenic aerosols through the air over relatively short distances. Long-distance airborne transmission, on the other hand, involves the spread of pathogenic aerosols through the air over long distances. Airborne transmission is often mistakenly considered to be only long-distance because epidemiologists are unable to distinguish short-distance airborne transmission from droplet transmission.

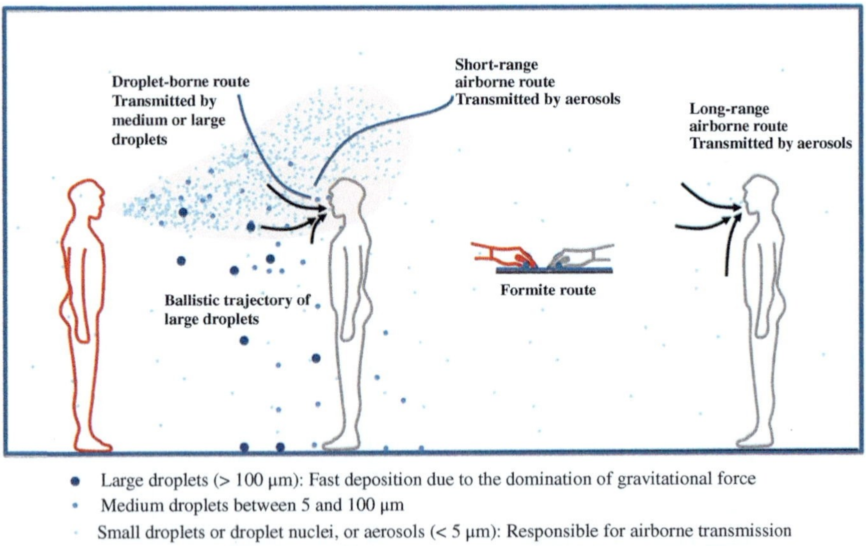

Figure 10.6 Illustration of different transmission routes for respiratory infectious diseases. Small liquid aerosols (<5 μm) are transmitted by the short-range airborne route, long-range airborne route, and indirect contact route; large droplets are transmitted by the direct spray route and indirect contact route. *Source:* Wei and Li (2016) / with permission of Elsevier.

Short-range airborne transmission occurs when an infected person exhales tiny droplets or droplet nuclei that are subsequently inhaled directly by an adjacent susceptible individual. When the distance between the susceptible individual and the infected person is less than 1.5 m, the exhaled airflow can easily reach the breathing zone of the susceptible individual. The probability that the susceptible individual will encounter the exhaled droplet nuclei is significant over such a short distance. As the distance increases, the impact of the infected subject's exhaled airflow on the airborne transmission of droplet nuclei decreases. When the distance exceeds 2 m, the concentration of droplet nuclei in the exhaled airflow quickly decays to the average concentration in the indoor air. This phenomenon is known as the proximity effect in short-range airborne transmission (Liu et al. 2017a). In addition, droplet nuclei can remain suspended in the air for extended periods and can travel further than droplets, and are thus capable of transmitting pathogens over greater distances. The spread of respiratory viruses such as SARS-CoV-2 and influenza virus (Liu et al. 2017a) has been attributed to long-distance airborne transmission.

In indoor environments, airflow patterns can strongly affect airborne transmission (Liu et al. 2022). Airflow patterns are mainly subject to the ventilation mode, ventilation rate, and supplementary systems (such as air filtration). Effective ventilation is key to controlling the transmission of respiratory diseases. Increasing the ventilation rate from 1.7 l/(s·person) to 24 l/(s·person) has been found to reduce the secondary morbidity rate of tuberculosis to zero (Du et al. 2020). Although ventilation can facilitate the removal of pathogen-laden aerosols, an appropriate ventilation strategy must be implemented to effectively reduce airborne transmission. When researchers investigated the dispersion of respiratory infectious aerosols in an elevator car, they found that increasing the ventilation rate may not significantly reduce the inhalation fraction (Zhu et al. 2024). A good ventilation strategy lowers the transmission risk more effectively than an increase in the ventilation rate in this case.

It is known that a vertical temperature gradient is created in the indoor environment when displacement ventilation is employed (Liu et al. 2023). Due to this thermal stratification, aerosols are very likely to become locked within regions at specific heights. If the aerosols are trapped in the human breathing zone, the risk of airborne transmission may increase. Proper placement of portable high-efficiency particulate air (HEPA) purifiers, which are capable of removing over 99.97% of aerosol particles, is effective in reducing exposure to infectious aerosols, especially when combined with ventilation and mask-wearing (Lindsley et al. 2021).

COVID-19 transmission events in a restaurant (Li et al. 2021a) and on a bus (Ou et al. 2022) have indicated that ventilation rates below 3 l/(s·person) may lead to long-distance transmission. If long-distance airborne transmission is possible, short-distance airborne transmission must also exist, and it poses an even higher infection risk. From another perspective, airborne transmission can be classified into two types: the first is preferential airborne transmission that occurs in indoor environments with adequate ventilation, and the second is opportunistic airborne transmission that occurs in environments with insufficient ventilation. The transmission of COVID-19 in the aforementioned restaurant and bus cases was opportunistic airborne transmission. In normal indoor environments, short-distance airborne transmission is predominant. In a poorly ventilated indoor environment, however, the concentration of pathogenic aerosols is high throughout the space, and long-distance airborne transmission is likely to occur. Short-distance transmission requires close gathering, and the number of people in close proximity to an infected individual within a certain period is limited. Therefore, short-distance transmission of COVID-19 is relatively unlikely to cause a super-spreading event in a poorly ventilated environment. Opportunistic long-distance airborne transmission is more likely to lead to super-spreading events.

10.3.4 Contact Transmission

Contact transmission refers to the spread of infectious pathogens among individuals through direct or indirect physical contact. This mode of transmission plays a critical role in the spread of numerous diseases, including viral, bacterial, fungal, and parasitic infections. Direct contact transmission occurs when an infected person comes into physical contact with a susceptible individual, allowing the pathogens to be transferred directly from one person to another. Direct exposure to infected body fluids transmits pathogens, particularly during medical procedures or childbirth. For respiratory diseases, activities like hugging, kissing, and handshaking facilitate pathogen transfer. Indirect contact transmission involves the transfer of pathogens through an intermediary object or surface (Tharayil et al. 2022). Fomites are non-living objects or surfaces that can harbor pathogens and serve as vehicles for their transmission; examples include doorknobs, towels, and electronic devices. Pathogens on contaminated surfaces can spread when touched and transferred to the face.

Direct contact includes mucous-membrane to mucous-membrane contact (such as kissing) and social contact (such as handshaking and hugging) (Zhang et al. 2020). The former involves the direct transfer of infectious pathogens from an infected patient to a susceptible individual without exposure to the environment. The latter involves contact through skin or clothing contamination, followed by self-inoculation via mucous membranes. Short airborne transmission can also take place during kissing. While mouth-to-mouth kissing leads to direct exchange of respiratory secretions, cheek or hand kissing or other skin contact may also lead to indirect contact transmission.

For centuries, respiratory infectious diseases have been understood to spread primarily through close contact with infected individuals or via airborne transmission, with limited attention given to contaminated surfaces (Cozad and Jones 2003). Surfaces can be contaminated through a variety of mechanisms, such as the deposition of droplets released by various respiratory activities, deposition of pathogen-laden aerosols, or the touching of surfaces with contaminated hands (Li et al. 2021b).

10.4 Infectious Dose and Infection Risk

10.4.1 Brief Introduction

The transmission of respiratory infectious diseases is impacted by various factors, such as individual respiratory activities, environmental factors, and the types, quantities, and viability of pathogens present in respiratory droplets. There are significant differences in the viral loads found in exhaled droplets for different respiratory diseases. During the transmission process, the activity of viruses in droplets may be diminished by the surrounding environment, and the viruses can even be completely inactivated.

Quantitative evaluation of infection risk requires a scientific and systematic assessment of the likelihood and potential severity of infectious diseases within a specific population. The Wells–Riley model and the dose–response model are the most commonly used models for quantitatively assessing the infection risk of respiratory diseases (Sze To and Chao 2010). In this section, the infectious dose is presented first, followed by a demonstration of the Wells–Riley model and the dose–response model, along with some typical airborne infection events.

10.4.2 Minimum Dose for Infection

The infectious dose is the minimum amount or concentration of a pathogenic microorganism required to cause infection and disease in a susceptible host. Different quantitative evaluation methods incorporate the infectious dose in different ways. In the Wells–Riley model, the infectious dose is a hypothetical value ("quanta") rather than a realistic physical unit. One quantum is defined as the minimum number of pathogens needed to reach a pathogenic level (Wells 1955). In other words, a susceptible person may become infected if he or she inhales one quantum of the pathogen. The quantum generation rate in the Wells–Riley equation cannot be obtained directly, but rather is inversely calculated from certain infection events. For instance, in the case of airborne transmission of COVID-19 on a bus, the quantum generation rate for SARS-CoV-2 was estimated to be 36.6 quanta/h (Luo et al. 2022). Quantum generation rates vary significantly across settings. For example, in the Prince of Wales Hospital in Hong Kong, the SARS-CoV-1 quantum generation rate was postulated to be as high as 4,680

Table 10.2 Viral load of different SARS-CoV-2 variants in saliva and nasopharyngeal fluid.

Virus variant	Saliva (gene copies/ml)	Nasopharyngeal fluid (gene copies/ml)
Wild type	10^4–10^9 (Imai et al. 2023)	10^5–10^{11} (Jones et al. 2020)
Delta	10^4–10^{10} (Imai et al. 2023)	10^3–10^{12} (de Michelena et al. 2023)
Omicron	10^4–10^{10} (Imai et al. 2023)	10^4–10^{11} (de Michelena et al. 2023)

quanta/h, whereas a much lower value of 28.8 quanta/h was reported in a Municipal Ho-Ping Hospital case in Taipei (Qian 2007).

In the dose–response model, the infectious dose is the true number of pathogens inhaled by the individual. The critical infectious dose is often obtained from experiments in which test animals (or human test tissues) are infected with pathogens. The critical infectious dose is affected by the type of virus and the particle size. Viruses vary in terms of their mechanism of infection as well as the dose required for infection. Different viruses must be present in different doses in order to infect the host. For example, influenza is one of the most contagious and rapidly spreading viruses, and a very low dose leads to infection. In contrast, for less virulent viruses (such as the respiratory syncytial virus), larger infectious doses are often required to overcome the host's immune defenses (Zhou et al. 2020). In addition, crossovers and mutations in viruses may influence their infectivity. Roughly 300–2,000 gene copies were required to cause an infection of COVID-19 in the early stages of the pandemic (Prentiss et al. 2020). For the Omicron variants, susceptible persons were infected if they inhaled 400 gene copies (Riediker et al. 2022) or even fewer. The viral loads in saliva and nasopharyngeal fluid vary significantly among different SARS-CoV-2 variants, as shown in Table 10.2. The wild type and Delta variants exhibit a wide range, with Delta showing the highest viral load in the nasopharyngeal fluid (up to 10^{12} gene copies/ml). The Omicron variant has similar concentrations to Delta in the saliva but has a slightly lower upper bound of the concentration in the nasopharyngeal fluid.

10.4.3 Wells–Riley Model

The Wells–Riley model was first proposed by Riley et al. (1978) in 1978 for investigating the airborne transmission of measles based on Wells's theory. The model assumes that there is a certain quantity of infectious pathogens, such as viruses and bacteria, suspended in the air. These pathogens are present in the form of aerosols, and infection occurs when susceptible individuals inhale aerosols containing sufficient doses of the pathogens. The model is based on the following assumptions: (i) droplet nuclei are evenly distributed in space, which means that the infection risk predicted by this equation is uniform within the space; (ii) changes in the viability and infectivity of the pathogens are negligible; (iii) the concentration of infectious agents is stable for the duration of the infection; and (iv) the amounts of infectious agents that might be removed by filtration or deposition are also negligible. The equation is as follows (Riley et al. 1978):

$$P = 1 - \exp\left(-\frac{Iqpt}{Q}\right) = 1 - \exp(-N_s) \tag{10.1}$$

where P is the infection probability for the susceptible population; I is the number of source patients; q represents the quanta produced by each source patient (quanta/h); p is the pulmonary ventilation rate of a susceptible person (m^3/h); t is the exposure time (h); Q is the ventilation rate of the supply air without the virus (m^3/h); and N_s is the quantity of quanta inhaled by the susceptible persons. Here N_s can be calculated by means of different equations, based on either aerosols or a tracer gas. For aerosols, N_s is defined as (Riley et al. 1978):

$$N_s(x, t_0) = C_v p \int_0^{t_0} C_d(t)\mathrm{d}t = C_v N_i \tag{10.2}$$

where C_v is the concentration of the virus in the exhaled droplets; $C_d(t)$ is the quantum concentration in an indoor environment at time t (quanta/m^3); t_0 is the exposure time (h); and N_i is the total number of droplets or droplet nuclei (that were exhaled by the index patient) inhaled by susceptible individuals.

For a tracer gas, N_s is defined as follows, according to the dilution-based evaluation method (Zhang and Lin 2021):

$$N_s = \int_0^{t_0} pC_{g,q}(t)\mathrm{d}t \tag{10.3}$$

$$C_{g,q} = \frac{qC_{g,p}}{p_s C_{g,s}} \tag{10.4}$$

where t_0 is the exposure time (h); $C_{g,q}$ is the airborne quantum concentration at the target position (quanta/m^3); p_s is the breathing rate of the source patient (m^3/h); and $C_{g,p}$ and $C_{g,s}$ are the concentrations of tracer gas (ppm) at the target and source positions, respectively.

The original Wells–Riley model assumed that the ventilation rate was the only factor determining the risk of infection. The model does not take into account inter-individual differences such as age, immunity, or exposure history. Many other factors and control measures can also affect infection risk, such as environmental conditions and the wearing of personal protection equipment (PPE). The later development of the Wells–Riley model attempted to extend the applicability of the Wells–Riley equation and relax some of the assumptions. For example, wearing a face mask reduces the number of inhaled droplets. It is feasible to add a parameter to the Wells–Riley equation to represent this reduction. The $\frac{Iqpt}{Q}$ term in Eqs. (10.1) and (10.5) is the inhalation dose in quanta. The effect of the mask can be considered by multiplying this term by a certain factor. The equation is as follows (Fennelly and Nardell 1998):

$$P = 1 - \exp\left(-\frac{Iqpt\theta}{Q}\right) \tag{10.5}$$

where θ (ranging from 0 to 1) is the fraction of particles that penetrate the mask. When $\theta = 1$, either there is no mask or the mask is completely ineffective.

Particle filtration and air disinfection, such as the use of ultraviolet (UV) irradiation, are additional airborne pathogen sinks other than removal by ventilation. The effects of these control measures can also be incorporated into the Wells–Riley model (Fisk et al. 2004):

$$P = 1 - \exp\left(-\frac{Iqpt}{Q + \lambda_{\mathrm{UV}}V + Q_{\mathrm{f}}\eta_{\mathrm{f}}}\right) \tag{10.6}$$

where λ_{UV} is the inactivation rate coefficient of UV irradiation (h^{-1}), V is the room volume (m^3), Q_{f} is the flow rate to the filter (m^3/h), and η_{f} is the filtration efficiency.

10.4.4 Demonstration of the Wells–Riley Model in a Classroom

The use of the Wells–Riley model to evaluate the impact of natural ventilation on the COVID-19 airborne infection probability was demonstrated in a classroom (Park et al. 2021), as shown in Figure 10.7. This classroom had dimensions of 8.18 m × 7.9 m × 2.6 m. Outside the classroom was a hallway with a width of 2.8 m. Openable windows were located on both the exterior wall of the classroom and the wall between the classroom and the hallway, as shown in Figure 10.7a. The hallway also had exterior windows.

Two different ventilation strategies were employed in the classroom: cross ventilation and single-sided ventilation. Figure 10.8 displays the window opening status and window opening ratios under the two scenarios. In the cross-ventilation mode, the exterior windows of the hallway were opened to 100%, while the windows of the classrooms on both sides were opened identically to 15%, 30%, and 100%, successively, as shown in Figure 10.8a. In the single-sided ventilation mode, the exterior windows of the hallway were opened to 100%, while the windows between the classroom and the hallway were closed, and the exterior windows of the classroom were opened to 15% and 30%, successively, as shown in Figure 10.8b. The above constituted a total of five different ventilation scenarios.

The ventilation rates were measured by the tracer-gas decay method for the five ventilation scenarios. The infection probabilities evaluated by the Wells–Riley model were compared among the different scenarios. It was assumed that there

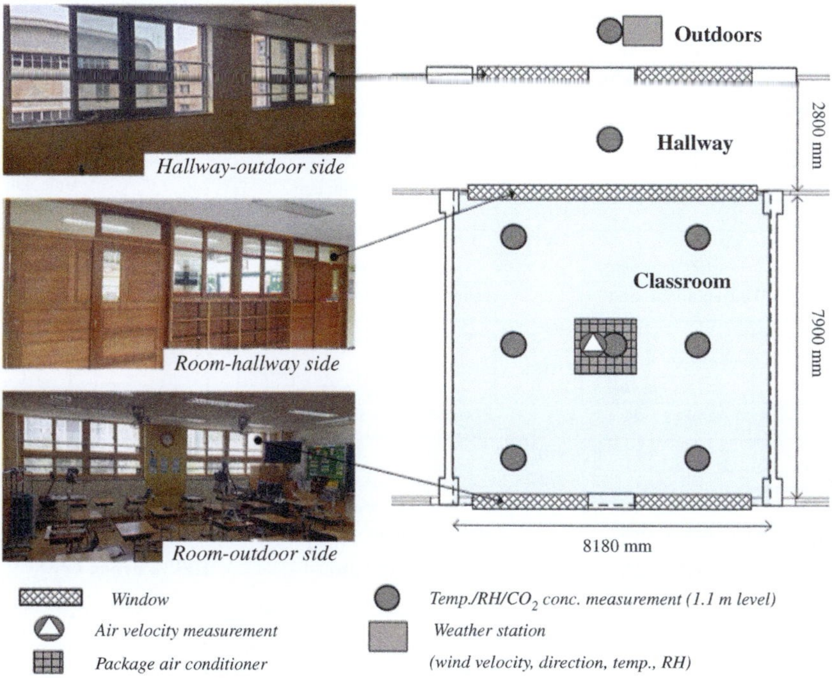

Figure 10.7 A classroom in which the impact of natural ventilation on the airborne infection probability was examined: (a) photographs of the studied classroom in which the ventilation conditions were measured; (b) planar view of the classroom and the sensor locations. *Source:* Park et al. (2021) / with permission of Elsevier.

was only one COVID-19 infector in the classroom. The pulmonary ventilation rate of the susceptible individual was set to $0.3\,m^3/h$, since he/she was sitting or engaging in light activity indoors. The quantum generation rate was defined as 48 quanta/h, and the exposure time was set successively to 0.5, 1, 2, and 3 hours. Additionally, the impact of mask-wearing on the infection probability was analyzed. The filtration efficiency of the surgical masks was set to 50% in light of the clearances between the mask and face that permitted air leakage.

Figure 10.9 presents the measured natural ventilation rates and the predicted COVID-19 infection probability. It can be seen in Figure 10.9a that the natural ventilation rates increased steadily with the window opening ratio. The natural ventilation rates in the cross-ventilation mode were much higher than in the single-sided ventilation mode. A prevailing east wind could slightly increase the ventilation rate as compared with a prevailing south wind.

Figure 10.9b presents the predicted COVID-19 infection probability under different ventilation rates, at different exposure times, and in situations with and

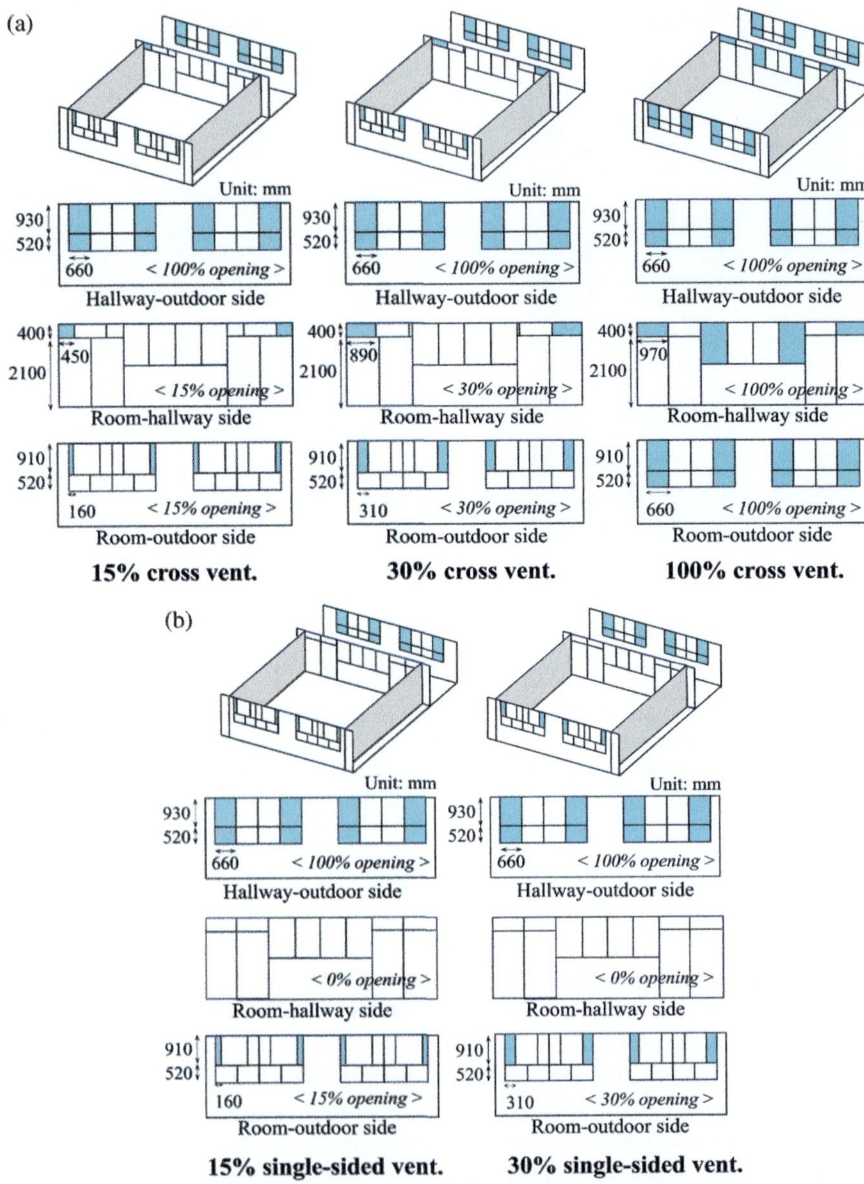

Figure 10.8 Schematics of five different ventilation scenarios and the window opening status, where blue shading indicates open windows: (a) cross-ventilation mode; (b) single-sided ventilation mode. *Source:* Park et al. (2021) / with permission of Elsevier.

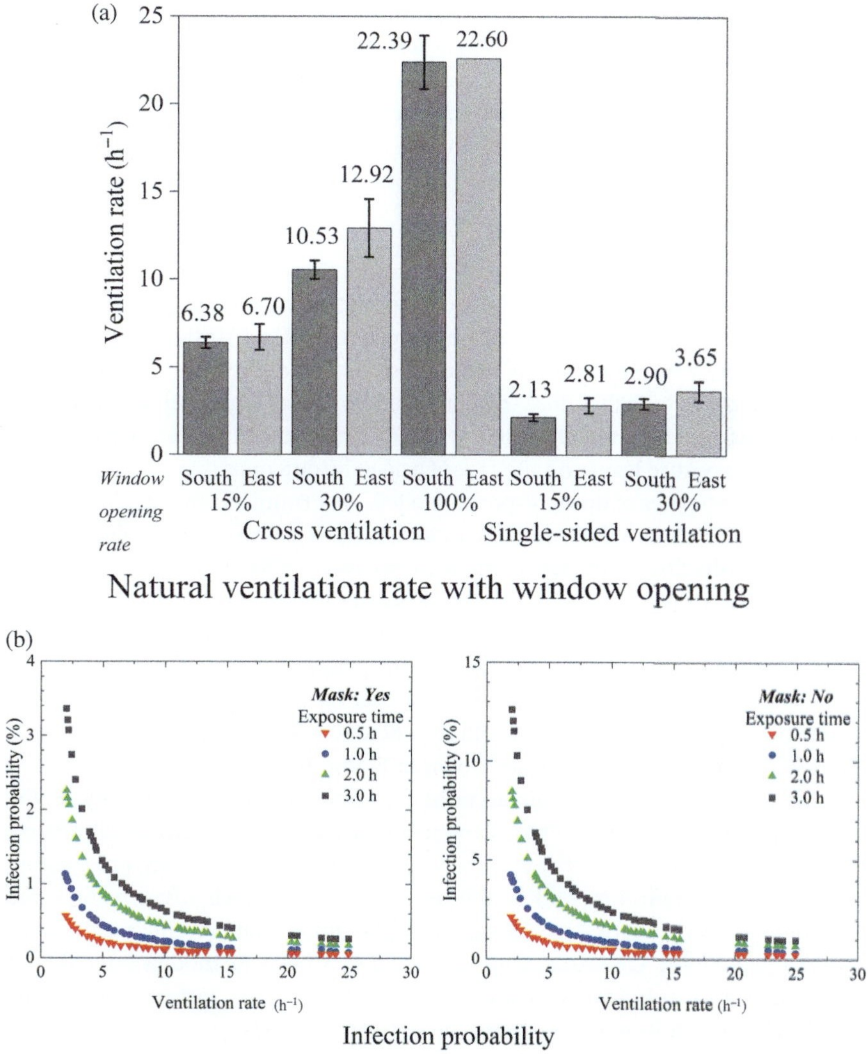

Figure 10.9 Measured natural ventilation rates and predicted COVID-19 infection probability: (a) natural ventilation rates in different ventilation scenarios; (b) COVID-19 infection probability under different ventilation rates. *Source:* Park et al. (2021) / with permission of Elsevier.

without face masks. It can be seen that the infection probability decreased notably as the ventilation rate increased. Longer exposure time apparently resulted in a higher infection rate. Mask-wearing could reduce the infection probability by approximately four times under the same ventilation rate. Specifically, if the windows were opened to 15% of the maximum openable area, then the natural ventilation rate was over 2 ACH, which was sufficient to confine the infection probability to a relatively low level. Consequently, wearing face masks, increasing the natural ventilation rate in transitional seasons, and shortening the exposure time were all effective means of reducing the airborne infection probability.

10.4.5 Dose–Response Model

The dose–response model is also widely used for assessing infection risk. This model integrates the principles of dose–response relationships, which describe how the quantity or concentration of a pathogen (the "dose") influences the infection probability. For infection risk evaluation, this model requires the infectious dose in order to establish a dose–response relationship (Rigaud et al. 2024). The dose–response relationship refers to the potential relationship between a dynamic variation in exposure level and the risk of a specific outcome.

There are two types of dose–response models: deterministic and stochastic (Sze To and Chao 2010). In a deterministic model, there is a clear, predictable relationship between the dose of the agent and the response or effect that is observed. This means that for a given dose, the same effect is expected to occur in all or a specified proportion of exposed individuals, as shown in Figure 10.10a. When each member of the susceptible population receives the same pathogenic dose, the infection risk is equal to the cumulative integration of the infection frequency at that dose. Deterministic models may adopt a threshold dose below which no infection occurs, and above which the effect emerges, as shown in Figure 10.10b.

In contrast, stochastic models account for variability and uncertainty in the dose–response relationship. They recognize that individuals may exhibit diverse responses to the same dose due to factors such as genetic differences, pre-existing conditions, and environmental conditions. Stochastic models often incorporate probabilities and statistics to describe the likelihood that an effect will occur within a population. Unlike deterministic models, stochastic models often do not assume a clear threshold dose below which no effect occurs; rather, they assume that the risk increases with the dose even at very low levels.

The dose–response model for assessing the risk of airborne infection for tuberculosis is formulated as (Nicas 1996):

$$P = 1 - \exp\left(-\frac{IGp\beta t}{Q}\right) \tag{10.7}$$

where P is the infection probability; I is the number of infectors; p is the pulmonary ventilation rate of a person (m^3/h); t is the exposure time (h); Q is the room ventilation rate with clean air (m^3/h); G is the number of airborne tuberculosis bacilli released per infector per unit time; and β is the deposition fraction of infectious particles in the alveolar region.

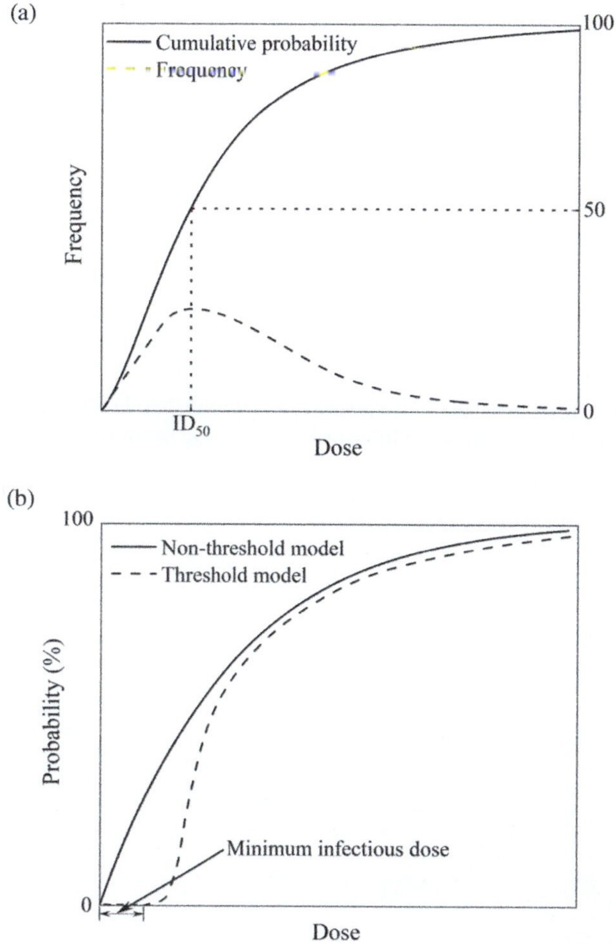

Figure 10.10 An illustration of the dose–response model: (a) cumulative probability and frequency distribution of the intake dose; (b) a non-threshold model versus a threshold model.

To establish the dose–response model for SARS-CoV-1, Watanabe et al. (2010) collected datasets for transgenic mice susceptible to SARS-CoV-1. The researchers then employed the maximum likelihood method to fit both the β-Poisson distribution and the exponential model. The general forms of the β-Poisson distribution model and the exponential model are shown in Eqs. (10.8) and (10.9), respectively:

$$P = 1 - \left[1 + \frac{d}{N_{50}}\left(2^{\frac{1}{\alpha}} - 1\right)\right]^{-\alpha} \tag{10.8}$$

$$P = 1 - \exp\left(-\frac{d}{k}\right) \tag{10.9}$$

where P is the infection risk; d is the inhalation dose (PFU); N_{50} and α are pathogen-specific parameters; and k is the reciprocal of the probability that a single pathogen will trigger a response.

10.4.6 Demonstration of the Dose–Response Model in an Aircraft Cabin

The Vietnam Airlines VN54 flight case involved a single index passenger (patient) in the business-class cabin of a B787 aircraft. The passenger transmitted COVID-19 to 12 susceptible passengers, resulting in an infection rate of around 60% within the business-class cabin (Figure 10.11). The flight was from London to Hanoi, with a total flight time of 10 hours 50 minutes. The index patient was seated in 5K, as

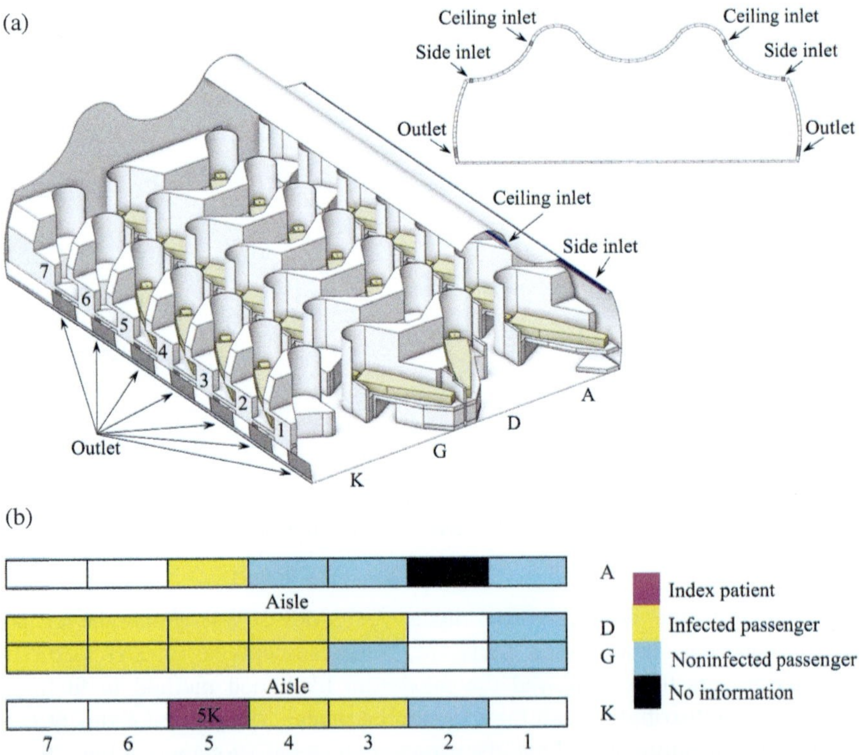

Figure 10.11 Schematic representation of the infection scenarios and seating in the VN54 aircraft cabin: (a) geometric model of the Boeing 787–9 business-class cabin; (b) planar view of the actual distribution of COVID-19 infection. *Source:* Wang et al. (2022) / with permission of John Wiley & Sons.

shown in Figure 10.11. In the B787 aircraft, conditioned air is supplied through slot inlets on the ceiling above and below the side stowage bins, respectively, on both sides of the aircraft.

CFD was used to numerically solve the airflow in the cabin, and then the spatial dispersion of respiratory aerosols released from passenger 5K was modeled (Wang et al. 2022). In this research, the coughing frequency was set to 15 times/h, while the breathing frequency was fixed at 720 times/h. The pulmonary ventilation rate followed a sinusoidal wave, and the total rate was approximately $0.3\,m^3/h$. The peak coughing jet speed was assumed to be 10 m/s. Table 10.3 presents the droplet size distributions and viral loads associated with droplets produced by various respiratory activities. For coughing, droplets less than 10 μm are mainly originated from the sputum in respiratory tracts, which have relatively higher gene copies

Table 10.3 Size distribution, release rate, and number of viral gene copies in the droplets produced by different respiratory activities.

Respiratory activity	Droplet diameter (μm)	Number of droplets from a single cough or breath	Gene copies per droplet
Coughing (Chao et al. 2009; Wang et al. 2022; Yang et al. 2007)	0.75	140,000	0.43
	1.32	71	5
	2.64	974	41
	5.28	362	80
	8.8	119	370
	12.32	44	1
	15.84	42	3
	19.8	36	6
	27.5	36	15
	38.5	25	42
	49.5	30	89
	60.5	28	163
	77	78	337
	99	44	716
	165	37	3,313
	330	25	26,507
Breathing (Fabian et al. 2008; Wang et al. 2022)	0.4	612	0.0045
	0.75	156	0.0294
	2.5	107	1.0904

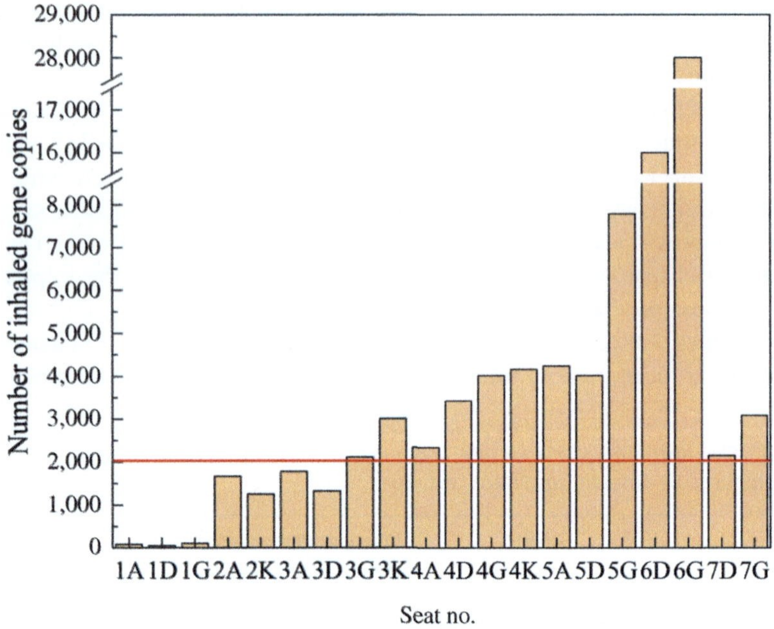

Figure 10.12 Evaluation of COVID-19 infection on flight VN54 by the dose–response model: predicted inhaled gene copies per passenger during the flight. *Source:* Wang et al. (2022) / with permission of John Wiley & Sons.

per droplet than those originated from the saliva in mouth, until the diameter exceeds 77 μm. Furthermore, it was assumed that the passengers did not talk at any time during the flight and that none of the passengers wore a face mask. A passenger was assumed to be infected if he/she inhaled a viral load of 2,000 or more gene copies. Further details of the CFD modeling can be found in Wang et al. (2022).

Figure 10.12 presents the number of gene copies inhaled by passengers during the entire flight. The red line indicates the threshold of 2,000 gene copies. Passengers were deemed to be infected when they inhaled gene copies at a quantity above this threshold. The infection status between the prediction and the actual situation was compared. There were actually 12 passengers infected with COVID-19, and the simulation model accurately predicted 11 of the infection cases, for an accuracy level of 91.7%.

10.5 Summary

Infectious aerosols in indoor environments have been found to transmit respiratory diseases such as COVID-19, MERS, and SARS. Developing strategies to control the indoor spread of respiratory diseases requires an understanding of the transmission mechanisms of pathogenic aerosols as well as an accurate prediction of the infection probability. This chapter briefly addresses the shedding and transmission of pathogenic aerosols and the resulting infection of susceptible individuals in indoor environments.

The aerosols shed during daily physiological activities such as breathing, coughing, sneezing, and speaking vary considerably in terms of their formation process, particle size distribution, and dispersion pattern, as well as the distance over which they can be transmitted. Respiratory pathogens are transmitted by three mechanisms: airborne transmission, droplet transmission, and contact transmission. Droplet transmission involves two key parameters: critical transport distance and critical droplet size. Airborne transmission can be either short-range or long-range, and is affected by the ventilation mode, ventilation rate, and particle removal efficiency (e.g., the ability to remove pathogen-laden particles). Meanwhile, contact transmission can be further classified as direct or indirect. Determining the inhalation dose that can lead to infection is a complex process. The viral load in the droplets exhaled by infected individuals can vary significantly with the type of respiratory disease or activity. The Wells–Riley model and the dose–response model are the most commonly used models for quantitatively assessing the infection risks of respiratory diseases.

Problems

10.1 How does relative humidity (RH) influence the critical droplet size for deposition versus evaporation?

10.2 Based on Table 10.3, why do larger droplets (>50 μm) from coughing pose a higher infection risk despite their shorter suspension time?

10.3 Explain the role of vocalization intensity (e.g., loud versus quiet talking) in aerosol emission rates and size distribution.

10.4 What are key limitations of the original Wells–Riley model?

10.5 Compare the size distributions, generation mechanisms, and transmission distances of aerosols produced by breathing, talking, coughing, and sneezing. How do factors like airflow velocity and relative humidity influence the behavior of these aerosols in indoor environments?

10.6 Explain the difference between airborne aerosol transmission and droplet transmission. How does the critical droplet size (e.g., 50 μm versus 5 μm) define these two transmission modes, and what role does ambient relative humidity play in determining the dominance of one mode over the other?

10.7 Using the classroom case study (Section 10.4.4), discuss the ways in which natural ventilation rates and mask-wearing influence COVID-19 infection probabilities. What are the key assumptions of the Wells–Riley model in predicting airborne transmission risks?

10.8 Analyze the roles of ventilation modes (e.g., cross ventilation versus single-sided ventilation) and supplementary systems (e.g., HEPA filters) in mitigating airborne transmission. Why might increasing ventilation rates alone sometimes fail to reduce infection risks significantly, as seen in the elevator car case?

10.9 Contrast the dose–response model with the Wells–Riley model in evaluating infection risks. How do stochastic and deterministic approaches differ in accounting for individual variability and environmental factors? Use the aircraft cabin case study (Section 10.4.6) to illustrate practical challenges in modeling real-world transmission events.

References

Abkarian, M. and Stone, H.A. (2020). Stretching and break-up of saliva filaments during speech: a route for pathogen aerosolization and its potential mitigation. *Physical Review Fluids* 5 (10): 102301.

Asadi, S., Wexler, A.S., Cappa, C.D. et al. (2019). Aerosol emission and superemission during human speech increase with voice loudness. *Scientific Reports* 9: 2348.

Asadi, S., Cappa, C.D., Barreda, S. et al. (2020). Efficacy of masks and face coverings incontrolling outward aerosol particle emission from expiratory activities. *Scientific Reports* 10: 15665.

Bourouiba, L. (2020). Turbulent gas clouds and respiratory pathogen emissions: potential implications for reducing transmission of COVID-19. *The Journal of the American Medical Association* 323 (18): 1837–1838.

Chao, C.Y.H., Wan, M.P., Morawska, L. et al. (2009). Characterization of expiration air jets and droplet size distributions immediately at the mouth opening. *Journal of Aerosol Science* 40 (2): 122–133.

Chen, W., Zhang, N., Wei, J. et al. (2020). Short-range airborne route dominates exposure of respiratory infection during close contact. *Building and Environment* 176: 106859.

Cozad, A. and Jones, R.D. (2003). Disinfection and the prevention of infectious disease. *American Journal of Infection Control* 31 (4): 243–254.

Dai, H. and Zhao, B. (2021). Movement and transmission of human exhaled droplets/droplet nuclei. *Chinese Science Bulletin* 66 (Z1): 493–500. (in Chinese).

Du, C.R., Wang, S.C., Yu, M.C. et al. (2020). Effect of ventilation improvement during a tuberculosis outbreak in underventilated university buildings. *Indoor Air* 30 (3): 422–432.

Duguid, J.P. (1945). The numbers and the sites of origin of the droplets expelled during expiratory activities. *British Medical Journal* 52 (11): 385–401.

Fabian, P., McDevitt, J.J., DeHaan, W.H. et al. (2008). Influenza virus in human exhaled breath: an observational study. *PLoS One* 3 (7): e2691.

Fennelly, K.P. (2020). Particle sizes of infectious aerosols: implications for infection control. *Lancet Respiratory Medicine* 8 (9): 914–924.

Fennelly, K.P. and Nardell, E.A. (1998). The relative efficacy of respirators and room ventilation in preventing occupational tuberculosis. *Infection Control and Hospital Epidemiology* 19 (10): 754–759.

Fisk, W.J., Seppänen, O., Faulkner, D. et al. (2004). Economic benefits of an economizer system: energy savings and reduced sick leave. *ASHRAE Transactions.* 111: 673–679.

Han, Z.Y., Weng, W.G., and Huang, Q.Y. (2013). Characterizations of particle size distribution of the droplets exhaled by sneeze. *Journal of the Royal Society Interface* 10 (88): 20130560.

Imai, K., Ikeno, R., Tanaka, H., and Takada, N. (2023). SARS-CoV-2 omicron variant in human saliva samples in cell-free form. *JAMA Network Open* 6 (1): e2250207.

Jayaweera, M., Perera, H., Gunawardana, B., and Manatunge, J. (2020). Transmission of COVID-19 virus by droplets and aerosols: a critical review on the unresolved dichotomy. *Environmental Research* 188: 109819.

Johnson, G.R. and Morawska, L. (2009). The mechanism of breath aerosol formation. *Journal of Aerosol Medicine and Pulmonary Drug Delivery* 22 (3): 229–237.

Jones, D.L., Baluja, M.Q., Graham, D.W. et al. (2020). Shedding of SARS-CoV-2 in feces and urine and its potential role in person-to-person transmission and the environment-based spread of COVID-19. *Science of the Total Environment* 749: 141364.

Kwon, S.-B., Park, J., Jang, J. et al. (2012). Study on the initial velocity distribution of exhaled air from coughing and speaking. *Chemosphere* 87 (11): 1260–1264.

Li, X. and Feng, B. (2023). Transmission of droplet aerosols in an elevator cabin: effect of the ventilation mode. *Building and Environment* 236: 110261.

Li, Y., Qian, H., Hang, J. et al. (2021a). Probable airborne transmission of SARS-CoV-2 in a poorly ventilated restaurant. *Building and Environment* 196: 107788.

Li, Y., Cheng, P., and Qian, H. (2021b). Dominant transmission route of SARS-CoV-2 and its implication to indoor environment. *Chinese Science Bulletin* 66 (Z1): 417–423. (in Chinese).

Lindsley, W.G., Derk, R.C., Coyle, J.P. et al. (2021). Efficacy of portable air cleaners and masking for reducing indoor exposure to simulated exhaled SARS-CoV-2 aerosols – United States, 2021. *Morbidity and Mortality Weekly Report* 70 (27): 972–976.

Liu, L., Li, Y., Nielsen, P.V. et al. (2017a). Short-range airborne transmission of expiratory droplets between two people. *Indoor Air* 27 (2): 452–462.

Liu, L., Wei, J., Li, Y., and Ooi, A. (2017b). Evaporation and dispersion of respiratory droplets from coughing. *Indoor Air* 27 (1): 179–190.

Liu, S., Zhao, X., Nichols, S.R. et al. (2022). Evaluation of airborne particle exposure for riding elevators. *Building and Environment* 207: 108543.

Liu, F., Luo, Z., and Qian, H. (2023). Impact of thermal stratification on airborne transmission risk of SARS-CoV-2 in various indoor environments. *Building Simulation* 16 (7): 1159–1172.

Luo, Q., Ou, C., Hang, J. et al. (2022). Role of pathogen-laden expiratory droplet dispersion and natural ventilation explaining a COVID-19 outbreak in a coach bus. *Building and Environment* 220: 109160.

de Michelena, P., Torres, I., Ferrando, E.-C. et al. (2023). RNA loads of severe acute respiratory syndrome coronavirus 2 in patients with breakthrough coronavirus disease 2019 caused by the Delta and omicron variants. *Clinical Microbiology and Infection* 29 (2): 256.e1–256.e4.

Morawska, L. (2006). Droplet fate in indoor environments, or can we prevent the spread of infection? *Indoor Air* 16 (5): 335–347.

Nicas, M. (1996). An analytical framework for relating dose, risk, and incidence: an application to occupational tuberculosis infection. *Risk Analysis* 16 (4): 527–538.

Ou, C., Hu, S., Luo, K. et al. (2022). Insufficient ventilation led to a probable long-range airborne transmission of SARS-CoV-2 on two buses. *Building and Environment* 207: 108414.

Park, S., Choi, Y., Song, D., and Kim, E.K. (2021). Natural ventilation strategy and related issues to prevent coronavirus disease 2019 (COVID-19) airborne transmission in a school building. *Science of the Total Environment* 789: 147764.

Prentiss, M., Chu, A., and Berggren, K.K. (2020). Superspreading events without superspreaders: using high attack rate events to estimate N_0 for airborne transmission of COVID-19. *MedRxiv* https://doi.org/10.1101/2020.10.21.20216895.

Qian, H. (2007). *Ventilation for Airborne Infection in Hospital Environment*. Hong Kong: Department of Mechanical Engineering of University of Hong Kong.

Riediker, M., Briceno-Ayala, L., Ichihara, G. et al. (2022). Higher viral load and infectivity increase risk of aerosol transmission for Delta and Omicron variants of SARS-CoV-2. *Schweizerische Medizinische Wochenschrift* 152: w30133.

Rigaud, M., Buekers, J., Bessems, J. et al. (2024). The methodology of quantitative risk assessment studies. *Environmental Health* 23 (1): 13.

Riley, E.C., Murphy, G., and Riley, R.L. (1978). Airborne spread of measles in a suburban elementary school. *American Journal of Epidemiology* 107 (5): 421–432.

Scharfman, B.E., Techet, A.H., Bush, J.W.M., and Bourouiba, L. (2016). Visualization of sneeze ejecta: steps of fluid fragmentation leading to respiratory droplets. *Experiments in Fluids* 57 (2): 24.

Stadnytskyi, V., Anfinrud, P., and Bax, A. (2021). Breathing, speaking, coughing or sneezing: what drives transmission of SARS-CoV-2? *Journal of Internal Medicine* 290 (5): 1010–1027.

Sze To, G.N. and Chao, C.Y. (2010). Review and comparison between the Wells-Riley and dose-response approaches to risk assessment of infectious respiratory diseases. *Indoor Air* 20 (1): 2–16.

Tang, J.W., Nicolle, A.D., Klettner, C.A. et al. (2013). Airflow dynamics of human jets: sneezing and breathing – potential sources of infectious aerosols. *PLoS One* 8 (4): e59970.

Tharayil, A., Rajakumari, R., Mozetic, M. et al. (2022). Contact transmission of SARS-CoV-2 on fomite surfaces: surface survival and risk reduction. *Interface Focus* 12 (1): 20210042.

Wang, C.C., Prather, K.A., Sznitman, J. et al. (2021). Airborne transmission of respiratory viruses. *Science* 373 (6558): eabd9149.

Wang, W., Wang, F., Lai, D., and Chen, Q. (2022). Evaluation of SARS-COV-2 transmission and infection in airliner cabins. *Indoor Air* 32 (1): e12979.

Watanabe, T., Bartrand, T.A., Weir, M.H. et al. (2010). Development of a dose-response model for SARS coronavirus. *Risk Analysis* 30 (7): 1129–1138.

Wei, J. and Li, Y.J.B. (2015). Environment, enhanced spread of expiratory droplets by turbulence in a cough jet. *Building and Environment* 93: 86–96.

Wei, J. and Li, Y. (2016). Airborne spread of infectious agents in the indoor environment. *American Journal of Infection Control* 44 (9): S102–S108.

Wells, W.F. (1934). On air-borne infection. Study II. Droplets and droplet nuclei. *The. American Journal of Epidemiology* 20 (3): 611–618.

Wells, W.F. (1955). *Airborne Contagion and Air Hygiene*, 117–122. Cambridge, MA: Harvard University Press.

WHO. (2014). Infection prevention and control of epidemic-and pandemic-prone acute respiratory infections in health care. WHO Guidelines. ISBN 9789241507134.

Xie, X., Li, Y., Chwang, A.T.Y. et al. (2007). How far droplets can move in indoor environments – revisiting the Wells evaporation–falling curve. *Indoor Air* 17 (3): 211–225.

Xu, C., Nielsen, P.V., Liu, L. et al. (2017). Human exhalation characterization with the aid of Schlieren imaging technique. *Building and Environment* 112: 190–199.

Yang, S., Lee, G.W.M., Chen, C.M. et al. (2007). The size and concentration of droplets generated by coughing in human subjects. *Journal of Aerosol Medicine* 20 (4): 484–494.

Zhang, S. and Lin, Z. (2021). Dilution-based evaluation of airborne infection risk – thorough expansion of Wells-Riley model. *Building and Environment* 194: 107674.

Zhang, N., Chen, W., Chan, P.T. et al. (2020). Close contact behavior in indoor environment and transmission of respiratory infection. *Indoor Air* 30 (4): 645–661.

Zhao, X., Liu, S., Yin, Y. et al. (2022). Airborne transmission of COVID-19 virus in enclosed spaces: an overview of research methods. *Indoor Air* 32 (6): e13056.

Zhou, P., Yang, X.L., Wang, X.G. et al. (2020). A pneumonia outbreak associated with a new coronavirus of probable bat origin. *Nature* 579 (7798): 270–273.

Zhu, S., Kato, S., and Yang, J.H. (2006). Study on transport characteristics of saliva droplets produced by coughing in a calm indoor environment. *Building and Environment* 41 (12): 1691–1702.

Zhu, L., Li, X., Feng, B., and Liu, F. (2024). Evaluation of transmission risk of respiratory particles under different ventilation strategies in an elevator. *Building Simulation* 17 (5): 771–784.

11

Particle Filtration and Indoor Air Cleaning

11.1 Introduction

In case the indoor aerosol concentration is above the prescribed limit, the concentration must be lowered through filtration. In general, four different filtration approaches and their combinations can be employed to remove airborne particles: filtration with the use of a fiber filter, electrostatic separation, mechanical separation by means of the particles' gravity and inertia, and washing to remove dust. In light of the relatively low concentration and small size of suspended particles in typical indoor spaces, filtration is usually accomplished by means of fiber filters, which are generally very efficient and reliable. Electrostatic separation is also becoming increasingly popular because of its relatively high filtration efficiency and low resistance. This chapter mainly addresses particle filtration by fiber filters and electrostatic separation, including their basic principles, performance evaluation models, and guidelines for operation.

11.2 Fiber Filters

11.2.1 Filtration Mechanisms

Fiber filters are widely used to remove airborne particles. The filters remove particles by means of interception, inertia impaction, Brownian diffusion, gravitational settling, and even electrostatic attraction. The following sections briefly introduce each of these mechanisms.

11.2.1.1 Interception

Within the fiber filters, the fibers are randomly distributed, forming numerous meshes. The meshes may be comprised of dozens or even hundreds of layers. Figure 11.1 is an illustration of particles of a certain size as they approach the

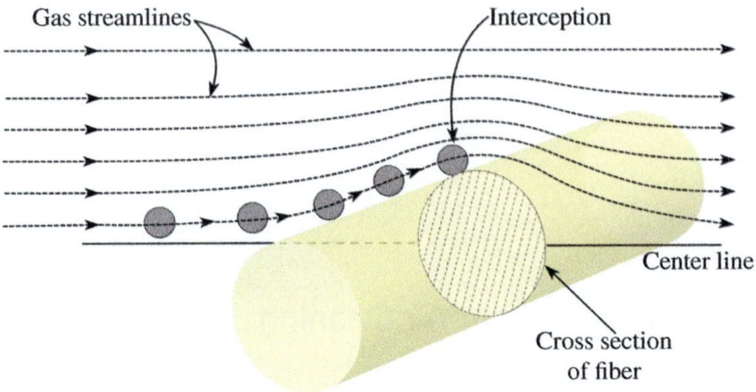

Figure 11.1 Schematics of single-fiber particle collection by interception. *Source:* Hinds and Zhu (2022) / with permission of John Wiley & Sons.

fiber surface along streamlines. If the distance between a given streamline and the fiber surface is equal to or less than the particle radius, the particles in the region between the two symmetric streamlines along the center line will be intercepted and subsequently deposited on the fiber surface. This phenomenon is referred to as the interception of particles. A sieve can also have an interception effect.

The particle collection efficiency of a single fiber resulting from interception varies with the dimensionless parameter R:

$$R = \frac{d_p}{d_f} \tag{11.1}$$

where d_p refers to the particle diameter, and d_f is the fiber diameter. The particle collection efficiency of a single fiber due to interception, E_R, can be given as (Kulkarni et al. 2011):

$$E_R = \frac{(1-\alpha)R^2}{\mathrm{Ku}(1+R)} \tag{11.2}$$

where Ku is the Kuwabara hydrodynamic factor, which is a dimensionless number that accounts for the flow distortion around a fiber. Ku is solely dependent on the solidity α (the ratio of the solid material's volume to the overall volume of the fiber filter in space) as follows:

$$\mathrm{Ku} = -\frac{\ln \alpha}{2} - \frac{3}{4} + \alpha - \frac{\alpha^2}{4} \tag{11.3}$$

Ku ranges from 1.9 to 0.25 for α from 0.005 to 0.2. For $d_f < 2\,\mu\mathrm{m}$, a slip correction is needed in Eq. (11.3).

11.2.1.2 Inertial Impaction

Due to the complex arrangement of fibers, streamlines change direction abruptly and frequently when air passes through the fibrous layers. If the size of the particles is relatively large or their velocity is high, they do not follow a given streamline closely as the air flows around the fibers, due to the particles' inertia. Consequently, particles deviate from the airflow streamline when approaching a fiber and may subsequently collide with the fiber, as shown in Figure 11.2. If particles do not collide with the fiber due to the inertial effect but do collide due to the interception effect, then particle capture can be ascribed to a combination of these two effects.

The Stokes number, defined in Eq. (11.4) as the ratio of the particle stopping distance to the fiber diameter, can be used as an indicator of the inertial impaction effect:

$$\mathrm{Stk} = \frac{\tau U_0}{d_\mathrm{f}} = \frac{\rho_\mathrm{p} d_\mathrm{p}^2 C_\mathrm{c} U_0}{18\eta d_\mathrm{f}} \tag{11.4}$$

Equation (11.4) describes the ratio of the "persistence" of a particle to the target, though the airflow direction has changed. The efficiency of single-fiber collection by impaction increases with the Stokes number due to: (i) greater particle inertia for a larger ρ_P and/or d_p, (ii) higher particle velocity, or (iii) more pronounced streamline curvature, resulting from a smaller fiber size. The particle collection efficiency of a single fiber for impaction, E_I, is expressed as (Yeh and Liu 1974):

$$E_\mathrm{I} = \frac{(\mathrm{Stk})J}{2\mathrm{Ku}^2} \tag{11.5}$$

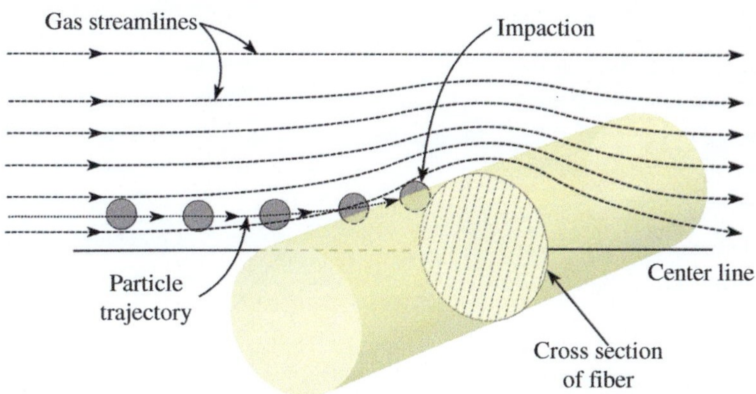

Figure 11.2 Schematics of single-fiber particle collection by impaction. *Source:* Hinds and Zhu (2022) / with permission of John Wiley & Sons.

where

$$J = (29.6 - 28\alpha^{0.62})R^2 - 27.5R^{2.8}, \quad \text{for } R < 0.4 \tag{11.6}$$

There is no simple equation for J when R is greater than 0.4. For approximate estimation, a value of 2.0 can be assigned to J when $R > 0.4$ (Hinds and Zhu 2022). As anticipated, inertial impaction is the predominant mechanism to collect large particles; however, such particles also typically undergo collection by interception. It should be noted that:

$$E_{\mathrm{I}} + E_{\mathrm{R}} \leq 1 + R \tag{11.7}$$

11.2.1.3 Diffusion

The Brownian motion of fine particles, due to random collisions by gas molecules, can be substantial. At normal pressure and temperature, 0.1 μm particles have an averaged Brownian displacement of 37 μm per second, as shown in Table 5.3, which is several times or even dozens of times the interspace distance between fibers (Xu 2018). As illustrated in Figure 11.3, Brownian diffusion may increase the probability of fine particle deposition on the fiber surface. However, for particles larger than 0.3 μm, Brownian diffusion is less pronounced and typically insufficient to cause significant deviation from the flow streamline or deposition onto the fiber surface.

The particle collection efficiency of a single fiber due to diffusion varies with the dimensionless Peclet number, Pe, which is defined as:

$$\mathrm{Pe} = \frac{d_{\mathrm{f}} U_0}{D} \tag{11.8}$$

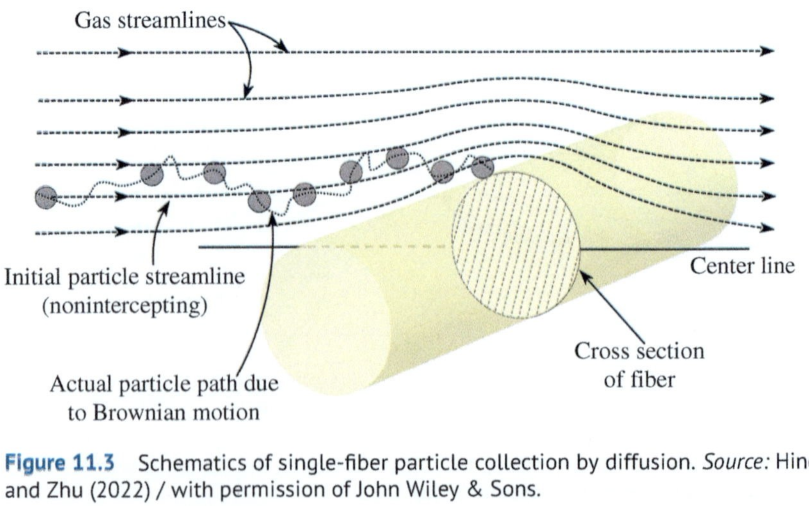

Figure 11.3 Schematics of single-fiber particle collection by diffusion. *Source:* Hinds and Zhu (2022) / with permission of John Wiley & Sons.

where D is the particle diffusion coefficient. The single-fiber efficiency due to diffusion, E_D, can then be predicted as (Kirsch and Fuchs 1968):

$$E_D = 2\,\text{Pe}^{-2/3} \tag{11.9}$$

Equation (11.9) is derived from experimental measurement data, and the coefficient 2 is an empirically determined value (Kirsch and Fuchs 1968). Single-fiber efficiency increases with the decrease of both Pe and particle size. E_D is the only deposition mechanism that enhances with the decrease of d_p. Theoretical expressions of E_D should incorporate the effect of Ku for the flow distortion around a fiber (Davies 1973; Brown 1993). When calculating the overall collection efficiency of a single fiber near the size of minimum efficiency, it is vital to account for the enhanced collection arising from the interception of the diffusing particles, E_{DR}, as (Hinds and Zhu 2022):

$$E_{DR} = \frac{1.24R^{2/3}}{(\text{Ku}\cdot\text{Pe})^{1/2}} \tag{11.10}$$

11.2.1.4 Gravitational Settling

Particles may deviate from the gas streamlines under the action of gravity when passing through fibrous layers, resulting in their deposition on the fiber surfaces by gravitational settling. The time required for air to pass through a fibrous filter, particularly a paper filter, is significantly less than 1 second. Consequently, particles smaller than 0.5 μm passing through the fibrous layers do not deposit on fiber surfaces as a result of gravitational settling because gravitational settling hardly occurs in such a short time. In such circumstances, it is allowable to omit the gravitational deposition effect.

The dimensionless number, G, which indicates the particle deposition due to gravitational settling can be expressed as:

$$G = \frac{V_{TS}}{U_0} = \frac{\rho_p d_p^2 C_c g}{18\eta U_0} \tag{11.11}$$

where V_{TS} is the terminal settling velocity. If U_0 and V_{TS} are in the same direction:

$$E_G \approx G(1 + R) \tag{11.12}$$

If U_0 and V_{TS} are in opposite directions:

$$E_G \approx -G(1 + R) \tag{11.13}$$

In horizontal flow scenarios, gravitational settling can generally be disregarded. Commonly, E_G is negligible compared to other mechanisms, except in cases where particle size is quite large and U_0 is low. If U_0 exceeds 10 cm/s, impaction becomes more significant than gravitational settling.

11.2.1.5 Electrostatic Attraction

Electrical charges may be present on both fibers and particles, resulting in an electrostatic field and particle attraction. Except for intentionally charged fibers or particles, neither the fiber charges created by friction during the manufacturing process nor the particle charges arising from induction persist for a long time. The electric field intensities produced by these charges are generally weak, and the resulting attractive forces are negligible.

Increasing the charge on either the particles or the fibers, or reducing the face velocity, can enhance the particle collection efficiency by electrostatic attraction. The theory of particle collection by charged fibers, charged particles, and their combination has been reviewed by Brown (1993). Particles with charges are attracted to oppositely charged fibers through coulombic attraction. When the electric field generated by the charged fibers induces dipoles on particles or when the charges are separated within the particle, an originally neutral particle can also be attracted to a charged fiber. In the non-uniform electric field surrounding the fiber, the proximal side of the particle undergoes an attractive force that exceeds the repulsive force on the distal side. Consequently, a net force arises in the direction of approaching the fiber, and the particle migrates accordingly. Moreover, a charged particle can be attracted to a neutral fiber at close range by image forces. The charged particle induces equal and opposite charges on the fiber surface, thereby creating an electric field between. Image forces are weaker than coulombic forces.

To predict the electrostatic attraction of particles to fiber filters, the charge distribution at a microscopic level is required. However, the microscopic charge distribution is typically not known and may significantly differ from the macroscopic charge state. The charges on a particle may be estimated from the conditions under which the charges are acquired. Based on experimental measurements of glass fiber filters, the particle collection efficiency by a single fiber due to electrostatic image forces, for a neutral fiber and a particle with charge q, was proposed as (Brown 1993):

$$E_{\mathrm{q}} = 1.5\left[\frac{(\varepsilon_{\mathrm{f}}-1)}{(\varepsilon_{\mathrm{f}}+1)}\frac{q^2}{12\pi^2\eta U_0\varepsilon_0 d_{\mathrm{p}} d_{\mathrm{f}}^2}\right]^{1/2} \tag{11.14}$$

where ε_{f} is the fiber's relative permittivity (dielectric constant), q is the number of charges on the particle, and ε_0 is the permittivity in a vacuum.

11.2.2 Performance Evaluation of Fiber Filters

11.2.2.1 Face Velocity and Filtration Velocity

Face velocity U_0 is defined as the velocity of air passing the cross section of an air filter (m/s), i.e.:

$$U_0 = \frac{Q}{A} \tag{11.15}$$

Table 11.1 Range of filtration velocities in typical air filters.

Type	Coarse filter	Fine filter, high-to-medium efficiency air filter	Sub-HEPA filter	HEPA filter
Magnitude of filtration velocity	m/s	dm/s	cm/s	cm/s

Source: Xu (2018) / with permission of Springer Nature.

where Q refers to the volume flow rate, and A is the cross-sectional area of the filter. The face velocity represents the capacity of the flow rate passing through the cross-sectional area of the air filter.

Filtration velocity U is defined as the flow velocity of the air passing through the expanded area of the filter medium, as follows:

$$U = \frac{Q}{A(1 - \alpha)} \tag{11.16}$$

where α refers to solidity or packing density. For fibrous filters, α is defined as:

$$\alpha = \frac{\text{fiber volume}}{\text{total volume}} = 1 - \text{porosity} \tag{11.17}$$

where α is typically between 0.01 and 0.3. The filtration velocity represents the capacity for airflow to pass through the filter medium, particularly in relation to the filtration efficacy of the medium. Within a filter, the air velocity U is marginally greater than U_0 due to the reduction in air passage volume caused by the presence of fibers. Table 11.1 illustrates the range of filtration velocities for some typical air filters. The magnitude of the filtration velocity across coarse filters is in m/s, whereas for fine filters and high-to-medium efficiency filters, the velocity is typically in dm/s. The range of filtration velocities for sub-HEPA (high-efficiency particulate air) filters is 5–7 cm/s, while that for HEPA filters and ULPA (ultra-low penetration air) filters is generally 2–3 cm/s.

11.2.2.2 Efficiency and Penetration Ratio

When filters are used for particle sampling, their particle collection capability is typically characterized by their collection efficiency, i.e., the fraction of entering particles retained by the filter. This efficiency can be expressed in terms of either particle number collection efficiency (E) or mass collection efficiency (E_m). The latter refers to the fraction of the entering particulate mass retained by the filter. Generally, the mass efficiency is higher than the number count efficiency because large particles possessing a higher mass ratio are easier to filter. The parameters E and E_m are defined as:

$$E = \frac{N_{in} - N_{out}}{N_{in}} \tag{11.18}$$

where N is the particle number concentration, and

$$E_m = \frac{C_{in} - C_{out}}{C_{in}} \tag{11.19}$$

where C is the particle mass concentration.

Single-fiber efficiencies can be predicted by Eq. (11.20) if each filtration mechanism acts independently and is less than 1.0:

$$E_\Sigma = 1 - (1 - E_R)(1 - E_I)(1 - E_D)(1 - E_{DR})(1 - E_G) \tag{11.20}$$

or

$$E_\Sigma = E_R + E_I + E_D + E_{DR} + E_G \tag{11.21}$$

where the equations for calculating the single-fiber efficiencies are provided in Section 11.2.1. Although Eq. (11.21) is theoretically imprecise, it is frequently used for rough calculations due to its simplicity. However, it may overestimate the particle collection efficiency, as different mechanisms can compete for capture of the same particle, potentially resulting in duplicate calculations of the efficiency. In many instances, one mechanism predominates, allowing the approximation of overall efficiency to be based solely on that mechanism. The values calculated by Eqs. (11.20) and (11.21) are in agreement within 5% when a single mechanism predominates and when the other mechanisms' contributions to the efficiency are less than 1%.

The overall efficiency of a filter, E, can be expressed as a function of the single-fiber efficiency, E_Σ. And E can be expressed as:

$$E = 1 - \exp\left(\frac{-4\alpha E_\Sigma t}{\pi d_f}\right) \tag{11.22}$$

where t is the thickness of the filter material, α is the filling ratio or the solidity, and d_f is the fiber diameter. When the performance of air filters is evaluated, it is standard practice to use the penetration ratio P, rather than the filtration efficiency.

$$P = 1 - E \tag{11.23}$$

The mass penetration ratio P_m is:

$$P_m = 1 - E_m \tag{11.24}$$

The particle collection efficiency of a filter varies with the aerosol particle size. As for fine particles, the primary filtration is by means of Brownian diffusion. Meanwhile, interception and inertial impaction play a major role in collecting large particles. There is an intermediate particle size at which the particle penetration through the filter peaks, and this size is defined as the most penetrating particle size (MPPS). An example of MPPS is shown in Figure 11.4. The MPPS is subject to the face velocity of a filter, the fiber packing density, and the fiber diameter (Lee and Liu 1980).

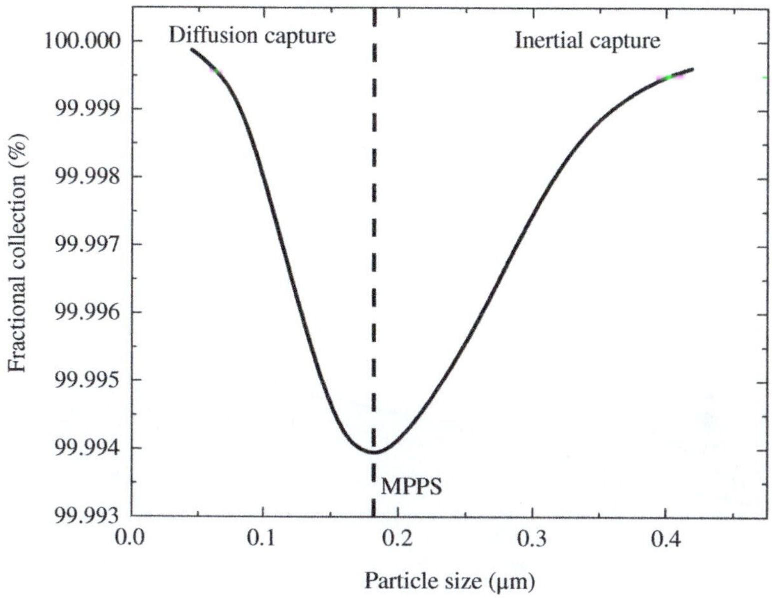

Figure 11.4 Particle filtration efficiency for a fiber filter, where the most penetrating particle size (MPPS) is around 0.2 μm because larger particles are collected due to impaction and interception, while smaller particles are captured by diffusion. *Source:* Xu (2018) / with permission of Springer Nature.

11.2.2.3 Pressure Drop

The pressure drop across an air filter involves two components: the pressure drop across the filter medium and the pressure drop across the supporting and enclosure structure of the air filter. The pressure drop across the filter medium exhibits a linear proportional relationship with the filtration velocity, whereas the total pressure drop across the air filter demonstrates an exponential relationship with the filtration velocity, as illustrated in Figure 11.5.

The pressure drop can be generally estimated as (Davies 1973):

$$\Delta p = \frac{\eta t U_0 f(\alpha)}{d_\text{f}^2} \tag{11.25a}$$

$$f(\alpha) = 64\alpha^{1.5}(1 + 56\alpha^3), \quad \text{for } 0.006 < \alpha < 0.3 \tag{11.25b}$$

It is evident that the pressure drop is directly proportional to the filter thickness t and inversely proportional to d_f^2. Additionally, pressure drop is directly proportional to U_0. For $d_\text{f} < 1$ μm, the aforementioned equation may slightly overestimate the pressure drop due to the slip effect.

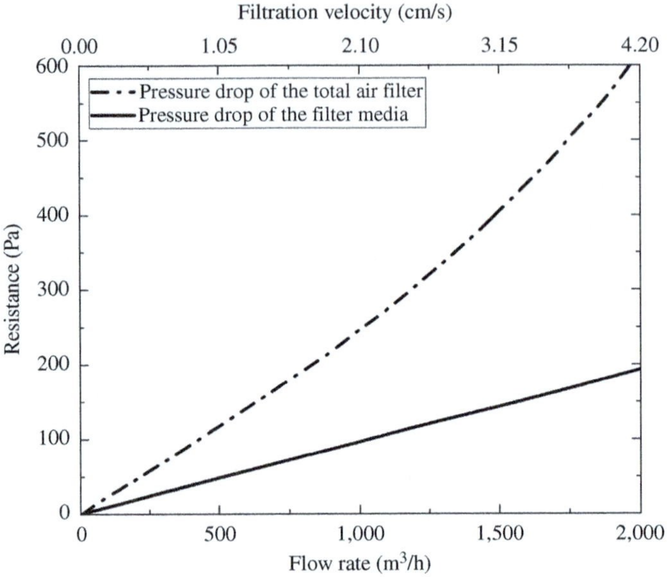

Figure 11.5 Variation of pressure drop with the airflow rate and the filtration velocity for a HEPA filter. *Source:* Xu (2018) / with permission of Springer Nature.

11.2.2.4 Quality Factor

An ideal filter is expected to have the highest filtration efficiency while maintaining the minimum pressure drop across the filter. The quality factor of a filter is defined as:

$$q_F = \frac{-\ln(1-E)}{\Delta p} = \frac{\ln(1/P)}{\Delta p} \tag{11.26}$$

where q_F is the quality factor of a filter, E is the particle filtration efficiency, P is the particle penetration ratio, and Δp is the pressure drop across the filter. Substituting Eqs. (11.22) and (11.25a) into Eq. (11.26), it yields:

$$q_F = \frac{4\alpha E_\Sigma d_f}{\pi \eta U_0 f(\alpha)} \tag{11.27}$$

The quality factor of a fiber filter does not depend on the filter medium's thickness. A higher q_F indicates better performance of a filter. It should be noted that the comparison of q_F among different filters should be based on the same face velocity and particle size.

11.2.2.5 Dust-holding Capacity

Dust-holding capacity is a parameter directly correlated with the operational lifetime of an air filter. Historically, the dust-holding capacity of an air filter was defined as the mass of dust deposited on the filter when the final pressure drop

Table 11.2 Relationship between the operational flow rate and the time to reach the operational flow rate.

Ratio between operational flow rate and nominal flow rate	0.5	0.7	0.75	0.8	1.0	1.25
Time to reach the operational flow rate	$3.5T_0$	$2.15T_0$	$1.91T_0$	$1.7T_0$	T_0	$0.59T_0$

Source: Xu (2018) / with permission of Springer Nature.

during operation reaches approximately twice the initial pressure drop. Table 11.2 illustrates the correlation between the operational flow rate and the time to reach the operational flow rate.

11.3 Electrostatic Precipitator

ESPs can employ either dry or wet technology. In the dry ESP, the adhered particulate on electrodes is collected in dry form after mechanical rapping, while in a wet ESP, the adhered particulate on electrodes is washed and collected in liquid. In this textbook, only the dry type is addressed.

11.3.1 Basic Principles of ESPs

ESPs utilize the forces exerted by an electric field on charged particles to separate the particles from a gas stream. These forces act exclusively on the particles and can be significantly greater than the gravitational or inertial force. Two basic steps are involved: charging the particles and subjecting them to an electric field, such that their electrostatic migration leads them to deposit on a collection surface. Removal of the deposited particles from the collection surface is crucial for the continuous operation of a precipitator. ESPs feature low pressure drop and high efficiency for small particles, as well as the capacity to handle high concentrations of dust. In an ESP, particles are typically charged by a corona discharge.

Figure 11.6 illustrates the components of a dry ESP (plate-type). Dust-laden gases flow horizontally through parallel grounded collecting plates. Between these plates, there are a number of high-voltage electrodes. This system is powered by a transformer rectifier (TR), and the assembly is referred to as a bus section. There are usually several independent bus sections along the length of the precipitator. For high gas flows, there may be multiple bus sections in parallel. One or several (parallel) casings enclose the bus sections to form the ESP plant. TR sets are commonly positioned on the roof of the electrostatic precipitator.

Figure 11.7 illustrates the basic principles of particle capture by an ESP. The high voltage generates a corona current that flows from the discharge electrodes

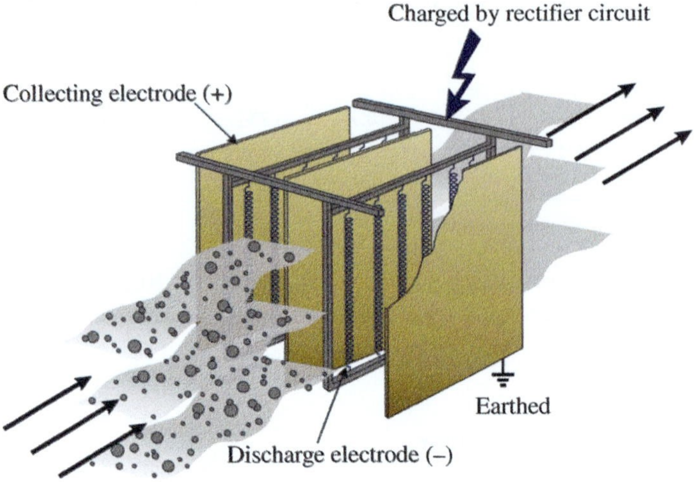

Figure 11.6 Schematic of the components of a dry electrostatic precipitator.

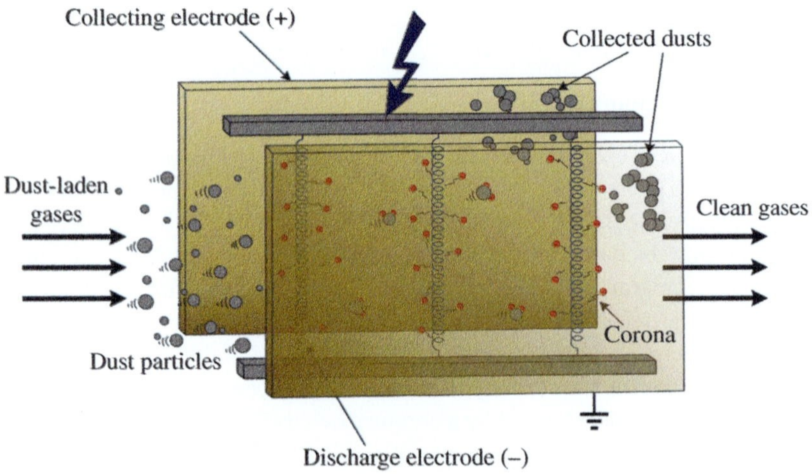

Figure 11.7 Principle of particle collection in an electrostatic precipitator.

to the collecting plates. As the particulate-laden gases enter the ESP, they become electrically charged. An electric field subsequently directs the charged particles toward the collecting plate. A layer of particulate matter accumulates on the plate. At predetermined intervals, this layer is dislodged, most commonly through mechanical rapping of the plate, causing the particulate matter to descend for collection.

11.3.2 Evaluation of ESP Performance

11.3.2.1 Electric Field

An electric field in space exerts an electrostatic force on a charged particle. The strength of this electric field is expressed in terms of the force $\boldsymbol{F}$ produced per unit charge on the particle. The electric field strength, or intensity, $\boldsymbol{E}_{\mathrm{e}}$ is defined as:

$$\boldsymbol{E}_{\mathrm{e}} = \frac{\boldsymbol{F}_{\mathrm{E}}}{q} \tag{11.28}$$

where q is the number of charges on the particle. The unit of the electric field strength is N/C. Electric field strength is a vector in the same direction as the force $\boldsymbol{F}_{\mathrm{E}}$. Commonly, the number of charges q can be expressed as n multiples of the smallest unit of charge, i.e., the charge of an electron e, and $e = 1.6 \times 10^{-19}$ C.

$$q = ne \tag{11.29}$$

Thus, the force on a particle with n elementary units of charge in an electric field $\boldsymbol{E}_{\mathrm{e}}$ is:

$$\boldsymbol{F}_{\mathrm{E}} = q\boldsymbol{E}_{\mathrm{e}} = ne\boldsymbol{E}_{\mathrm{e}} \tag{11.30}$$

Equation (11.30) represents the fundamental equation for the electrostatic force exerted on an aerosol particle. The implementation of this equation is simple and direct once the values of n and $\boldsymbol{E}_{\mathrm{e}}$ are known.

11.3.2.2 Particle Migration Velocity

When a charged particle is subjected to an electric field, it undergoes a force $\boldsymbol{F}_{\mathrm{E}}$ (Eq. 11.30). The resultant particle velocity can be determined through a method analogous to that employed for terminal settling velocity. In the case of particle motion within the Stokes region, equating the electrostatic force to the Stokes drag yields the terminal electrostatic velocity as:

$$\boldsymbol{V}_{\mathrm{TE}} = \frac{ne\boldsymbol{E}_{\mathrm{e}}C_{\mathrm{c}}}{3\pi\eta d_{\mathrm{p}}} \tag{11.31}$$

The aforementioned equation can be rewritten in terms of the particle mechanical mobility B as:

$$\boldsymbol{V}_{\mathrm{TE}} = ne\boldsymbol{E}_{\mathrm{e}}B \tag{11.32}$$

To determine $\boldsymbol{V}_{\mathrm{TE}}$ when the particle motion is outside the Stokes region, it is necessary to employ a procedure analogous to that utilized for calculating $\boldsymbol{V}_{\mathrm{TS}}$ (settling velocity due to gravity) for Re > 1. A value of Re > 1 is more prevalent for electrostatic motion than for gravitational settling, as the electrostatic force can significantly exceed the gravitational force. Consequently, we have:

$$neE_{\mathrm{e}} = C_{\mathrm{D}}\frac{\pi}{8}\rho_{\mathrm{g}}d_{\mathrm{p}}^2V_{\mathrm{TE}}^2 \tag{11.33}$$

Multiplying both sides of Eq. (11.33) by Re^2 and reorganizing, we obtain the following equation:

$$C_\text{D}\ \text{Re}^2 = \frac{8neE_\text{e}\rho_\text{g}}{\pi\eta^2} \tag{11.34}$$

$C_\text{D}\ \text{Re}^2$ can be calculated without information on the velocity or the Reynolds number at the terminal electrostatic velocity. According to the value of $C_\text{D}\ \text{Re}^2$, the terminal electrostatic velocity can subsequently be determined.

It should be noted that the migration velocity cannot be measured directly; it must be calculated from the measured collection efficiency, the measured or calculated gas flow, and the actual collecting area, as will be explained below.

11.3.2.3 Particle Collection Efficiency

The collection efficiency can be determined through various methods. A widely used approach is the modified Deutsch equation developed by Matts and Ohnfeldt (1963):

$$E = 1 - \exp\left[-\left(\frac{V_\text{TE}A}{Q}\right)^k\right] \tag{11.35}$$

where E represents the particle collection efficiency, A denotes the collection area, Q stands for the gas volume flow rate, and k is a constant ranging from 0.5 to 1.0. A significant parameter is A/Q, which facilitates comparison (normalization) between different ESPs. This parameter indicates the collection area required per m^3/s of gas, which is also called the specific collection area (SCA).

11.4 Operation of Air Cleaners

11.4.1 PM Exposure Limits

We shall understand the allowed PM (particulate matter) exposure levels before carrying out particle filtration. Over-filtration may not provide any benefits; rather, it may waste energy and filtration materials. Meanwhile, under-filtration may result in unacceptable indoor PM concentrations.

The WHO, the US EPA, and the European Commission of the EU have recommended various particle exposure concentration limits, as listed in Tables 11.3–11.5, respectively. The PM exposure concentration limits are commonly evaluated in terms of annual mean or 24-hour mean. The allowed daily exposure level is higher than the annual mean. The maximum exposure limit for PM_{10} is higher than that for $\text{PM}_{2.5}$, given that $\text{PM}_{2.5}$ can reach the alveoli. The PM exposure limits set by the WHO or the so-called air quality guideline (AQG) levels are the strictest, and the exceedance of these limits is associated with risks to public health. The US EPA recommends different exposure limits for sensitive populations and for the

Table 11.3 Particle exposure concentration limits recommended by the WHO.[a)]

Particle size range	Annual mean	24 h mean
$PM_{2.5}$	5 μg/m³	15 μg/m³
PM_{10}	15 μg/m³	45 μg/m³

a) Adapted from WHO global air quality guidelines: particulate matter ($PM_{2.5}$ and PM_{10}), ozone, nitrogen dioxide, sulfur dioxide and carbon monoxide, 2021. https://www.who.int/publications/i/item/9789240034228/, last accessed on 22 October 2025.

Table 11.4 Particle exposure concentration limits recommended by the US EPA.[a)]

Particle size range	Primary/ secondary	Averaging period	Concentration limit	Algorithm for averaging calculation
$PM_{2.5}$	Primary	1 yr	9.0 μg/m³	Annual mean, averaged over 3 yr
	Secondary	1 yr	15.0 μg/m³	Annual mean, averaged over 3 yr
	Primary and secondary	24 h	35 μg/m³	98th percentile, averaged over 3 yr
PM_{10}	Primary and secondary	24 h	150 μg/m³	Not to be exceeded more than once per year on average over 3 yr

a) Note: Primary standards provide public health protection, including protection of "sensitive" populations such as children, older adults, and people with asthma. Secondary standards provide public welfare protection, including protection against decreased visibility and damage to animals, crops, vegetation, and buildings. Adapted from NAAQS Table, https://www.epa.gov/criteria-air-pollutants/naaqs-table, last accessed on 22 October 2025.

general public. Furthermore, the EPA accounts for the PM concentrations over three consecutive years, and this approach yields the most complicated evaluation among the three organizations. The EU recommendations are more or less similar to those of the EPA, but the evaluation process is much simpler.

Although Tables 11.3–11.5 address the general PM standards for atmospheric air, the recommended exposure limits can also be considered in the evaluation of indoor particle exposure. Small amounts of PM may be removed or deposited when PM of outdoor origin migrates into an indoor space. In this case, the ratio of the indoor PM concentration to the outdoor concentration, i.e., the I/O ratio, will generally be less than 1.0. However, I/O ratios vary greatly with numerous factors. Consequently, it is generally safe to use the values in Tables 11.3–11.5 as a criterion to determine the need for operating indoor PM filtration devices, although this approach may be somewhat conservative.

Table 11.5 Particle exposure concentration limits recommended by the European Commission of the European Union.[a]

Particle size range	Averaging period	Concentration limit
$PM_{2.5}$	1 yr	20 μg/m^3
PM_{10}	24 h	50 μg/m^3
	1 yr	40 μg/m^3

a) Adapted from EU air quality standards, https://environment.ec.europa.eu/topics/air/air-quality/eu-air-quality-standards_en, last accessed on 22 October 2025.

11.4.2 Air Cleaner Operation

11.4.2.1 Clean Air Delivery Rate (CADR)

The method for testing air cleaners was initially developed by the Association of Home Appliance Manufacturers (AHAM) in the United States. This method lacked theoretical derivation and instead relied on the concept of the clean air delivery rate (CADR), denoted as Q_{CADR}, to characterize the "purification performance for the target pollutant". This metric was employed to assess the operational properties of air cleaners. The CADR, Q_{CADR}, is defined as:

$$Q_{\text{CADR}} = kV \tag{11.36}$$

where V is the volume of the well-enclosed test chamber (m^3), and k is the decay constant (h^{-1}). This constant, in turn, is defined as:

$$k = nE \tag{11.37}$$

where n is the equivalent air change rate (h^{-1}) of the test chamber provided by the air cleaner, and E is the total efficiency (%) of the air filters inside the air cleaner.

The efficacy of an air cleaner in reducing indoor pollutant concentrations is contingent upon the equivalent air change rate of the room provided by the air cleaner, the room's volume, as well as the efficiency of the air filters installed in the cleaner. Furthermore, the placement of the air cleaner, along with the configuration and allocation of supply and return air openings, can influence air distribution, subsequently affecting purification performance. This phenomenon can be quantified in terms of the non-uniform coefficient in cleanroom environments. For air cleaners in test chambers, the value of k does not reflect this influence, which is not addressed at present. Consequently, the well-mixed condition inside the test chamber should be ensured. Substituting Eq. (11.37) into Eq. (11.36) yields:

$$Q_{\text{CADR}} = nEV \tag{11.38}$$

Since the nominal flow rate $Q_0 = nV$, Eq. (11.38) can be rewritten as:

$$Q_{\text{CADR}} = Q_0 E \tag{11.39}$$

A higher CADR implies greater air purification capability in an air cleaner. Either an increase in the nominal flow rate of the air cleaner or the total efficiency of the air filters can increase the CADR. Consequently, for instance, if we have an air cleaner with $E = 0.95$ and wish to achieve the same CADR and purification capability as an air cleaner with $E = 0.995$, then we need to increase Q_0 by a factor of $0.995/0.95 = 1.05$.

11.4.2.2 Operation of Air Cleaners and Replacement of Air Filters

If the indoor particle emission rates are known, it is not difficult to use the multi-zone model or CFD model to predict either the transient or steady-state particle concentrations, assuming that all the related boundary conditions are known. The CADR that will provide the expected well-mixed particle concentrations can be roughly estimated from the indoor particle emission rate, the outdoor particle infiltration rate, the air change rate of the room, the room's volume, the particle deposition rate inside the room, and so on. However, the particle concentration is only one of a number of critical parameters in the selection and operation of an air cleaner. The noise level when running the air cleaner, the flow distribution created inside the room, the energy consumption rate, the initial investment cost, and the readiness and expense of maintenance are also factors in the use of indoor air cleaners. A detailed analysis of air cleaner operation is beyond the scope of this book. Interested readers are advised to find this information in other reference materials.

It is known that the resistance of a fiber filter increases as dust accumulates on the filter. Consequently, the fan pressure must increase to ensure the same operating flow rate. In most situations, the fan output is fixed, and hence the operational flow rate decreases as the operating time increases. To prolong the lifetime of air filters, ASHRAE Standard 52.2-2017 specifies that they not be replaced until the final pressure drop reaches a value that is twice the designed initial pressure drop.

11.5 Summary

This chapter addresses the fundamental principles of fiber filters and dry ESPs. Fiber filters capture particles by means of interception, inertia impaction, Brownian diffusion, gravitational settling, and even electrostatic attraction. The performance of fiber filters can be evaluated in terms of face velocity, filtration velocity, efficiency, penetration ratio, pressure drop, quality factor, dust-holding capacity, etc. The particle collection efficiency of a filter varies with the size of the aerosol particles. For fine particles, the primary filtration mechanism is Brownian diffusion. In contrast, large particles are mainly captured by interception and inertial impaction. An intermediate particle size at which the particle penetration through the filter reaches the peak is defined as the MPPS. For polypropylene fibrous media, the MPPS is around 0.2 μm.

ESPs utilize the forces exerted on charged particles by an electric field to separate the particles from a gas stream. ESPs feature low pressure drop and high efficiency for small particles, and the capacity to handle high concentrations of dust. The performance of ESPs can be evaluated in terms of the electric field they create, the particle migration velocity, the particle collection efficiency, and so on.

The WHO, the US EPA, and the European Commission of the EU have published various PM exposure concentration limits to ensure the health of the general public. Most of these concentration limits are evaluated in terms of an annual mean or 24-hour mean. The recommended exposure limits can also be used as criteria to determine the need for indoor filtration, although this approach may be somewhat conservative.

The CADR, defined as the product of the operational flow rate of an air cleaner with its particle filtration efficiency, is widely used to assess air cleaners. The CADR reflects the rate at which pure particle-free air is delivered. To prolong the lifetime of air filters, it is recommended that they not be replaced until the final pressure drop reaches a value that is twice the designed initial pressure drop.

Problems

11.1 What are the main mechanisms by which fiber filters capture particles? How does the dominance of these mechanisms change with particle sizes? Explain this phenomenon in relation to the most penetrating particle size (MPPS).

11.2 Which mechanism is dominant for filtration of nanoscale particles (e.g., $PM_{0.1}$)? How can the filter design be optimized to enhance the capture efficiency of ultrafine particles?

11.3 Explain how an electrostatic precipitator (ESP) utilizes electric field forces to capture particulate matter. Describe the process of particle charging, migration, and deposition.

11.4 What key parameters do influence the efficiency of electrostatic precipitators? How can adjustments to these parameters enhance dust removal performance?

11.5 What metrics are typically used to determine when a filter should be replaced? How do these metrics reflect the operational status of the filter?

11.6 Based on Figure 11.5 and Eq. (11.25a), explain how the pressure drop across a HEPA filter changes with filtration velocity (U). Why does the total pressure drop exhibit an exponential relationship with U in practical applications?

11.7 For particles in the size range of 1–5 μm, which filtration mechanisms (interception, impaction, diffusion) are most effective? Justify your answer by referencing the governing equations and particle behavior described in Section 11.2.1.

11.8 According to ASHRAE Standard 52.2-2017, air filters should be replaced when the final pressure drop reaches twice the initial value. Discuss the practical implications of this guideline for maintaining indoor air quality while balancing energy consumption and operational costs.

References

Brown, R.C. (1993). *Air Filtration: An Integrated Approach to the Theory and Applications of Fibrous Filters*. New York: Pergamon Press.

Davies, C.N. (1973). *Air Filtration*. London: Academic Press.

Hinds, W.C. and Zhu, Y. (2022). *Aerosol Technology: Properties, Behavior, and Measurement of Airborne Particles*, 3e. New York: Wiley.

Kirsch, A.A. and Fuchs, N.A. (1968). Studies on fibrous aerosol filters-III diffusional deposition of aerosols in fibrous filters. *Annals of Occupational Hygiene* 11 (4): 299–304.

Kulkarni, P., Baron, P.A., and Willeke, K. (2011). *Aerosol Measurement: Principles, techniques, and applications*. Hoboken, NJ: Wiley.

Lee, K.W. and Liu, B.Y.H. (1980). On the minimum efficiency and the most penetrating particle size for fibrous filters. *Journal of the Air Pollution Control Association* 30 (4): 377–381.

Matts, S. and Öhnfeldt, P.O. (1963). Efficient gas cleaning with SF electrostatic precipitators. *Flakten Rev* 6: 93–110.

Xu, Z. (2018). *Air Purifier: Property, Assessment and Applications*. Singapore: Springer.

Yeh, H.C. and Liu, B.Y.H. (1974). Aerosol filtration by fibrous filters – I: theoretical. *Journal of Aerosol Science* 5 (2): 191–204.

12

Inactivation of Indoor Microbial Aerosols

12.1 Introduction

Indoor aerosols may contain harmful pathogens, such as viruses, bacteria, and fungi, exposure to which may cause adverse health outcomes. For a healthful indoor environment, the concentrations of bioaerosols must be maintained within acceptable limits. However, there is no universal concentration limit that can guide indoor bioaerosol control worldwide. The Chinese national standard (GB/T 18883 2022) sets the upper limit of total airborne culturable bacteria as 1,500 CFU/m^3 in indoor environments. In Finland, the indoor airborne fungal spore concentration must not exceed 500 CFU/m^3 in urban housing (WHO 2009). It is claimed that maintaining a total microbial concentration below 1,000 CFU/m^3 indoors is generally sufficient (Chawla et al. 2023).

To reduce occupants' exposure to pathogenic aerosols in indoor spaces and mitigate the associated adverse health outcomes, indoor bioaerosols may require inactivation. Inactivation can be accomplished through the spraying of chemical agents, irradiation with ultraviolet (UV) light, plasma treatment, and other approaches. This chapter addresses the basic principles, as well as the pros and cons, of some typical methods for inactivating indoor bioaerosols.

12.2 Chemical Disinfectants

Chemical disinfectants are fairly effective in inactivating microorganisms in the air, water, or on solid surfaces. Most chemical disinfectants are manufactured into liquid-phase solutions. When a chemical disinfectant is sprayed on a surface, the microorganisms on the surface are effectively sterilized. Meanwhile, the evaporated chemical disinfectant vapor or suspending droplets of the disinfectant can also inactivate some airborne microorganisms. Common chemical disin-

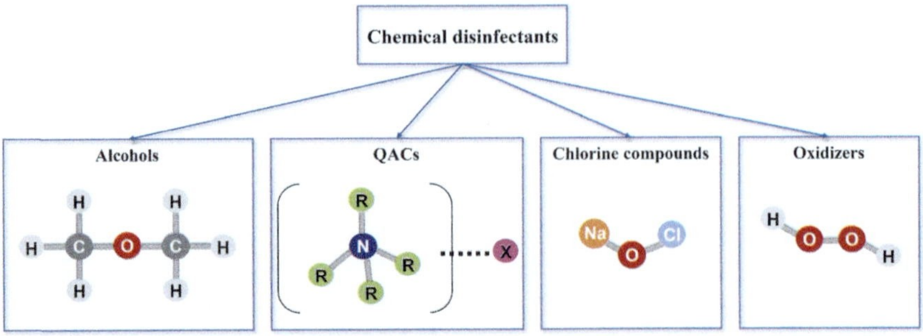

Figure 12.1 Molecular structure of example chemical disinfectants.

fectants include alcohols, quaternary ammonium compounds (QACs), chlorine compounds, and oxidizers, as shown in Figure 12.1.

12.2.1 Alcohol-Based Disinfectants

The antimicrobial effect of alcohol-based disinfectants mainly involves lipid membrane disruption and protein denaturation, which disrupts the membrane integrity and metabolic processes of pathogens (Golin et al. 2020). Alcohols are both hydrophilic and lipophilic and thus can effectively penetrate pathogens. Ethanol also inhibits protein synthesis in pathogens by affecting ribosomes and RNA polymerase (Haft et al. 2014). Alcohols are highly volatile, and their concentration after spraying is difficult to control. Therefore, alcohols are rarely used for bioaerosol disinfection but are often applied for surface disinfection. The optimal bactericidal concentration of alcohol ranges from 60% to 90% v/v solution. Concentrations below 50% v/v are usually ineffective. A few minutes of contact time is typically sufficient for achieving a 3–4 log reduction in the pathogen count (Boyce 2018).

Alcohol-based disinfectants are fast-acting, easy to use, and inexpensive, and they do not leave behind toxic residues. However, because alcohols are flammable and explosive, they must be used with caution. In addition, alcohols can irritate the eyes and skin and can even cause damage to the skin with prolonged use (Goh et al. 2020). Some people are sensitive to alcohol exposure, such that they cannot use alcohol-based disinfectants.

12.2.2 Quaternary Ammonium Compounds

QACs are among the most commonly used disinfectants in the healthcare and food-processing industries. The mechanism for sterilizing microorganisms with QACs is as follows: (i) adsorption and penetration of QACs into the cell walls of microorganisms, (ii) reaction with cytoplasmic membranes (lipids or proteins) followed by membrane disintegration, (iii) leakage of intracellular low-molecular-weight substances, (iv) degradation of proteins and nucleic acids, and (v) autolysis of the enzymes, leading to lysis of the cell walls. Thus, QACs are effective as cationic detergents against bacteria, yeasts, and lipid-containing viruses. Depending on the product formulation, QACs are also effective in sterilizing non-lipid-containing viruses and spores because the QACs can interact with intracellular targets and bind to DNA (Zinchenko et al. 2004).

The QACs must be used at the proper dosage, and excessive dilution, insufficient exposure time, or overdosing should be avoided. In addition, QACs' wide range of chemical structures facilitates consistent improvement in the efficacy of specific applications and target organisms while toxicity is progressively reduced. These structural variations can affect the dose and performance of QACs against different microorganisms. For example, QACs with methyl groups from C_{12} to C_{16} usually have the strongest antimicrobial activity (Jones and Joshi 2021). QACs are used primarily for surface disinfection.

Some concerns remain in regard to the potential effects of QACs on human health, animal health, the environment, and the ecological impacts. Improper selection and use of QACs may lead to additional public health and environmental concerns.

12.2.3 Chlorine Compounds

A large number of antimicrobial chlorine compounds, including sodium hypochlorite and hypochlorous acid, are used for inactivating microorganisms. When hypochlorous acid is produced by hypochlorite disinfectants, it can destroy membrane lipids and nucleic acids due to its ability to permeate the membranes and its strong oxidizing capability. In addition, hypochlorous acid inhibits key enzymatic reactions and results in protein damage within cells (McDonnell and Russell 1999).

Chlorine-based disinfectants are commonly used for efficient sterilization of indoor surfaces. A 1:10–1:100 dilution of 5.25%–6.15% by weight sodium hypochlorite (i.e., household bleach) is commonly used for surface disinfection (Rutala and Weber 2024). The hypochlorous acid in a concentration of 100–550 ppm is optimal for wound management and environmental disinfection, while posing minimal environmental toxicity (Sulaiman 2024).

It should be noted that high concentrations of chlorine can lead to corrosion of materials and irritation of skin or mucous membranes, as well as the emission of a pungent, unpleasant odor. Excess quantities of chlorinated disinfectants take time to degrade when they enter the natural environment, and they may be acutely toxic to aquatic organisms.

12.2.4 Hydrogen Peroxide

Hydrogen peroxide is a strong oxidizing agent with broad-spectrum efficacy against a wide range of microorganisms. This chemical compound produces hydroxyl radicals, which in turn react with lipids, proteins and nucleic acids, cleaving and oxidizing RNA and DNA, denaturing proteins, and disrupting biological membranes and sulfhydryl bonds in proteins and enzymes (Linley et al. 2012).

Hydrogen peroxide is environmentally benign because it can rapidly degrade into harmless products (water and oxygen) during dissolution. It is also non-toxic, making it safe to use as a disinfectant for medical devices and surfaces (even skin). For routine disinfection, a concentration of 3–6% by weight is typically used (Jones and Joshi 2021), although the presence of catalase or other peroxidases in some microorganisms may increase their tolerance to low concentrations of disinfectants. Hydrogen peroxide is widely employed not only for surface disinfection but also for aerosol disinfection, in the form of evaporation by fumigation.

Table 12.1 Pros, Cons, and Effectiveness of Different Chemical Disinfectants.

Chemical disinfectant	Pros	Cons	Recommended concentration	References
Alcohol	• Convenient and quick action • Inexpensive • No residue • Acceptable odor	• Not recommended for bioaerosol sterilization • Flammable and explosive • Can irritate eyes and skin	60–90% v/v	Boyce (2018)
QACs	• Low corrosivity • Non-carcinogenic • Inexpensive	• Less effective against biofilms • Strain-specific efficacy • Efficacy varies with temperature	0.04–0.05%	Tezel and Pavlostathis (2015)
Hypochlorous acid	• High inactivation efficiency • Inexpensive • Easily decomposed	• Corrosion of materials • Irritation of skin or mucous membranes • Potential side effects • Unpleasant odor	0.01–0.055%	Sulaiman (2024)
Hydrogen peroxide	• Environmentally friendly • Broad-spectrum efficacy	• Unstable and difficult to store • Effectiveness varies with the application method	3–6%	Jones and Joshi (2021)

Since hydrogen peroxide vapor can decontaminate hard-to-reach surfaces, it may be used for decontaminating an entire space, such as a laboratory and hospital ward. However, care should be taken to avoid excessive inhalation of hydrogen peroxide by humans.

The pros and cons of each type of chemical disinfectant are summarized in Table 12.1, and the recommended concentrations for indoor use are listed. Except for alcohol, the recommended concentrations are expressed by means of the mass ratio. The use of chemical disinfectants has long been an effective method of indoor bioaerosol inactivation and control. However, the disinfectant formulations are complex and may include adjuvants that influence the effectiveness of the disinfectant. Therefore, it is imperative to compare the efficacy of disinfectant products through appropriate testing based on the standards in different countries or regions. Finally, care should be taken to avoid improper disinfectant use that may affect human and animal health, and to minimize environmental outcomes and the ecological burden.

12.3 Ultraviolet Germicidal Irradiation (UVGI)

UV light (100–400 nm) is divided into four main bands based on wavelength and energy intensity: vacuum UV (100–200 nm), UVC (200–280 nm), UVB (280–315 nm), and UVA (315–400 nm) (Khan et al. 2022). The UVC band is the most widely used and is effective for the inactivation of microorganisms. Unlike chemical disinfection, UVC irradiation does not create harmful chemical residues. Consequently, UV irradiation has become more and more popular for indoor bioaerosol sterilization.

12.3.1 Mechanism of UVC Inactivation of Microbes

UVC irradiation inactivates microbes primarily through photochemical reactions with DNA or RNA, as shown in Figure 12.2. UVC rays break adenine-thymine bonds in the nucleic acid chains and form dimers. These dimers interfere with nucleic acid assembly, disrupting both transcription and replication of RNA or DNA. The highest microbial inactivation efficiency occurs at wavelengths between 250 and 270 nm, as the UV light in this range is strongly absorbed by microbial nucleic acids (Bhardwaj et al. 2021).

12.3.2 UVC Light Sources

A variety of UVC light sources are available, including mercury-vapor lamps, light-emitting diodes (LEDs), and excimer lamps, among others. The major pros and cons of different UVC light sources are summarized in Table 12.2.

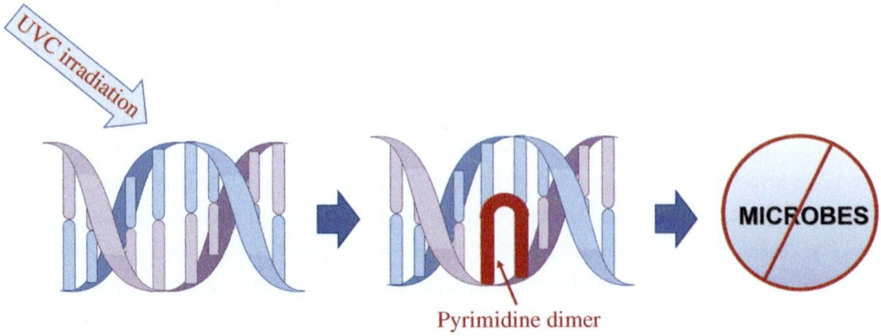

Figure 12.2 Mechanism for inactivation of microbes with the UVC irradiation, which produces pyrimidine dimer and subsequently prevents their reproduction.

Table 12.2 Pros and cons of different UVC light sources.

UVC light source	Pros	Cons
Low-pressure mercury lamp (254 nm)	• Low cost • Technical maturity	• Damage to skin and eyes • Hazards of mercury • Warm-up required • Ozone production
UVC-LED (255–275 nm)	• Diversity in wavelengths • No warm-up time • Long lifespan • Compact size	• Relatively low technological maturity • Relatively expensive
Far-UVC (222 nm) excimer lamp	• Reduced damage to skin and eyes • Long service life	• High energy consumption (Sun et al. 2023) • Ozone production • Expensive

12.3.2.1 Mercury-Vapor Lamps

Mercury-vapor UV lamps are classified into three types: low-pressure (254 nm), medium-pressure (220–580 nm), and high-pressure (220–1,000 nm) (Demeersseman et al. 2023). Low-pressure mercury-vapor lamps, which generate the conventional 254 nm UVC, are the most widely used UVC light source due to their affordability.

12.3.2.2 Light-Emitting Diodes (LEDs)

To address the environmental and health concerns associated with mercury, UVC-LEDs (wavelength between 255 and 275 nm) have been developed as an alternative (Demeersseman et al. 2023). UVC-LEDs are typically made from aluminum gallium nitride (AlGaN). They offer benefits such as a diversity in wavelengths, a longer lifespan, and a compact size, although they are more expensive than traditional UVC lamps (Nyangaresi et al. 2018).

12.3.2.3 Excimer Lamps

Excimer lamps, such as xenon-bromide (XeBr, 282 nm), krypton-bromide (KrBr, 206 nm), and krypton-chlorine (KrCl, 222 nm) (Erofeev and Tarasenko 2006), can effectively inactivate a broad range of pathogens, in a similar manner to that of 254 nm UVC. The KrCl lamps, emitting rays at 222 nm (also known as far-UVC), are the most commonly used far-UVC light source for disinfection. Far-UVC is a relatively new and effective method of disinfection. Due to its poor penetration rate, far-UVC light is much safer for the skin and eyes than that of 254 nm UVC (Barnard et al. 2020). However, ozone production is considered a side effect of far-UV lamp usage.

12.3.3 Factors Affecting UVC Inactivation Performance

12.3.3.1 Microbial Species

UVC susceptibility varies with the species of microorganism. Generally, the susceptibility of airborne microorganisms to UVC irradiation follows this order: viruses > bacteria > fungi (Kowalski 2009). Differences in susceptibility are likely due to variations in the cell envelope composition and cell size, which affect the UVC irradiation to penetrate the cells. In addition, the DNA repair mechanism may play a role in microorganisms' susceptibility to UV irradiation.

12.3.3.2 Environmental Conditions

Environmental condition parameters also influence the UVC dose required for microbial inactivation. The microbial susceptibility to UVC irradiation may be impacted by the thermal conditions. This is because the combination of UVC and heat promotes cell membrane destabilization, which subsequently compromises membrane integrity. In addition, at higher relative humidity, UVC-based inactivation of microorganisms is less effective (Kowalski 2009). This reduction in susceptibility is due to the formation of a water layer around the microorganisms, protecting them from UVC-induced disruption of DNA or RNA.

12.3.3.3 Medium of Microbial Attachment

The UVC dose required for microbial inactivation varies depending on the medium in which microorganisms reside (e.g., air, liquid, or surface). The disinfection of bio-contaminated air and surfaces with UV irradiation is generally more practical and predictable than the inactivation of the microorganisms in a liquid. Moreover, the inactivation of airborne microorganisms is more effective than the inactivation of microorganisms on surfaces, as the latter is subject to

Table 12.3 Examples of studies evaluating the inactivation of microorganisms by UVC lamps.

UVC lamp type (wavelength)		Log reduction UVC dose (mJ/cm^2)				References
		1	2	3	4	
Mercury-vapor UVC (254 nm)	*Staphylococcus aureus*	N/A	N/A	N/A	6	Narita et al. (2020)
	Murine hepatitis virus (MHV)	0.66	N/A	N/A	N/A	Walker and Ko (2007)
	MS2	0.34–0.42	0.80–0.91	N/A	N/A	Tseng and Li (2005)
	Porcine reproductive and respiratory syndrome virus (PRRSV)	N/A	N/A	1.21	N/A	Cutler et al. (2012)
	Influenza A virus	0.67	N/A	N/A	N/A	McDevitt et al. (2012)
UVC-LED (255–275 nm)	*Salmonella typhimurium*	N/A	N/A	4.6	N/A	Kim and Kang (2018)
	Aspergillus flavus	N/A	N/A	N/A	23	
	Phi X174	N/A	N/A	N/A	4.6	
Far-UVC lamp (222 nm)	*Escherichia coli*	N/A	N/A	N/A	10.6	Clauß et al. (2005)
	Staphylococcus aureus	N/A	N/A	N/A	6	Narita et al. (2020)
	Human coronavirus HCoV-OC43	0.39	0.78	1.2	N/A	Buonanno et al. (2020)
	Human coronavirus HCoV-229E	0.56	1.1	1.7	N/A	

surface heterogeneity and roughness and to the potential presence of biofilms (Gora et al. 2019). Typical examples of indoor bioaerosol inactivation by means of UVC irradiation are provided in Table 12.3.

12.3.4 Environmental and Health Safety Issues

When mercury lamps are used to generate UVC rays, it is imperative to address the environmental and health concerns associated with mercury. Some UVC light sources produce ozone or other harmful gases. The US Environmental Protection Agency (EPA) warns against the use of ozone-emitting UVC lamps in enclosed spaces without proper ventilation.

Table 12.4 Threshold limit values (TLVs) for exposure of eyes and skin to UVC light at typical wavelengths.

Wavelength (nm)	TLV for eyes (mJ/cm^2)	TLV for skin (mJ/cm^2)
222	160	479
254	6.0	10
270	3.0	10

Exposure limits for UVC irradiation vary with the wavelength, as do exposure limits for the skin and eyes. The American Conference of Governmental Industrial Hygienists (ACGIH) has published tentative guidelines for limiting human exposure, in terms of threshold limit values (TLVs) (ACGIH 2021). TLVs are defined as doses to which nearly any healthy worker can be exposed over an eight-hour period without acute adverse effects or increased risk of delayed symptoms. TLVs for the eyes and skin at several typical UVC wavelengths are shown in Table 12.4. The TLVs for the far-UVC exposure at 222 nm are at least one order of magnitude higher than the values for the traditional UVC exposure at 254 or 270 nm.

12.4 Plasma Treatment

Plasma treatment is used in hospitals, public places, and enclosed spaces, providing a means of inactivating airborne microorganisms, removing particles, oxidizing certain VOCs, and so on. Plasma can be categorized by temperature as: high-temperature plasma, thermal plasma, or non-thermal plasma. High-temperature plasma and thermal plasma are not appropriate for use in indoor environments. However, the temperature of non-thermal plasma remains low, and thus it is suitable for microbial treatment indoors. The non-thermal plasma is typically classified according to the adopted discharge type as: glow discharge, corona discharge, atmospheric-pressure plasma jet (APPJ), dielectric barrier discharge (DBD), micro-hollow cathode discharge (MHCD), DC discharge, pulse discharge, or high/low-frequency discharge (Sakudo et al. 2019). The type of discharge depends on factors such as the mode and frequency of the power source (e.g., direct current [DC] or alternating current [AC]), ambient gas pressure (low-pressure or atmospheric-pressure plasma), and the specific shape and configuration of the electrodes.

12.4.1 Mechanisms of Plasma Treatment

Plasma ionizes the gas under the excitation of a high electric field to produce a large number of reactive species (e.g., hydroxyl radicals, oxygen radicals, and

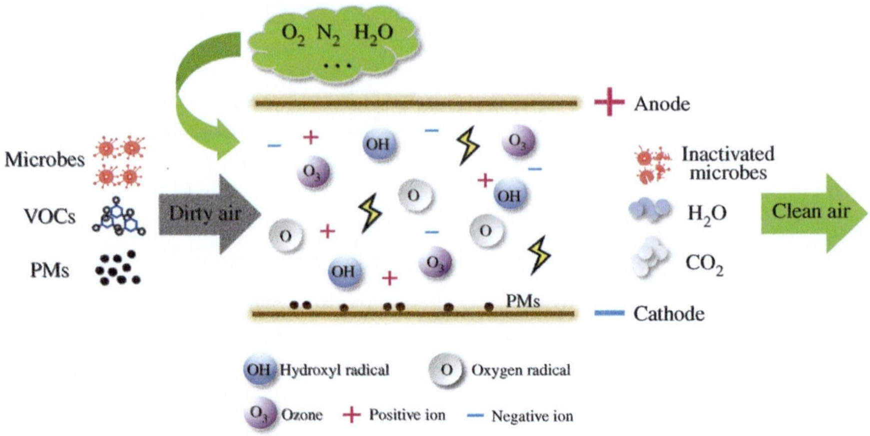

Figure 12.3 Mechanism for inactivation of microorganisms, oxidation of VOCs, and removal of PMs by plasma treatment.

ozone), UV rays, and charged particles. The reactive substances are capable of inactivating microorganisms and degrading VOCs, and the electrostatic action settles particles, as shown in Figure 12.3.

The inactivation of microorganisms by plasma consists of three main aspects: etching effect on cells, cell membrane perforation and electrostatic disruption, and oxidation of macromolecules.

12.4.1.1 Etching Effect on the Cell

Etching refers to the breaking of the chemical bonds of certain substances in the microorganisms in the presence of plasma, which ultimately results in the production of volatile substances such as CO_2 and C_xH_y. Reactive oxygen radicals and UV light generated by plasma are considered to be the main causes of etching (Moisan et al. 2001).

12.4.1.2 Cell Membrane Perforation and Electrostatic Disruption

When high-energy electrons generated by plasma act on the surface of a microorganism, the cell membrane is locally corroded, and the chemical bonds of phospholipids and other substances are destroyed (Sreedevi and Suresh 2023). The plasma-generated electrons and ions are enriched on the surface of the cell membrane, so that the curvature of the microbial surface changes, which triggers local irregularity. Consequently, the total electrical tension across the microbial membrane exceeds the total tensile strength of the membrane, resulting in the formation of small pores on the membrane surface and hence disruption.

12.4.1.3 Oxidation of Intracellular Macromolecules

Reactive oxygen species (ROS) and reactive nitrogen species (RNS) produced by plasma can oxidize cellular macromolecules. Lipids on the surface of the cell membrane are the primary targets of ROS (Joshi et al. 2011). The ROS adheres to polyunsaturated fatty acids in the cell membrane, causing a chain reaction in the lipids. The ROS can also interact with proteins, resulting in the modification of amino acid side chains and changes in the tertiary structure of proteins. Another target of the ROS reaction is DNA. The ROS reacts with DNA bases and sugar moieties to cause various types of oxidative damage to the DNA.

12.4.2 Effectiveness of Plasma Treatment

The effectiveness of plasma in inactivating indoor microorganisms depends on its air treatment capability, the air circulation rate, the room size, and the duration of the action. Among these factors, the air treatment capability can be evaluated in terms of the single-pass sterilization efficiency, which depends mainly on the concentration of reactive substances that are generated and the exposure time of the microorganisms. The typical single-pass sterilization efficiency for low-temperature plasma devices is shown in Table 12.5. It can be seen in the table that the devices achieve good inactivation of airborne viruses, bacteria, and fungi across a range of exposure times, from a few milliseconds to several hundred milliseconds.

Table 12.5 Single-pass sterilization efficiency of low-temperature plasma treatment on microbial aerosols.

Test microorganism	Duration of action	Log reduction	References
Staphylococcus epidermidis	<0.2 s	3.7	Bisag et al. (2020)
Porcine reproductive and respiratory syndrome virus	Milliseconds	3.5	Nayak et al. (2020)
Pseudomonas fluorescens	0.12 s	Total elimination	Liang et al. (2012)
Bacillus subtilis	0.12 s	1.80	
Escherichia coli	Milliseconds	1.5	Gallagher et al. (2007)

Table 12.6 Cyclic sterilization efficiency of low-temperature plasma treatment on indoor microbial aerosols.

Scenario	Microorganism	Duration of action	Log reduction	References
Chamber of 17.28 m^3	Total microorganisms	90 min	>2	Jangra et al. (2023)
Chamber of 0.216 m^3	*Escherichia coli*	90 min	5	
	MS2	30 min	>5	
Chamber of 0.060 m^3	*Escherichia coli*	10 min	Total elimination	Baek et al. (2022)
	Staphylococcus epidermidis	10 min		
	Mycobacterium smegmatis	60 min		

The cyclic sterilization effects of low-temperature plasma on indoor microbial aerosols are shown in Table 12.6. Numerous studies have shown that low-temperature plasma treatment is effective in inactivating airborne microorganisms, and that the smaller the operating environment, the shorter the time required to achieve the same sterilization rate. In addition, the sterilization rate may vary with the target microorganism. Low-temperature plasma is more effective in inactivating viruses than Gram-negative bacteria, followed by Gram-positive bacteria.

Although the low-temperature plasma treatment shows great potential for inactivating indoor airborne microbes, its engineering application still faces many challenges. The main challenge is its limited reaching depth, which restricts the sterilization effect in complex indoor environments, especially when porous materials or inaccessible spaces are present. In addition, the byproduct ozone poses a health hazard to indoor occupants, and therefore the risk of occupants' exposure to indoor ozone needs to be balanced with the effectiveness of microbial control.

In summary, low-temperature plasma treatment is effective and broad-spectrum, making it suitable for use in hospitals, laboratories, and other facilities with high cleanliness requirements. However, the technology is limited by the potential for harmful by-products, inadequate spatial coverage, and equipment costs. When plasma treatment is used, the specific needs and risks of the indoor setting should be taken into account. The plasma generation device should also be appropriately selected, and all necessary safety measures should be taken to prevent adverse impacts.

12.5 Other Disinfection Technologies

12.5.1 Ozone Oxidation

As a strong oxidizing gas, ozone provides highly efficient, broad-spectrum, and thorough inactivation of microorganisms. As ozone generation technology has matured, ozone sterilization has been gradually applied in many fields, for air purification, food processing, medical treatment, and other purposes.

Ozone reacts with the cell walls and cell membranes of microorganisms, changing the permeability of the microbial membranes and denaturing the microorganisms with the released endolysins. Meanwhile, the ozone oxidizes and decomposes the glucose oxidase enzymes required for the oxidation of glucose inside the microorganisms, thereby destroying microbial organelles and nucleic acids and interfering with cellular growth, metabolism, and reproduction (Brodowska et al. 2018).

The effectiveness of ozone in sterilizing viruses, bacteria, and fungi is well-established. However, it should be noted that the protocols for using ozone and its effects vary with the microbes with which aerosols are laden and the types of surfaces are to which they adhere. The disinfection effect of ozone on various types of microorganisms in the air is shown in Table 12.7. It can be seen that when the concentration of ozone is increased or the action time is extended, the inactivation rate of ozone on microorganisms can exceed 99%. In addition, microbial species, ozone concentration, humidity, and action time all have a significant influence on the sterilization effect.

Although ozone is highly effective, excessive exposure to ozone should be avoided. The primary adverse health effects of ozone exposure include lung damage, asthma, and the worsening of chronic obstructive pulmonary disease (COPD). As a result, some governments have imposed restrictions on the widespread use of ozone generators for sanitizing public and/or private spaces. According to the

Table 12.7 Sterilization efficiency of ozone on different microorganisms.

	Log reduction under different ozone doses (ppm.min)					
Microorganism	**1**	**2**	**3**	**4**	**5**	**References**
Candida famata	N/A	17	N/A	N/A	N/A	Huang et al. (2012)
Influenza A virus	N/A	N/A	N/A	2,100	3,000	Tanaka et al. (2009)
SARS-CoV-2	N/A	N/A	60	N/A	N/A	Yano et al. (2020)
Respiratory syncytial virus A2	N/A	N/A	N/A	800	N/A	Blanchard et al. (2020)

World Health Organization (WHO), the eight-hour average ozone concentration in indoor environments should not exceed 0.1 mg/m^3 (WHO 2021).

12.5.2 Photocatalytic Oxidation (PCO)

In recent years, PCO-related technologies have garnered significant interest. PCO is considered a relatively safe, stable, and economical strategy for sterilizing various microorganisms in indoor air. Common photocatalysts that generate ROS include titanium dioxide, zinc oxide, copper oxide, zirconium dioxide, and iron trioxide. The electronic band structure consists of a valence band (VB) and a conduction band (CB). When exposed to light, a semiconductor absorbs photons, exciting electrons (e–) from the VB to the CB and leaving corresponding holes (h+) in the VB. The electrons and holes can interact with water and OH– on the surface of the photocatalyst to generate highly reactive species such as hydroxyl radicals (•OH) and superoxide anions (Mamaghani et al. 2017). These ROS induce oxidative stress on microbial cells and biomolecules (proteins, lipids, carbohydrates, DNA, and amino acids), damaging microbial cell-wall structures. Additionally, semiconductor molecules and ROS can disrupt intracellular electron transport, leading to protein denaturation, ATP depletion, DNA replication disruption, and ultimately cell death (Ray et al. 2018).

Table 12.8 provides an overview of the photocatalysis technologies employed to inactivate microorganisms. Among photocatalysts, titanium dioxide is the most commonly used due to its well-known reaction mechanism. Titanium dioxide can be applied as a coating on air filtration media to provide antimicrobial capabilities to filters. In addition, titanium dioxide has been immobilized on metal nanoparticles for use in filters.

Table 12.8 Performance of photocatalytic oxidation for inactivating microorganisms.

Substrate/laden material	Photocatalyst	Microorganism	Log reduction	References
Glass rings	TiO_2	*Bacillus subtilis*	2 for vegetative cells	Zacarías et al. (2021)
Glass fiber	Cu-TiO2	*Escherichia coli*	0.92	Pham and Lee (2014)
Non-woven spunbond polypropylene	ZnO in polyacrylonitrile (PAN)	*Escherichia coli* *Staphylococcus aureus*	>2	Pardo-Figuerez et al. (2021)
Activated carbon nanofiber	ZnO	*Escherichia coli* *Staphylococcus epidermidis*	0.84 1	Kang et al. (2022)

While PCO shows great potential for air purification, there are several issues that hinder its frequent usage. A key issue is that the short-lived free radicals generated during the purification process react with the airborne contaminants, producing secondary chemicals such as aldehydes and ketones. These by-products may negatively influence indoor air quality. Furthermore, the use of PCO for pathogen inactivation is still being explored. Although some studies have shown that PCO is effective against various pathogens, the efficiencies can vary significantly with the type of pathogen, environmental conditions, the design of the PCO system, and other parameters.

12.6 Summary

This chapter summarizes various indoor microbial aerosol inactivation technologies and some key factors affecting the choice of appropriate strategies for disinfecting indoor environments.

There is no single indoor bioaerosol abatement method that is universally optimal. Chemical disinfectants are effective and inexpensive, but the disinfectant residues can have deleterious effects on human and animal health, corrode materials, and may lead to significant environmental and ecological burdens. Compared to the chemical disinfection methods, UVC irradiation is an affordable and easy-to-deploy technique that effectively inactivates microorganisms without generating harmful chemical residues. However, there are shadowed regions for UVC disinfection and additional health and safety concerns regarding human exposure to leaked UVC irradiation. Plasma treatment is also highly effective and broad-spectrum, but there is a risk that harmful by-products will be emitted. Ozone is a broad-spectrum biocide that may inactivate a wide range of microorganisms, but excessive exposure of humans and materials to ozone should be avoided. Meanwhile, although PCO shows promise for sterilizing microbial aerosols, PCO may produce secondary chemicals that pollute indoor air environments.

The selection of an appropriate method for microbial inactivation is subject to the type of pathogen to be controlled and the specific indoor air quality requirements. In addition, factors such as temperature, humidity, and airflow distribution may significantly affect the performance of a microbial inactivation device.

Problems

12.1 What are the primary mechanisms by which alcohol-based disinfectants inactivate microorganisms?

12.2 How does UVC irradiation disrupt microbial DNA or RNA, and which wavelength range is most effective?

12.3 List the five sequential steps in the sterilization mechanism for quaternary ammonium compounds (QACs).

12.4 Describe the three main mechanisms by which plasma treatment inactivates microorganisms.

12.5 What are the two key pathways through which ozone achieves microbial sterilization?

12.6 What are the primary photocatalysts used in photocatalytic oxidation (PCO) technology, and how do they generate reactive oxygen species (ROS)?

12.7 Compare the advantages and limitations of chemical disinfectants versus UVC irradiation for indoor bioaerosol inactivation. Consider perspectives such as residue, safety, and ease of use.

12.8 Plasma treatment shows promise for microbial inactivation, but challenges to its application include ozone by-products and limited spatial coverage. Explain how these issues can be mitigated in future engineering applications.

12.9 Photocatalytic oxidation (PCO) is considered generally effective but may produce secondary pollutants like aldehydes. Debate the feasibility of widespread PCO application in residential settings in light of this drawback.

12.10 Ozone is highly effective for sterilization but poses health risks at elevated concentrations. Propose strategies to safely integrate ozone-based devices into occupied indoor environments.

References

ACGIH (2021). *Documentation on Threshold Limit Values and Biological Exposure Indices – Ultraviolet*. Cincinnati, OH: American Conference of Governmental Industrial Hygienists.

Baek, K.H., Kim, S.J., Kim, T. et al. (2022). Evaluation of zone disinfection by air-passable plasma filter in a novel chamber model. *Journal of Hospital Infection* 128: 1–7.

Barnard, I.R.M., Eadie, E., Wood, K. et al. (2020). Further evidence that far-UVC for disinfection is unlikely to cause erythema or pre-mutagenic DNA lesions in skin. *Photodermatology, Photoimmunology & Photomedicine* 36: 476–477.

Bhardwaj, S.K., Singh, H., Deep, A. et al. (2021). UVC-based photoinactivation as an efficient tool to control the transmission of coronaviruses. *Science of the Total Environment* 792: 148548.

Bisag, A., Isabelli, P., Laurita, R. et al. (2020). Cold atmospheric plasma inactivation of aerosolized microdroplets containing bacteria and purified SARS-CoV-2 RNA to contrast airborne indoor transmission. *Plasma Processes and Polymers* 17 (10): 2000154.

Blanchard, E.L., Lawrence, J.D., Noble, J.A. et al. (2020). Enveloped virus inactivation on personal protective equipment by exposure to ozone. *medRxiv*.

Boyce, J.M. (2018). Alcohols as surface disinfectants in healthcare settings. *Infection Control & Hospital Epidemiology* 39 (3): 323–328.

Brodowska, A.J., Nowak, A., and Śmigielski, K. (2018). Ozone in the food industry: principles of ozone treatment, mechanisms of action, and applications: an overview. *Critical Reviews in Food Science and Nutrition* 58 (13): 2176–2201.

Buonanno, M., Welch, D., Shuryak, I. et al. (2020). Far-UVC light (222 nm) efficiently and safely inactivates airborne human coronaviruses. *Scientific Reports* 10 (1): 1–21.

Chawla, H., Anand, P., Garg, K. et al. (2023). A comprehensive review of microbial contamination in the indoor environment: sources, sampling, health risks, and mitigation strategies. *Frontiers in Public Health* 23 (11): 1285393.

Clauß, M., Mannesmann, R., and Kolch, A. (2005). Photoreactivation of *Escherichia coli* and *Yersinia enterolytica* after irradiation with a 222 nm excimer lamp compared to a 254 nm low-pressure mercury lamp. *Acta Hydrochimica et Hydrobiologica* 33 (6): 579–584.

Cutler, T.D., Wang, C., Hoff, S.J., and Zimmerman, J.J. (2012). Effect of temperature and relative humidity on ultraviolet (UV 254) inactivation of airborne porcine respiratory and reproductive syndrome virus. *Veterinary Microbiology* 159: 47–52.

Demeersseman, N., Saegeman, V., Cossey, V. et al. (2023). Shedding a light on ultraviolet-C technologies in the hospital environment. *Journal of Hospital Infection* 132: 85–92.

Erofeev, M.V. and Tarasenko, V.F. (2006). XeCl-, KrCl-, XeBr- and KrBr-excilamps of the barrier discharge with the nanosecond pulse duration of radiation. *Journal of Physics D: Applied Physics* 39 (16): 3609–3614.

Gallagher, M.J., Vaze, N., Gangoli, S. et al. (2007). Rapid inactivation of airborne bacteria using atmospheric pressure dielectric barrier grating discharge. *IEEE Transactions on Plasma Science* 35 (5): 1501–1510.

GB/T 18883 (2022). *Standards for Indoor Air Quality*. China: National Health Commission of the People's Republic of China.

Goh, C.F., Ming, L.C., and Wong, L.C. (2020). Dermatologic reactions to disinfectant use during the COVID-19 pandemic. *Clinics in Dermatology* 39: 314–322.

Golin, A.P., Choi, D., and Ghahary, A. (2020). Hand sanitizers: a review of ingredients, mechanisms of action, modes of delivery, and efficacy against coronaviruses. *American Journal of Infection Control* 48 (9): 1062–1067.

Gora, S.L., Rauch, K.D., Ontiveros, C.C. et al. (2019). Inactivation of biofilm-bound Pseudomonas aeruginosa bacteria using UVC light emitting diodes (UVC LEDs). *Water Research* 151: 193–202.

Haft, R.J.F., Keating, D.H., Schwaegler, T. et al. (2014). Correcting direct effects of ethanol on translation and transcription machinery confers ethanol tolerance in bacteria. *Proceedings of the National Academy of Sciences of the United States of America* 111: E2576–E2585.

Huang, H.L., Lee, M.G., and Tai, J.H. (2012). Controlling indoor bioaerosols using a hybrid system of ozone and catalysts. *Aerosol and Air Quality Research* 12 (1): 73–82.

Jangra, R., Ahlawat, K., Dixit, A., and Prakash, R. (2023). Efficient deactivation of aerosolized pathogens using a dielectric barrier discharge based cold-plasma detergent in environment device for good indoor air quality. *Scientific Reports* 13 (1): 10295.

Jones, I.A. and Joshi, L.T. (2021). Biocide use in the antimicrobial era: a review. *Molecules* 26 (8): 2276.

Joshi, S.G., Cooper, M., Yost, A. et al. (2011). Nonthermal dielectric-barrier discharge plasma-induced inactivation involves oxidative DNA damage and membrane lipid peroxidation in *Escherichia coli*. *Antimicrobial Agents and Chemotherapy* 55 (3): 1053–1062.

Kang, S., Park, D.H., and Hwang, J. (2022). Hierarchical ZnO nanospines grown on a carbon fiber seed layer for efficient VOC removal and airborne virus and bacteria inactivation. *Journal of Hazardous Materials* 424: 127262.

Khan, M., McDonald, M., Mundada, K., and Willcox, M. (2022). Efficacy of ultraviolet radiations against coronavirus, bacteria, fungi, fungal spores and biofilm. *Hygiene* 2: 120–131.

Kim, D.K. and Kang, D.H. (2018). UVC LED irradiation effectively inactivates aerosolized viruses, bacteria, and fungi in a chamber-type air disinfection system. *Applied and Environmental Microbiology* 84 (17): e00944–18.

Kowalski, W. (2009). *Ultraviolet Germicidal Irradiation Handbook: UVGI for Air and Surface Disinfection*. Berlin, Heidelberg: Springer.

Liang, Y., Wu, Y., Sun, K. et al. (2012). Rapid inactivation of biological species in the air using atmospheric pressure nonthermal plasma. *Environmental Science & Technology* 46 (6): 3360–3368.

Linley, E., Denyer, S.P., McDonnell, G. et al. (2012). Use of hydrogen peroxide as a biocide: new consideration of its mechanisms of biocidal action. *Journal of Antimicrobial Chemotherapy* 67 (7): 1589–1596.

Mamaghani, A.H., Haghighat, F., and Lee, C.S. (2017). Photocatalytic oxidation technology for indoor environment air purification: the state-of-the-art. *Applied Catalysis B: Environmental* 203: 247–269.

McDevitt, J.J., Rudnick, S.N., Radonovich, L.J. et al. (2012). Aerosol susceptibility of influenza virus to UV-C light. *Applied and Environmental Microbiology* 78 (6): 1666–1669.

McDonnell, G. and Russell, A.D. (1999). Antiseptics and disinfectants: activity, action, and resistance. *Clinical Microbiology Reviews* 12: 147–179.

Moisan, M., Barbeau, J., Moreau, S. et al. (2001). Low-temperature sterilization using gas plasmas: a review of the experiments and an analysis of the inactivation mechanisms. *International Journal of Pharmaceutics* 226 (1–2): 1–21.

Narita, K., Asano, K., Naito, K. et al. (2020). Ultraviolet C light with wavelength of 222 nm inactivates a wide spectrum of microbial pathogens. *Journal of Hospital Infection* 105: 459–467.

Nayak, G., Andrews, A.J., Marabella, I. et al. (2020). Rapid inactivation of airborne porcine reproductive and respiratory syndrome virus using an atmospheric pressure air plasma. *Plasma Processes and Polymers* 17 (10): 1900269.

Nyangaresi, P.O., Qin, Y., Chen, G. et al. (2018). Effects of single and combined UV-LEDs on inactivation and subsequent reactivation of *E. Coli* in water disinfection. *Water Research* 147: 331–341.

Pardo-Figuerez, M., Chiva-Flor, A., Figueroa-Lopez, K. et al. (2021). Antimicrobial nanofiber based filters for high filtration efficiency respirators. *Nanomaterials* 11: 900.

Pham, T.D. and Lee, B.K. (2014). Cu doped TiO_2/GF for photocatalytic disinfection of *Escherichia coli* in bioaerosols under visible light irradiation: application and mechanism. *Applied Surface Science* 296: 15–23.

Ray, S.K., Dhakal, D., Regmi, C. et al. (2018). Inactivation of *Staphylococcus aureus* in visible light by morphology tuned α-$NiMoO_4$. *Journal of Photochemistry and Photobiology* 350: 59–68.

Rutala, W.A. and Weber, D.J. (2024). *Guideline for disinfection and sterilization in healthcare facilities, 2008. Update: June 2024*. Atlanta, GA: Centers for Disease Control and Prevention (CDC).

Sakudo, A., Yagyu, Y., and Onodera, T. (2019). Disinfection and sterilization using plasma technology: fundamentals and future perspectives for biological applications. *International Journal of Molecular Sciences* 20 (20): 5216.

Sreedevi, P.R. and Suresh, K. (2023). Cold atmospheric plasma mediated cell membrane permeation and gene delivery – empirical interventions and pertinence. *Advances in Colloid and Interface Science* 320: 102989.

Sulaiman, S. (2024). *Hypochlorous Acid (HOCl) for Antisepsis, Disinfection, and Wound Care in Core Categories 15.1, 15.2, and 13 (Proposal for the Addition and Inclusion of Hypochlorous Acid to the 2025 WHO Essential Medicines List in the Categories of Disinfection, Antisepsis, and Wound Care).* Khartoum, Sudan: Sudanese National Academy of Sciences.

Sun, W., Jing, Z., Zhao, Z. et al. (2023). Dose–response behavior of pathogens and surrogate microorganisms across the ultraviolet-C spectrum: inactivation efficiencies, action spectra, and mechanisms. *Environmental Science & Technology* 57 (29): 10891–10900.

Tanaka, H., Sakurai, M., and Ishii, K. (2009). Inactivation of influenza virus by ozone gas. *IHI Engineering Review* 42: 108–111.

Tezel, U. and Pavlostathis, S.G. (2015). Quaternary ammonium disinfectants: microbial adaptation, degradation and ecology. *Current Opinion in Biotechnology* 33: 296–304.

Tseng, C.C. and Li, C.S. (2005). Inactivation of virus-containing aerosols by ultraviolet germicidal irradiation. *Aerosol Science and Technology* 39 (12): 1136–1142.

Walker, C.M. and Ko, G. (2007). Effect of ultraviolet germicidal irradiation on viral aerosols. *Environmental Science & Technology* 41 (15): 5460–5465.

World Health Organization (WHO) (2009). *WHO Guidelines for Indoor Air Quality: Dampness and Mould.* Geneva: World Health Organization.

World Health Organization (WHO) (2021). *WHO Global Air Quality Guidelines: Particulate Matter ($PM_{2.5}$ and PM_{10}), Ozone, Nitrogen Dioxide, Sulfur Dioxide and Carbon Monoxide.* Geneva: World Health Organization.

Yano, H., Nakano, R., Suzuki, Y. et al. (2020). Inactivation of severe acute respiratory syndrome coronavirus 2 (SARS-CoV-2) by gaseous ozone treatment. *Journal of Hospital Infection* 106: 837–838.

Zacarías, S.M., Manassero, A., Pirola, S. et al. (2021). Design and performance evaluation of a photocatalytic reactor for indoor air disinfection. *Environmental Science and Pollution Research* 28: 23859–23867.

Zinchenko, A.A., Sergeyev, V.G., Yamabe, K. et al. (2004). DNA compaction by divalent cations: structural specificity revealed by the potentiality of designed quaternary diammonium salts. *Chembiochem* 5: 360–368.

Appendices

Appendix A1 Some physical parameters of air and water at standard conditions.

Dry air at 293 K (20 °C) and 101 kPa (1 atm)

Molecular weight = 28.96 g/mol
Density = 1.205 kg/m^3
Dynamic viscosity = 1.81×10^{-5} Pa · s
Mean free path = 0.066 μm
Specific heat ratio = 1.40
Diffusion coefficient = 2.0×10^{-5} m^2/s
Composition of dry air (excluding water vapor) by volume

Gas	Volume fraction (%)	Molecular weight (g/mol)
N_2	78.1	28.01
O_2	20.9	32.00
Ar	0.93	39.95
CO_2	0.039	44.01
Other	0.031	N/A

Water at 293 K (20 °C) and 101 kPa (1 atm)

Dynamic viscosity = 1.002×10^{-3} Pa · s
Surface tension = 0.0727 N/m
Saturation vapor pressure = 2.34 kPa

Water vapor at 293 K (20 °C) and 101 kPa (1 atm)

Density = 0.75 kg/m^3
Diffusion coefficient = 2.4×10^{-5} m^2/s

Appendix A2 Physical parameters of dry air at standard atmospheric pressure.

Temperature (°C)	Density (kg/m^3)	Thermal conductivity (W/m · K)	Specific heat capacity (kJ/kg · K)	Thermal diffusivity (m^2/s)	Dynamic viscosity (Pa · s)	Prandtl number
−50	1.584	0.0204	1.013	1.27E-05	1.46E-05	0.728
−40	1.515	0.0212	1.013	1.38E-05	1.52E-05	0.728
−30	1.453	0.022	1.013	1.49E-05	1.57E-05	0.723
−20	1.395	0.0228	1.009	1.62E-05	1.62E-05	0.716
−10	1.342	0.0236	1.009	1.74E-05	1.67E-05	0.712
0	1.293	0.0244	1.005	1.88E-05	1.72E-05	0.707
10	1.247	0.0251	1.005	2.0E-05	1.76E-05	0.705
15	1.226	0.0255	1.005	2.07E-05	1.79E-05	0.704
20	1.205	0.0259	1.005	2.14E-05	1.81E-05	0.703
25	1.185	0.0263	1.005	2.22E-05	1.84E-05	0.702
30	1.165	0.0267	1.005	2.29E-05	1.86E-05	0.701
40	1.128	0.0276	1.005	2.43E-05	1.91E-05	0.699
50	1.093	0.0283	1.005	2.57E-05	1.96E-05	0.698
60	1.06	0.029	1.005	2.72E-05	2.01E-05	0.696
70	1.029	0.0296	1.009	2.86E-05	2.06E-05	0.694
80	1	0.0305	1.009	3.02E-05	2.11E-05	0.692
90	0.972	0.0313	1.009	3.19E-05	2.15E-05	0.69
100	0.946	0.0321	1.009	3.36E-05	2.19E-05	0.688
120	0.898	0.0334	1.009	3.68E-05	2.28E-05	0.686
140	0.854	0.0349	1.013	4.03E-05	2.37E-05	0.684
160	0.815	0.0364	1.017	4.39E-05	2.45E-05	0.682
180	0.779	0.0378	1.022	4.75E-05	2.53E-05	0.681
200	0.746	0.0393	1.026	5.14E-05	2.6E-05	0.68

Appendix A3 Physical parameters of saturated water at different temperatures.

Temperature (°C)	Density (kg/m^3)	Specific heat capacity (kJ/kg · K)	Thermal conductivity (W/m · K)	Dynamic viscosity (Pa · s)	Prandtl number
0	999.9	4.212	0.551	1.79E-03	13.67
5	999.8	4.202	0.563	1.55E-03	11.6
10	999.7	4.191	0.574	1.31E-03	9.52
15	999.0	4.187	0.587	1.16E-03	8.27
20	998.2	4.183	0.599	1.0E-03	7.02
25	997.0	4.179	0.609	9.03E-04	6.22
30	995.7	4.174	0.618	8.02E-04	5.42
35	994.0	4.174	0.627	7.27E-04	4.87
40	992.2	4.174	0.635	6.53E-04	4.31
45	990.2	4.174	0.642	6.01E-04	3.93
50	988.1	4.174	0.648	5.5E-04	3.54
60	983.1	4.179	0.659	4.7E-04	2.99
70	977.8	4.187	0.668	4.06E-04	2.55
80	971.8	4.195	0.674	3.55E-04	2.21
90	965.3	4.208	0.68	3.15E-04	1.95
100	958.4	4.22	0.683	2.83E-04	1.75
120	943.1	4.25	0.686	2.37E-04	1.47
140	926.1	4.287	0.685	2.01E-04	1.26
160	907.0	4.346	0.683	1.74E-04	1.10
180	886.9	4.417	0.674	1.53E-04	1.00
200	863.0	4.505	0.663	1.36E-04	0.93

Appendix A4 Properties of gases and vapors at 293 K (20 °C) and 101 kPa (1 atm).

Gas or vapor	Molecular weight (g/mol)	Density (kg/m³)	Dynamic viscosity (Pa · s)	Diffusion coefficient (m²/s)
Air (dry)	29.0	1.205	1.81E-05	2.0E-05
Air (saturated)	28.7	1.193	1.79E-05	2.0E-05
N_2	28.0	1.165	1.75E-05	2.0E-05
O_2	32.0	1.326	2.03E-05	2.0E-05
CO_2	44.0	1.832	1.46E-05	1.6E-05
CO	28.0	1.165	1.75E-05	1.6E-05
H_2O	18.0	0.747	9.60E-06	2.4E-05
CH_4	16.0	0.668	1.09E-05	1.6E-05

Source: Hinds and Zhu (2022) / with permission of John Wiley & Sons.

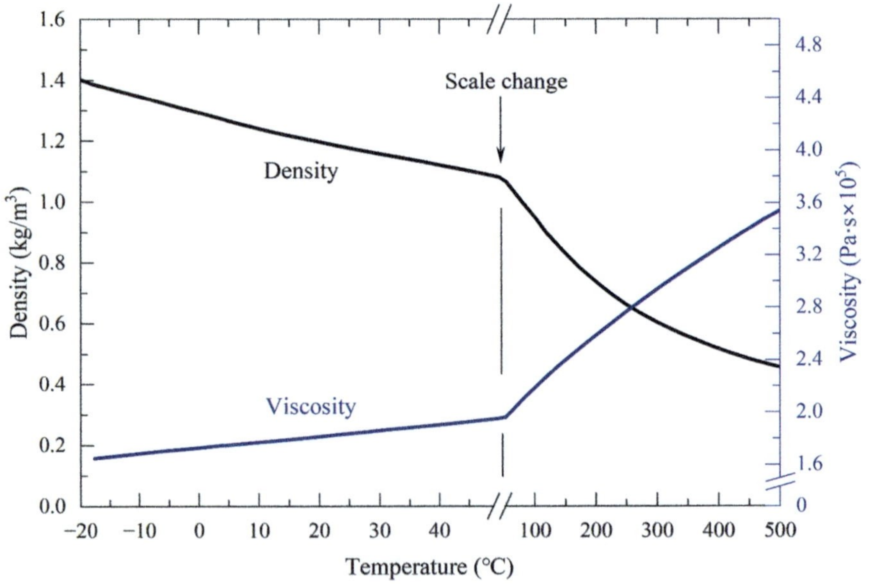

Figure A5 Variations of air viscosity and density with temperature. *Source:* Hinds and Zhu (2022) / with permission of John Wiley & Sons.

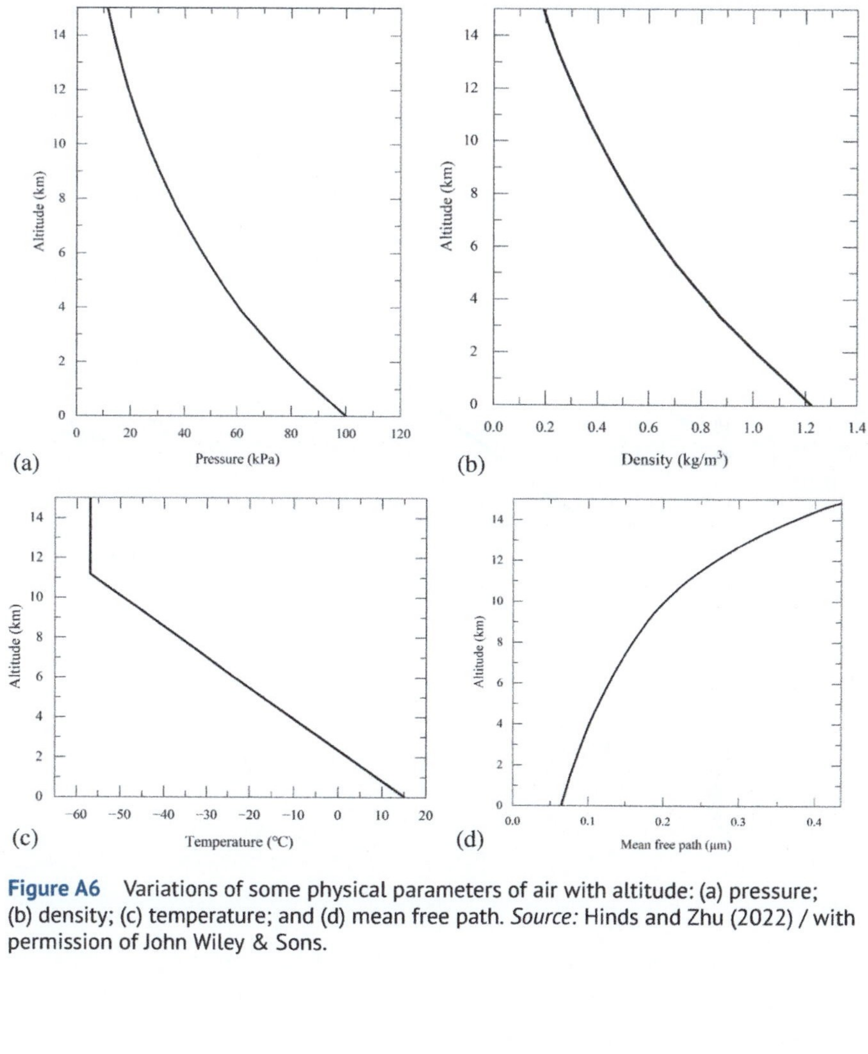

Figure A6 Variations of some physical parameters of air with altitude: (a) pressure; (b) density; (c) temperature; and (d) mean free path. *Source:* Hinds and Zhu (2022) / with permission of John Wiley & Sons.

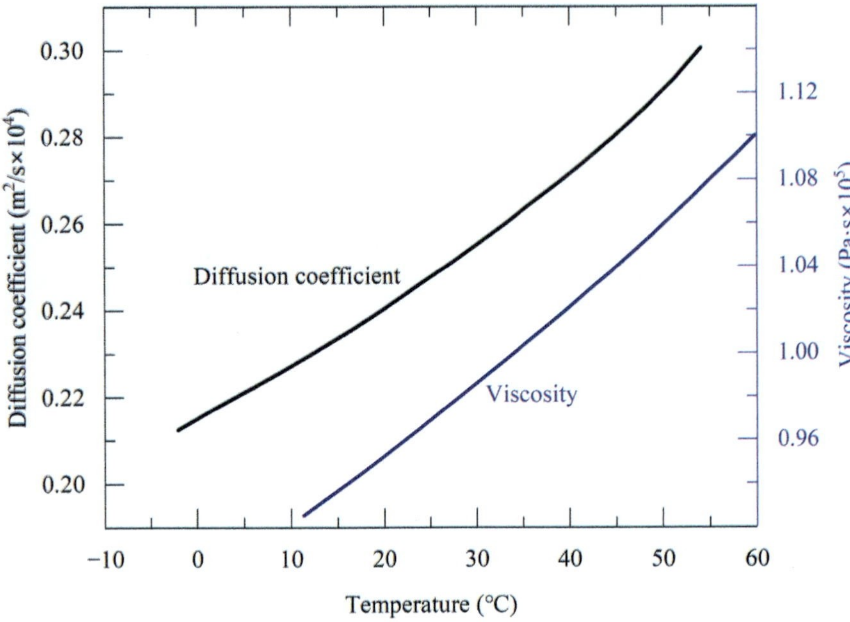

Figure A7 Diffusion coefficient and viscosity of water vapor versus temperature. *Source:* Hinds and Zhu (2022) / with permission of John Wiley & Sons.

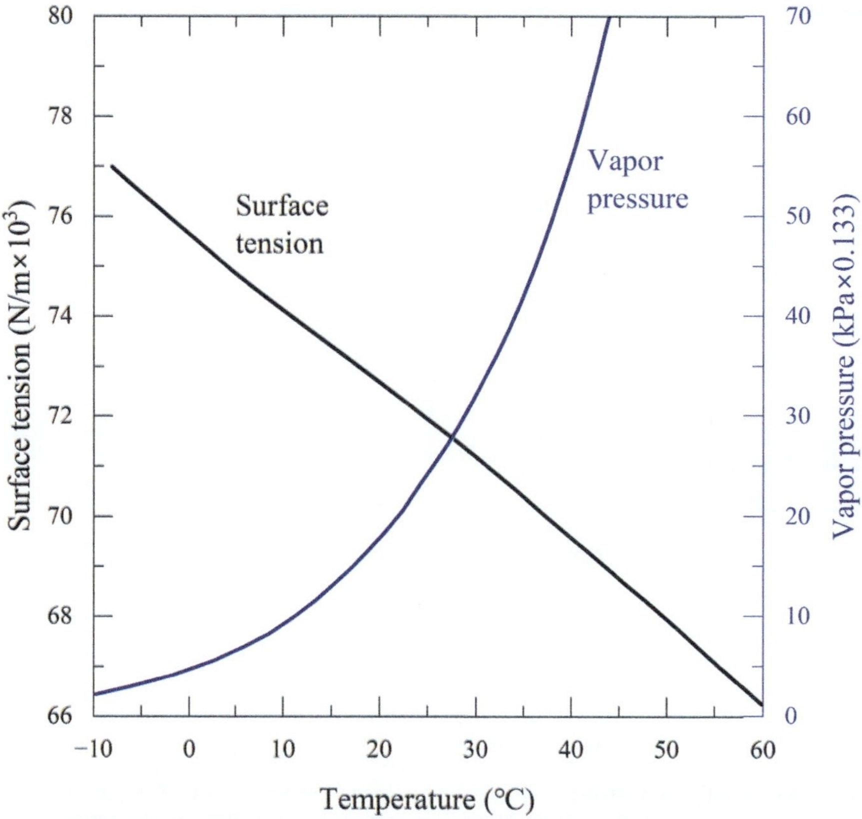

Figure A8 Surface tension and vapor pressure of water versus temperature. *Source:* Hinds and Zhu (2022) / with permission of John Wiley & Sons.

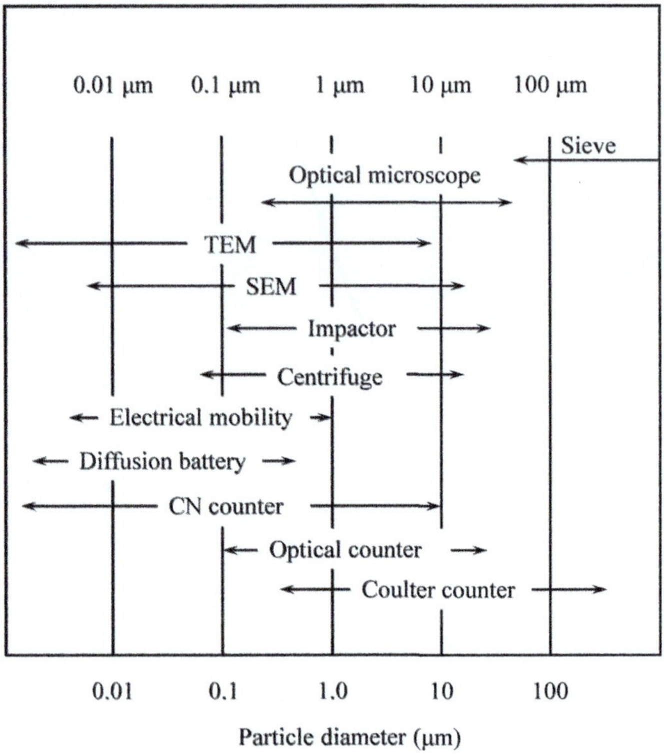

Figure A9 Application ranges of different particle measuring devices (TEM: transmission electron microscope; SEM: scanning electron microscope; CN counter: condensation nuclei counter). *Source:* Hinds and Zhu (2022) / with permission of John Wiley & Sons.

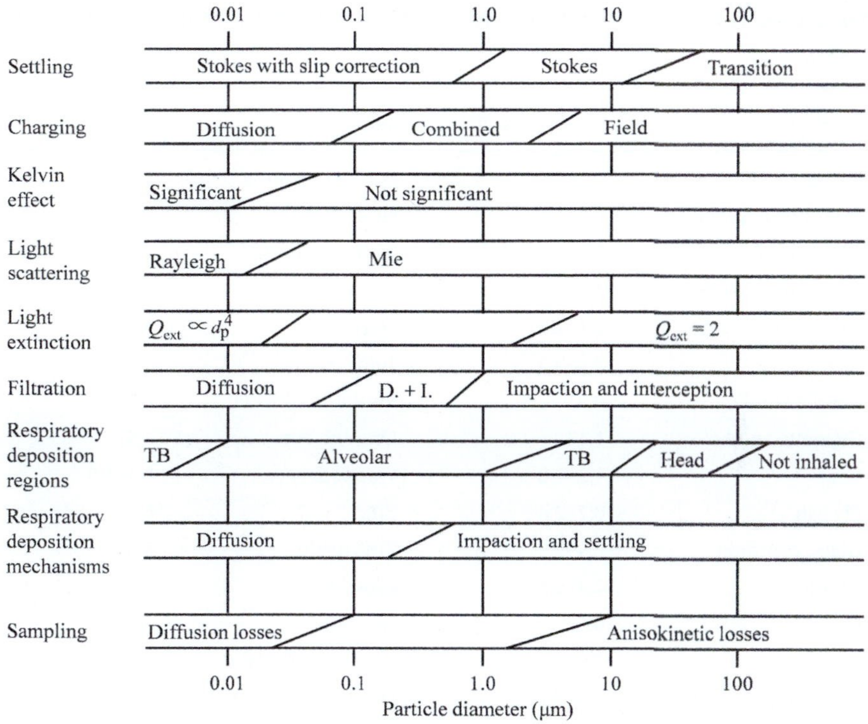

Figure A10 Size range of aerosol properties (Q_{ext}: extinction coefficient of particles; D.: abbreviation for diffusion; I.: abbreviation for impaction; TB: abbreviation for tracheobronchial). *Source:* Hinds and Zhu (2022) / with permission of John Wiley & Sons.

Appendix A11 Density of common aerosol materials (Multiply values by 1,000).

Substance	Density (kg/m^3)	Substance	Density (kg/m^3)
Solids		Portland cement	3.2
3D printing particles	1.1	Quartz	2.6
Aluminum	2.7	Sodium chloride	2.2
Aluminum oxide	4.0	Sulfur	2.1
Arizona test dust	2.5–2.7	Starch	1.5
Asbestos	2.4–3.3	Talc	2.6–2.8
Candle smoke	0.91	Titanium dioxide	4.3
Cigarette smoke	1.18	Uranine dye	1.53
Coal	1.2–1.8	Wood (dry)	0.4–1.0
Fly ash	0.7–2.6	Zinc	6.9
Glass, common	2.4–2.8	Zinc oxide	5.6
Granite	2.4–2.7	***Liquids***	
Ice	0.92	Alcohol	0.79
Iron	7.9	Dibutyl phthalate	1.05
Iron oxide	5.2–5.7	Dioctyl phthalate	0.98
Limestone	2.1–2.9	Dioctyl sebacate	0.92
Lead	11.3	Hydrochloric acid	1.19
Paraffin	0.9	Mercury	13.6
Natural fiber	1–1.6	Petroleum	0.88–0.94
Plastics	1–1.6	Polyethylene glycol	1.13
Pollens	0.45–1.05	Sulfuric acid	1.84
Polystyrene	1.05	Water	1.00
Polyvinyl toluene	1.03		

Source: Hinds and Zhu (2022) / with permission of John Wiley & Sons.

Appendix A12 Properties of standard-density airborne particles at standard conditions.

Particle diameter d_p (μm)	Slip correction factor C_c	Settling velocity V_{TS} (m/s)	Relaxation time τ (s)	Mobility B (m/N · s)	Diffusion coefficient D (m²/s)
0.001	224.332	6.75E-09	6.89E-10	1.32E+15	5.32E-06
0.002	112.463	1.35E-08	1.38E-09	3.30E+14	1.33E-06
0.004	56.530	2.72E-08	2.78E-09	8.28E+13	3.35E-07
0.006	37.888	4.11E-08	4.19E-09	3.70E+13	1.50E-07
0.008	28.568	5.51E-08	5.61E-09	2.09E+13	8.46E-08
0.01	22.976	6.92E-08	7.05E-09	1.35E+13	5.45E-08
0.02	11.801	1.42E-07	1.45E-08	3.46E+12	1.40E-08
0.04	6.229	3.00E-07	3.06E-08	9.13E+11	3.69E-09
0.06	4.384	4.75E-07	4.84E-08	4.28E+11	1.73E-09
0.08	3.470	6.69E-07	6.82E-08	2.54E+11	1.03E-09
0.1	2.928	8.82E-07	8.99E-08	1.72E+11	6.94E-10
0.2	1.878	2.26E-06	2.31E-07	5.51E+10	2.23E-10
0.3	1.554	4.21E-06	4.29E-07	3.04E+10	1.23E-10
0.4	1.402	6.76E-06	6.89E-07	2.06E+10	8.31E-11
0.5	1.316	9.91E-06	1.01E-06	1.54E+10	6.24E-11
0.6	1.261	1.37E-05	1.39E-06	1.23E+10	4.98E-11
0.8	1.194	2.30E-05	2.35E-06	8.75E+09	3.54E-11
1.0	1.155	3.48E-05	3.54E-06	6.77E+09	2.74E-11
2.0	1.077	1.30E-04	1.32E-05	3.16E+09	1.28E-11
3.0	1.051	2.85E-04	2.90E-05	2.05E+09	8.31E-12
4.0	1.039	5.00E-04	5.10E-05	1.52E+09	6.15E-12
5.0	1.031	7.76E-04	7.91E-05	1.21E+09	4.89E-12
6.0	1.026	1.11E-03	1.13E-04	1.00E+09	4.05E-12
8.0	1.019	1.96E-03	2.00E-04	7.47E+08	3.02E-12
10.0	1.015	3.06E-03	3.12E-04	5.95E+08	2.41E-12

Source: Hinds and Zhu (2022) / with permission of John Wiley & Sons.

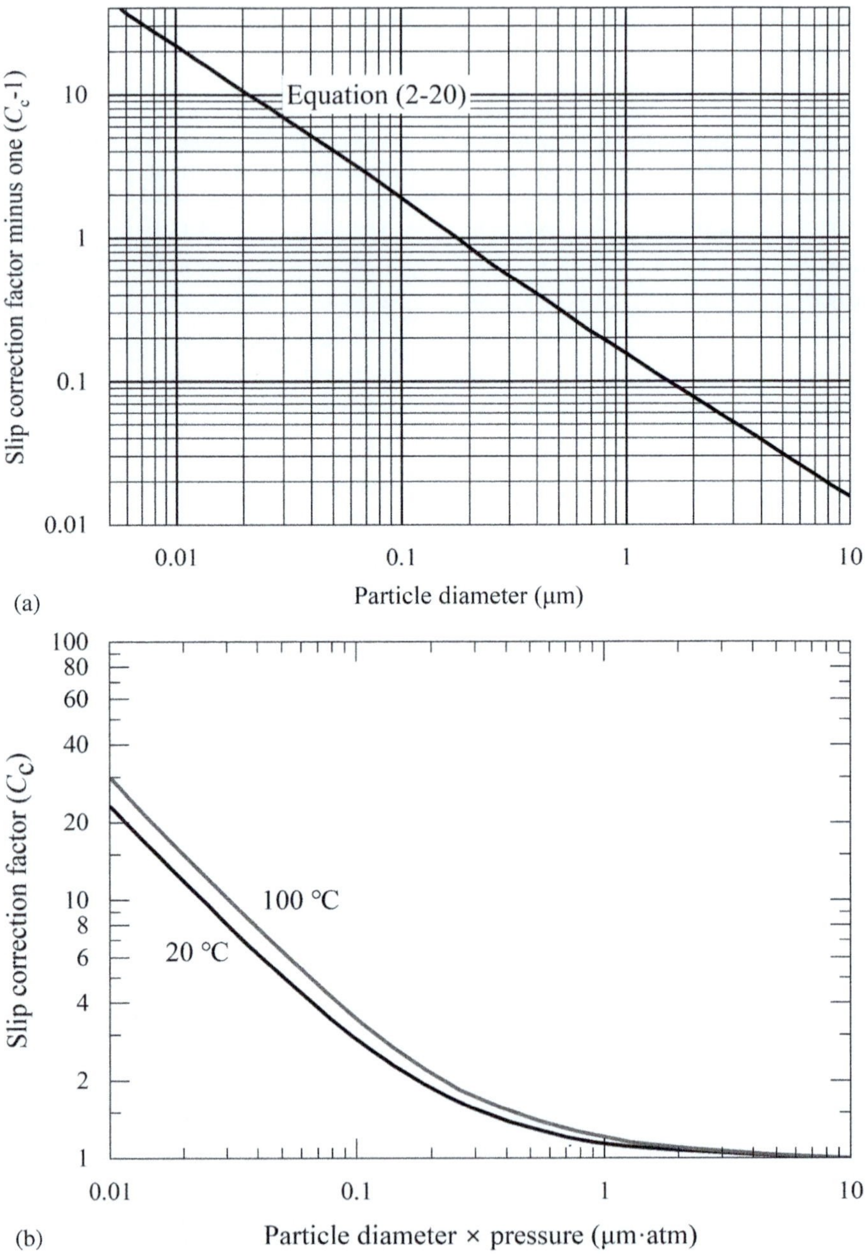

Figure A13 Slip correction factor under standard and nonstandard conditions: (a) variation of the slip correction factor minus one with particle diameter under standard conditions; (b) slip correction factor versus particle diameter multiplying by pressure for two different temperatures. *Source:* Hinds and Zhu (2022) / with permission of John Wiley & Sons.

Appendix A14 Commonly used constants.

Name	Symbol	Value	Unit
Boltzmann constant	k	1.381E-23	J/K
Elementary charge	e	1.602E-19	C
Gravitational acceleration	g	9.807	m/s^2
Planck constant	h	6.626E-34	$J \cdot s$
Stefan–Boltzmann constant	σ	5.670E-08	$W/m^2/K^4$
Universal gas constant	R_u	8.314	J/mol/K
Vacuum permittivity	ε_0	8.854E-12	F/m

Appendix A15 Important dimensionless numbers.

Dimensionless number	Symbol	Expression	Note
Knudsen number	Kn	$\mathrm{Kn} = \lambda/d_p$	Ratio of molecular mean free path to particle diameter
Peclet number	Pe	$\mathrm{Pe} = UL/D$	Ratio of convective to diffusive transport
Prandtl number	Pr	$\mathrm{Pr} = \nu/\alpha$	Ratio of momentum diffusivity to thermal diffusivity
Reynolds number	Re	$\mathrm{Re} = Vd_p/\nu$	Ratio of inertial forces to viscous forces
Schmidt number	Sc	$\mathrm{Sc} = \nu/D$	Ratio of momentum diffusivity to mass diffusivity
Stokes number	Stk	$\mathrm{Stk} = \tau U/L$	Ratio of particle relaxation time to flow characteristic time

Reference

Hinds, W.C., Zhu, Y. (2022). *Aerosol Technology: Properties, Behavior, and Measurement of Airborne Particles,* 3e. Wiley.

Index

h

i

Printed and bound by CPI Group (UK) Ltd, Croydon, CR0 4YY
24/08/2026
14947830-0002